J. O. Bird
B.Sc.(Hons), A.F.I.M.A., T.Eng.(CEI), M.I.T.E.

A. J. C. May
B.A., C.Eng., M.I.Mech.E., F.I.T.E., M.B.I.M.

Technician mathematics

Level 1

Second edition

Longman London and New York

Longman Group Limited,
Longman House,
Burnt Mill, Harlow, Essex,
United Kingdom.

Published in the United States of America
by Longman Inc., New York

First published 1977
Second edition 1982

British Library Cataloguing in Publication Data

Bird, J. O.
Technician mathematics level 1 – 2nd ed.
1. Shop mathematics
I. Title II. May, A. J. C.
510'.246 TJ1165

ISBN 0–582–41256–0

Library of Congress Cataloging in Publication Data

Bird, J. O.
Technician mathematics, level 1.

(Longman technician series. Mathematics and sciences sector)
Includes index.
1. Engineering mathematics. I. May, A. J. C.
II. Title. III. Series.
TA330.B532 1982 510'.2462 81–11724
ISBN 0–582–41256–0 AACR2

Printed in Great Britain by The Pitman Press Ltd, Bath

Sector Editor:

D. R. Browning, B.Sc., F.R.S.C., C.Chem., A.R.T.C.S.
Principal Lecturer and Head of Chemistry, Bristol Polytechnic

Books already published in this sector of the series:

Technician mathematics Level 2 Second edition **J. O. Bird and A. J. C. May**
Technician mathematics Level 3 **J. O. Bird and A. J. C. May**
Technician mathematics Levels 4 and 5 **J. O. Bird and A. J. C. May**
Mathematics for science technicians Level 2 **J. O. Bird and A. J. C. May**
Mathematics for electrical and telecommunications technicians Level 2 **J. O. Bird and A. J. C. May**
Mathematics for electrical technicians Level 3 **J. O. Bird and A. J. C. May**
Mathematics for electrical technicians Levels 4 and 5 **J. O. Bird and A. J. C. May**
Calculus for technicians **J. O. Bird and A. J. C. May**
Statistics for technicians **J. O. Bird and A. J. C. May**
Algebra for technicians **J. O. Bird and A. J. C. May**
Physical sciences Level 1 **D. R. Browning and I. McKenzie Smith**
Engineering science for technicians Level 1 **D. R. Browning and I. McKenzie Smith**
Safety science for technicians **W. J. Hackett and G. P. Robbins**
Fundamentals of chemistry **J. H. J. Peet**
Further studies in chemistry **J. H. J. Peet**
Technician chemistry Level 1 **J. Brockington and P. J. Stamper**

Mathematics for scientific and technical students
H. G. Davies and G. A. Hicks
Mathematical formulae for TEC courses **J. O. Bird and A. J. C. May**
Science formulae for TEC courses **D. R. Browning**

Contents

Section III – Geometry and Trigonometry 199

Section IV – Statistics 365

Chapter 12 An introduction to statistics 366

Preface

This textbook is the first of a series which deal simply and carefully with the fundamental mathematics essential in the development of technicians.

Technician Mathematics Level 1 provides coverage of the Technician Education Council level I Mathematics Unit U80/683 (formerly U75/005).

The aim of the book is to consolidate basic mathematical principles and establish a common base for further progress, taking into account the wide variety of approaches to mathematics previously encountered at earlier stages of education.

Each topic considered in the text is presented in a way that assumes in the reader little previous knowledge of that topic. This practical mathematics book contains over 350 detailed worked problems followed by some 700 further problems with answers. Although specifically written for the TEC Level 1 syllabus, the book may also be suitable for use by CSE final year students who are likely to move into engineering/science disciplines.

The authors would like to thank Mr David Browning, Principal Lecturer, Bristol Polytechnic, for his valuable assistance in his capacity as General Editor of the Mathematics and Sciences Sector of the Longman Technician Series. They would also like to express their appreciation for the friendly cooperation and helpful advice given to them by the publishers.

Thanks are also due to Mrs Elaine Woolley for the excellent typing of the manuscript.

Finally, the authors would like to add a word of thanks to their wives, Elizabeth and Juliet, for their patience during the preparation of this book.

J. O. Bird
A. J. C. May

Highbury College of Technology,
Portsmouth, 1982

Acknowledgements

We are indebted to the City of Portsmouth, Passenger Transport Department for our Table 3.9 from *Leigh Park Area Timetable* March 1979.

Section I

Arithmetic operations

Chapter 1

Revision of basic arithmetic

1.1 Arithmetic operations

Whole numbers such as +5, +16, and +7 are called **positive integers,** while whole numbers such as −3, −8, and −29, are called **negative integers.** The four basic arithmetic operations associated with these numbers are shown in Table 1.1.

Table 1.1

Operation	Symbol	Example
Addition	+	$12 + 4 = 16$
Subtraction	−	$12 - 4 = 8$
Multiplication	×	$12 \times 4 = 48$
Division	÷	$12 \div 4 = 3$

When positive and negative integers occur in the same calculation the following rules apply.
Unlike signs give a negative overall sign.
Like signs give a positive overall sign.
The addition of +4 to −2, is written $4 + -2$ since, if no sign is shown in front of an integer, a positive sign is assumed. The + and − are unlike signs and give a negative overall sign. Thus $4 + -2 = 4 - 2 = 2$.

Similarly, subtracting −2 from 4 is written $4 - -2$. The − and − are like signs and give a positive overall sign, hence $4 - -2 = 4 + 2 = 6$.

In multiplication $+ \times +$ gives $+$ and $+ \times -$ gives $-$,
$- \times -$ gives $+$ and $- \times +$ gives $-$.
Thus $4 \times -2 = -8$, whereas $-4 \times -2 = +8$, which is just written as 8.

For division $4 \div -2 = -2$ (unlike signs) and
$-4 \div -2 = 2$ (like signs).

Positive and negative integers can be represented by points marked on a line. The number 0 is the centre of the line and 1, 2, 3, . . . are represented by spaces to the right of 0. To find the value of 3 + 2 involves moving 3 places along the line to the right and then a further two places to the right since addition of positive integers is obtained by movement to the right. Figure 1.1 shows that doing this results in finishing by the number 5.

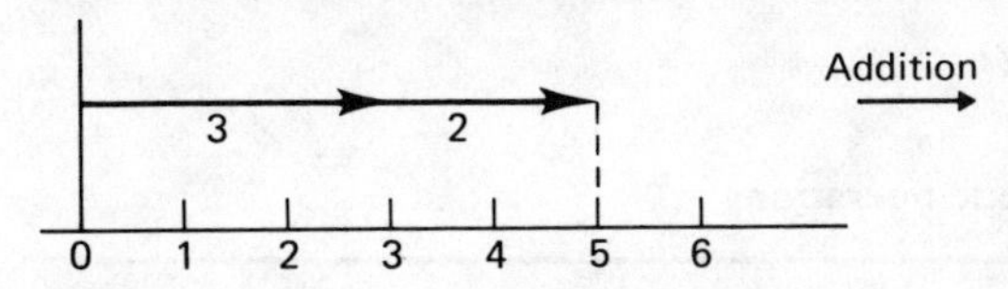

Fig. 1.1

The value of 3 − 2 is found by moving 3 places to the right and then 2 places to the left, since subtraction of positive integers is obtained by movement to the left. Figure 1.2 shows that the value of 3 − 2 is 1.

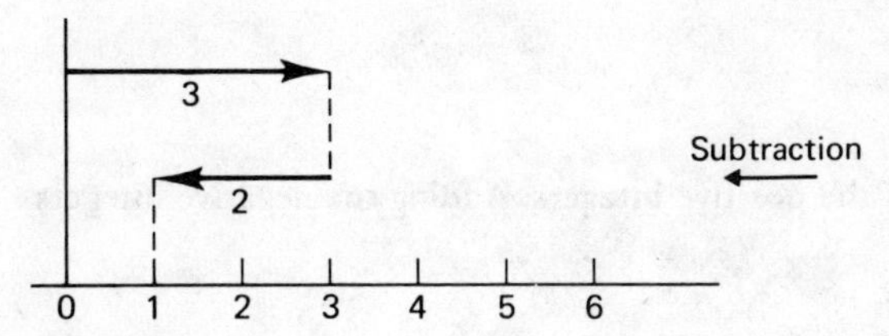

Fig. 1.2

Negative integers −1, −2, −3 . . . are represented by spaces to the left of 0. The value of 3 − 5 can be found by starting at 0 and moving 3 places to the right and 5 places to the left. Figure 1.3 shows that 3 − 5 is −2.

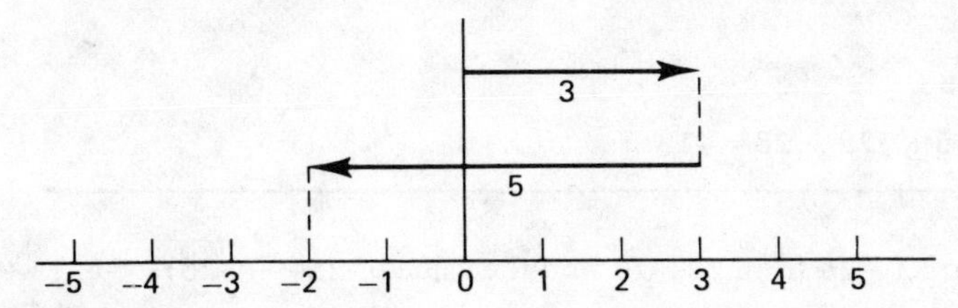

Fig. 1.3

When adding positive and negative integers, the overall sign in front of a number governs the direction of movement. For example, $-6 + -10$ becomes $-6 -10$ or -16 and can be indicated by movement to the left. Also $-4 - -10$ becomes $-4 + 10$ or $+6$ and can be shown by movement to the right.

The number 234 means $(2 \times 100) + (3 \times 10) + (4 \times 1)$ and the number 561 means $(5 \times 100) + (6 \times 10) + (1 \times 1)$. Adding these numbers gives
$(2 \times 100) + (3 \times 10) + (4 \times 1) + (5 \times 100) + (6 \times 10) + (1 \times 1)$
and grouping like terms gives
$(2 + 5) \times 100 + (3 + 6) \times 10 + (4 + 1) \times 1$
or
$700 \quad + 90 \quad + 5$
that is, 795.

When adding or subtracting integers, the units, tens, hundreds, . . . columns must be kept underneath one another. Thus

$$\begin{array}{lr} & 234 \\ & 561 \\ \hline \text{Adding} & 795 \\ \hline \end{array}$$

Worked problems on arithmetic operations

Problem 1. Add 263, − 146, 329 and −28.

This would be written mathematically as $263 - 146 + 329 - 28$.

Adding the positive integers

$$\begin{array}{lr} & 263 \\ & 329 \\ \hline \text{Adding} & 592 \\ \hline \end{array}$$

592 is known as the **sum of the positive integers.** Adding the negative integers

$$\begin{array}{lr} & 146 \\ & 28 \\ \hline \text{Adding} & 174 \\ \hline \end{array}$$

174 is known as the **sum of the negative integers.** Taking the sum of the negative integers from the sum of the positive integers

$$\begin{array}{lr} & 592 \\ & 174 \\ \hline \text{Subtracting} & 418 \\ \hline \end{array}$$

Hence $263 - 146 + 329 - 28 =$ **418**

Problem 2. Subtract −26 from 284 (or find the value of $284 - -26$)

When dealing with subtraction care must be taken in deciding what is to be taken from what. One way of determining this is to use the words **from** and **take**. This question then becomes **from 284 take** −26 and is written $284 - -26$. The like signs give a positive overall sign, hence $284 - -26 = 284 + 26$

$$\begin{array}{lr} & 284 \\ & 26 \\ \hline \text{Adding} & 310 \\ \hline \end{array}$$

Hence 284 − −26 = **310**

Problem 3. Subtract 143 from 72 (or find the vlaue of 72 − 143)

To depict the number 72, a line 72 units long is drawn to the right. To subtract 143 from this number, a line starting at point 72 and 143 units long is drawn to the left. The answer, with reference to Fig. 1.4 is **−71 units** from 0.

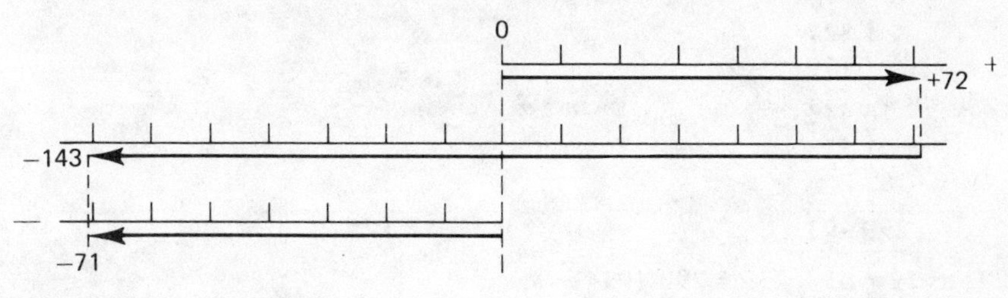

Fig. 1.4

Problem 4. Multiply 73 by 13 (or find the value of 73 × 13)

	73	
	13	
Multiply 73 by 3.	219	The units column figure is written under the 3.
Multiply 73 by 1.	73	The units column figure is written under the 1.
Adding	949	

Thus 73 × 13 = **949**

Problem 5. Find the value of 215 × 63

	215	
	63	
Multiply 215 by 3.	645	The units column figure is written under the 3.
Multiply 215 by 6.	12 90	The units column figure is written under the 6 of the 63.
Adding	13 545	

Thus 215 × 63 = **13 545**

Problem 6. Find the value of 215 × −63

The signs are unlike and will give a negative answer. Remembering this, 215 is multipled by 63 and gives 13 545 (from problem 5) and 215 × −63 = **−13 545** due to the unlike signs.

Problem 7. Find the value of $29 \times 63 \times -159$

$$\begin{array}{lr} & 29 \\ & \underline{63} \\ 29 \times 3 & 87 \\ 29 \times 6 & \underline{1\,74} \\ \text{Adding} & \underline{1\,827} \end{array}$$

The problem now becomes $1\,827 \times -159$. Treating these as positive numbers for the multiplication process, but remembering that the unlike signs will give a negative answer overall:

$$\begin{array}{r} 1\,827 \\ \underline{159} \\ 16\,443 \\ 91\,35 \\ 182\,7 \\ \hline 290\,493 \\ \hline \end{array}$$

Thus $29 \times 63 \times 159 = 290\,493$

and $29 \times 63 \times -159 =$ **−290 493**

Problem 8. Divide 9 417 by 73 (or find the value of $9\,417 \div 73$)

Yet another way of expressing this problem mathematically is to use a horizontal or oblique line, then

$$9\,417 \div 73 = \frac{9\,417}{73} \text{ or } 9\,417/73$$

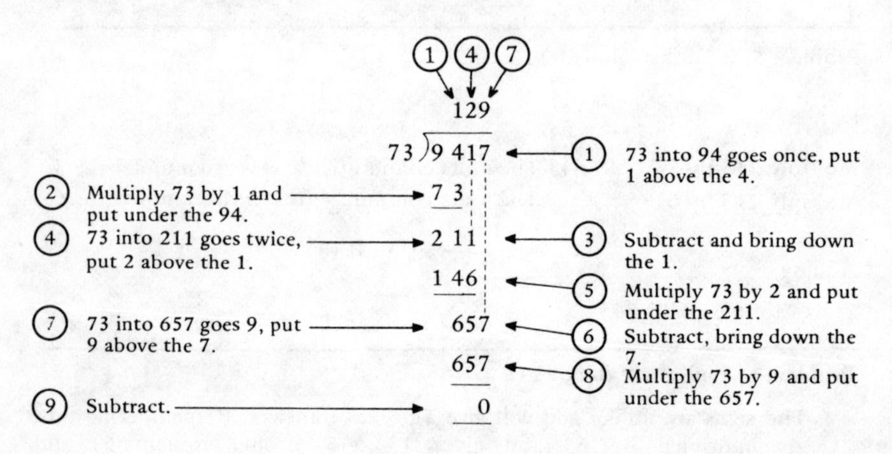

If, when dividing one number by another, there is still a number left resulting from the last subtraction this is called the **remainder.** In this case, the

0 on the final subtraction shows that there is no remainder, that is 73 goes an exact number of times into 9 417.

Thus 9 417 ÷ 73 = **129.**

Problem 9. Find the value of 6 738 ÷ 46

$$\begin{array}{r} 146 \\ 46\,\overline{)6\,738} \end{array}$$

Step		Working	Step	
(2)	1 × 46. →	4 6	← (1)	46 into 67 goes once, put 1 up above the 7.
(3)	Subtract. →	2 13	← (4)	Bring down the 3; 46 into 213 goes 4 times. Put up the 4 above the 3.
(5)	4 × 46. →	1 84		
(6)	Subtract. →	298	← (7)	Bring down the 8; 46 into 298 goes 6 times. Put up the 6 above the 8.
(8)	6 × 46. →	276		
(9)	Subtract. →	22	← (10)	Since there are no more numbers to bring down, 22 is the remainder.

Thus 6 738 ÷ 46 = **146, remainder 22.**

Problem 10. Find the value of $\dfrac{127\,364}{-123}$

The unlike signs show that the answer will be negative. Both numbers are treated initially, for the purpose of long division, as if they are positive.

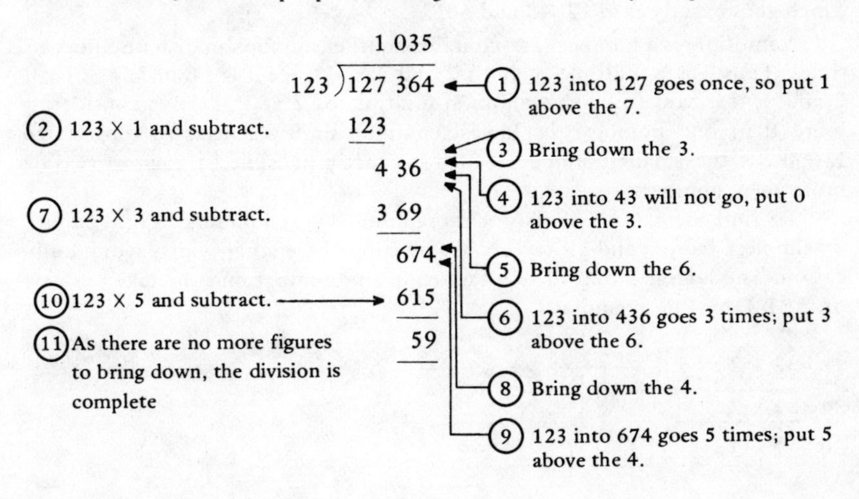

Thus 127 364 ÷ 123 = 1 035, remainder 59

And 127 364 ÷ −123 = **−1 035, remainder 59.**

Further problems on arithmetic operations may be found in Section 1.8 (Problems 1–15), page 31.

1.2 Highest common factor and lowest common multiple

A **factor** is a number which divides into another number exactly. Thus both 2 and 3 are factors of 6. The numbers 2 and 3 are also factors of 12, 18 and 24 and are called common factors of all these numbers. The **highest common factor,** usually abreviated H.C.F., is the largest number that divides into two or more numbers exactly.

To find the H.C.F. of two or more numbers, each number is broken into its simplest factors, and those which are common to all the numbers constitute the H.C.F. Numbers which have no factors apart from 1 are called **prime numbers.** Since 2, 3, 5, 7, 11, 13 and 17 . . . have no factors apart from 1, they are prime numbers. When looking for factors, numbers are expressed in terms of factors comprising these prime numbers. The simplest ones are found first; that is 2, 3, 5, 7 and so on. For example:

$$\left.\begin{aligned} 12 &= 2 \times 2 \times 3 \\ 48 &= 2 \times 2 \times 2 \times 2 \times 3 \\ 60 &= 2 \times 2 \times 3 \times 5 \end{aligned}\right\} \text{or} \quad \begin{array}{|l|l} 2 \times 2 \times 3 & \\ 2 \times 2 \times 3 & \times 2 \times 2 \\ 2 \times 2 \times 3 & \times 5 \end{array}$$

The factors contained in each of the numbers 12, 48 and 60 are 2 × 2 × 3 = 12, that is the H.C.F. of 12, 48 and 60 is 12, this being the largest number which goes exactly into 12, 48 and 60.

A **multiple** is a number that contains another number an exact number of times. Thus 6 is a multiple of both 2 and 3 and since it is a multiple of both 2 and 3, it is said to be the common multiple of 2 and 3. When considering more than one number, the **lowest common multiple**, usually abbreviated L.C.M., is the smallest number which is exactly divisible by each of two or more given numbers.

To find the L.C.M. of two or more numbers each number is broken into its simplest factors and all factors are multiplied together which avoid duplication. The largest group of like factors in any number must be taken as part of the L.C.M. For example:

$$\begin{aligned} 12 &= 2 \times 2 \times \boxed{3} \\ 24 &= \boxed{2 \times 2 \times 2} \times 3 \\ 60 &= 2 \times 2 \times 3 \times \boxed{5} \end{aligned}$$

The largest group of twos is in the number 24, and this covers the 2×2 factors in 12 and 60, so that they will not be required to contribute towards the L.C.M. As there is one 3 factor in each of the numbers only one of these will be required. Finally the 5 factor in 60 will also be required. The L.C.M. is $2 \times 2 \times 2 \times 3 \times 5 = 120$, the lowest number which can be exactly divided by 12, 24 and 60.

The method of finding the L.C.M. of a group of numbers one or more of which are negative is treated in a similar manner. The L.C.M. is usually expressed in terms of positive numbers. This is because, for example, the numbers 6 and 9 have factors 2×3 and 3×3 and their L.C.M. is 18. Also the numbers 6 and -9 have factors 2×3 and $-1 \times 3 \times 3$ and at first sight have an L.C.M. of $-1 \times 2 \times 3 \times 3$ or -18. However, since $\frac{18}{6} = 3$ and $\frac{18}{-9} = -2$, this shows that 18 is the L.C.M. of both 6 and -9 and the positive value of 18 is the one taken.

Worked problems on H.C.F. and L.C.M.

Problem 1. Find the H.C.F. of the numbers 42, 56 and 70

To determine the simplest factors of a number a check should first be made to see if it can be divided by 2. If it can be, this should be done and an attempt made to divide by 2 again, repeating until an odd number is left. Then a check should be made to see if it will divide by 3. This check should also be repeated as often as division is possible. This process is repeated with the prime numbers 5, 7, 11, 13, 17, 19 etc. until a prime number remains. Then

$42 = 2 \times 3 \times 7$

$56 = 2 \times 2 \times 2 \times 7$

$70 = 2 \times 5 \times 7$

The factors which are common to each of the numbers 42, 56 and 70 are $2 \times 7 = 14$. Thus **14 is the H.C.F. of these three numbers.**

Problem 2. Find the H.C.F. of 110, 286 and 330

$110 = 2 \times 5 \times 11$

$286 = 2 \times 11 \times 13$

$330 = 2 \times 3 \times 5 \times 11$

The H.C.F. is $2 \times 11 = 22$

Problem 3. Find the H.C.F. of 182, $-1\,547$ and $4\,199$

$182 = 2 \times 7 \times 13$

$-1\,547 = -1 \times 7 \times 13 \times 17$

$4\,199 = 13 \times 17 \times 19$

Thus the H.C.F. of 182, $-1\,547$ and $4\,199$ is **13**.

Problem 4. Find the L.C.M. of the numbers 42, 56 and 140

Determining the simplest factors of each of these numbers gives

$42 = 2 \times 3 \times 7$

$56 = 2 \times 2 \times 2 \times 7$

$140 = 2 \times 2 \times 5 \times 7$

The L.C.M. is $2 \times 2 \times 2 \times 3 \times 5 \times 7 = 840$.

Problem 5. Find the L.C.M. of the numbers 24, 56, 128 and 39

Determining the simplest factors of each of these numbers gives

$24 = 2 \times 2 \times 2 \times 3$

$56 = 2 \times 2 \times 2 \times 7$

$128 = 2 \times 2 \times 2 \times 2 \times 2 \times 2 \times 2$

$39 = 3 \times 13$

Selecting the largest groups of each of the prime numbers results in $2 \times 2 \times 2 \times 2 \times 2 \times 2 \times 2 \times 3 \times 7 \times 13$, or 34 944.

Then the L.C.M. of 24, 56, 128 and 39 is **34 944.**

Problem 6. Find the L.C.M. of −442, 663 and 1 105

$-442 = -1 \times 2 \times 13 \times 17$

$663 = 3 \times 13 \times 17$

$1\,105 = 5 \times 13 \times 17$

The L.C.M. of these three numbers is **$2 \times 3 \times 5 \times 13 \times 17$ or 6 630, the −1 factor being ignored.**

Further problems on H.C.F. and L.C.M. may be found in Section 1.8 (Problems 16–25), page 31.

1.3 Brackets and basic laws

Brackets are used in arithmetic to indicate that the operation inside the bracket must be done first. For instance, take the sum of 3 and 4 from 9 is written as

$9 - (3 + 4) = 9 - 7 = 2$

Since the (3 + 4) is in a bracket, this operation must be done before subtracting the result from 9.

Another use of brackets is to indicate multiplication. An integer which is next to a bracket means multiply the contents of that bracket by this integer. For instance, take twice the sum of 3 and 4 from 9 is written as

$9 - 2(3 + 4) = 9 - 2(7) = 9 - 2 \times 7 = 9 - 14 = -5$

Other examples are

$3(4 + 5) = 3 \times 9 = 27$

$-2(6 - 3) = -2 \times 3 = -6$

$(8 - 3)7 = 5 \times 7 = 35$

This last example shows that the integer can be written either in front or behind the bracket, although it is usual to write it in front of the bracket.

When there is no integer next to the bracket, it is inferred that the integer is 1. Thus

$(4 + 5) = 1 \times (4 + 5) = 1 \times 9 = 9$

but

$-(4 + 5) = -1 \times (4 + 5) = -1 \times 9 = -9$

Adjacent brackets with no + or − sign between them indicate multiplication.

$(2 + 3)(4 + 5) = (5)(9) = 5 \times 9 = 45$

There are certain laws which govern basic arithmetic operations and the use of brackets. In both addition and multiplication the order of writing integers does not affect the value. For example

$2 + 3 = 5$

Also $3 + 2 = 5$

Thus $2 + 3 = 3 + 2$ [I]

For multiplication $2 \times 3 = 6$

And $3 \times 2 = 6$

Thus $2 \times 3 = 3 \times 2$ [II]

However, the same laws must *not* be applied for the operation of subtraction and division. $2 - 3$ is *not* equal to $3 - 2$ and $2 \div 3$ is *not* equal to $3 \div 2$. The position of brackets in problems containing only addition signs *or* mutliplication signs does not alter the final value. For example

$2 + (3 + 4) = 2 + 7 = 9$

Also $(2 + 3) + 4 = 5 + 4 = 9$

Hence $2 + (3 + 4) = (2 + 3) + 4$ [III]

In multiplication $2 \times (3 \times 4) = 2 \times 12 = 24$

Also $(2 \times 3) \times 4 = 6 \times 4 = 24$

Then $2(3 \times 4) = (2 \times 3)4$ [IV]

These laws do *not* apply in the case of subtraction and division. $2 - (3 + 4)$ is *not* equal to $(2 - 3) + 4$. Also $(2 \div 3) + 4$ is *not* equal to $2 \div (3 + 4)$. Another way of determining the value of the expression in a bracket is given in law [V], and is sometimes helpful in evaluating brackets containing difficult numbers. Using the method of dealing with the bracket first:

$2\,(3 + 4) = 2 \times 7 = 14$

Alternatively $2\,(3 + 4) = (2 \times 3) + (2 \times 4)$

$$= 6 + 8 = 14 \qquad [V]$$

Frequently, in a problem, more than one operator is involved. Different answers to the problem can be obtained depending on the order in which operations are carried out. For example, to find the value of $2 \times 3 + 7$ can give an answer of $2 \times 3 = 6$ and $6 + 7 = 13$. Another way of getting an answer would be to add 3 and 7 and multiply by 2. Then $2 \times 3 + 7 = 2 \times 10 = 20$. Both of these answers cannot be correct and an **order of precedence** of operations is used. This order of operations is: **B**rackets **O**f **D**ivision **M**ultiplication **A**ddition and **S**ubtraction often remembered by the initials B O D M A S. Applying the order of precedence to $2 \times 3 + 7$, shows that since multiplication comes before addition, the only correct solution to this problem is $2 \times 3 + 7 = 6 + 7 = 13$, and 20 is incorrect.

Worked problems on brackets and basic laws

Problem 1. Find the value of $(6 + 7) + 8$

By law [III], the position of the brackets does not affect the answer. Thus

$(6 + 7) + 8 = \mathbf{6 + 7 + 8 = 21}$

Problem 2. Find the value of $6 - (7 + 8)$

Applying the rule that bracket terms must be dealt with first

$6 - (7 + 8) = \mathbf{6 - 15 = -9}$

Problem 3. Find the value of $6 - (7 - 8)$

$6 - (7 - 8) = 6 - (-1)$

and, since like signs give +,

$6 - (-1) = 6 + 1 = 7$

Alternatively $(7 - 8) = -1$

and $-1\,(-1) = (-1) \times (-1) = +1$

Hence $6 - (7 - 8) = 6 + 1 = 7$

Problem 4. Find the value of $(6 + 7)\,(8 + 9)$

Adjacent brackets mean multiply and operations within brackets must be done first. Thus

$(6 + 7)\,(8 + 9) = \mathbf{(13)\,(17) = 221}$

Problem 5. Find the value of $(16 - 7) \div 3$

This problem involves both **B**rackets and **D**ivision and remembering BODMAS, the brackets must be dealt with before dividing. Thus

$(16 - 7) \div 3 = 9 \div 3 = 3$

Problem 6. Find the value of $6 \times 4 \div 2\,(4 + 7)$

The operations are **M**ultiplication, **D**ivision and **B**rackets. From **BODMAS** the order is **B**rackets, **D**ivision and finally **M**ultiplication, thus

$$6 \times 4 \div 2\,(4 + 7) = 6 \times 4 \div 2 \times 11 \quad \text{(B)}$$
$$= 6 \times 2 \times 11 \quad \text{(D)}$$
$$= \mathbf{132} \quad \text{(M)}$$

Problem 7. Find the value of $(19 - 43) \div 6 \times 40 \div (3 + 17)$

In this problem there are two lots of brackets and two ÷ operators. Provided the overall order dictated by BODMAS is adhered to, the order in which repeated operators are determined does not affect the value of the answer. Thus

$$(19 - 43) \div 6 \times 40 \div (3 + 17) = -24 \div 6 \times 40 \div 20 \quad \text{(B)}$$
$$= -4 \times 2 \quad \text{(D)}$$
$$= \mathbf{-8} \quad \text{(M)}$$

Problem 8. Find the value of $(2 + 3) + (4 + 5) \div (10 - 7)\,(-3 + 2) - 16$

Applying the order of operations:

$$(2 + 3) + (4 + 5) \div (10 - 7)\,(-3 + 2) - 16 = 5 + 9 \div 3 \times (-1) - 16 \quad \text{(B)}$$
$$= 5 + 3 \times (-1) - 16 \quad \text{(D)}$$
$$= 5 - 3 - 16 \quad \text{(M)}$$
$$= \mathbf{-\,14} \quad \text{(S)}$$

The (-1) term is usually kept in a bracket when preceded by an operator, to prevent any confusion. Thus to take -3 from 4 is written as $4 - (-3)$ and to multiply 6 by -4 is written as $6 \times (-4)$ or $6\,(-4)$.

Problem 9. Find the value of $64 \div (-16) + (-7 - 12) - (-29 + 36)\,(-2 + 9)$

This expression becomes

$$64 \div (-16) + (-19) - (7)\,(7) \quad \text{(B)}$$
$$= (-4) + (-19) - (7)\,(7) \quad \text{(D)}$$

$= (-4) + (-19) - 49$ (M)

$= -23 - 49$ (A)

$= -72$ (S)

Further problems on brackets and basic laws may be found in Section 1.8 (Problems 26–40), page 32.

1.4 Fractions

Dividing 3 by 5 can be written as $3 \div 5$ or as $\frac{3}{5}$. In the latter form, $\frac{3}{5}$ is called a fraction and can be defined as one or more equal parts of something. For example, one-fifth of 100 metres means that the 100 metres must be divided into 5 equal parts and one of these parts is taken. Then one-fifth of 100 metres is 20 metres, obtained by dividing 100 into 5 equal parts of 20 metres and taking one of these parts. Three-fifths of 100 metres means divide 100 metres into 5 equal parts and then take 3 of these parts, that is $20 \times 3 = 60$ metres.

In every fraction the number above the line is called the **numerator** and that below the line the **denominator.** In the fraction $\frac{3}{5}$, 3 is the numerator and 5 the denominator. It is called a **proper fraction.** The number in the numerator can be larger than the number in the denominator, for example, the fraction $\frac{7}{5}$. This fraction is then called an **improper fraction** because it can be expressed as an integer plus a proper fraction. Seven-fifths comprises five-fifths plus two-fifths, and five-fifths is one. Then, mathematically:

$$\frac{7}{5} = \frac{5}{5} + \frac{2}{5} = 1\frac{2}{5}$$

To keep calculations to a minimum, it is usual to express fractions using the lowest possible numbers for both numerator and denominator. This can be achieved by dividing both numerator and denominator by the highest common factor of these two numbers when it is possible.

Now $\frac{3}{5}$, $\frac{6}{10}$, $\frac{9}{15}$ and $\frac{12}{20}$ are all the same fraction.

For $\frac{9}{15} = \frac{3 \times 3}{3 \times 5}$ and, dividing by the H.C.F. of the numerator and denominator,

$\frac{9}{15} = \frac{\not{3}^1 \times 3}{\not{3}_1 \times 5} = \frac{3}{5}$. This process is called cancelling. The fraction $\frac{32}{48}$ can be reduced to its simplest form by cancellation as follows:

$$\frac{32}{48} = \frac{\not{2}^1 \times \not{2}^1 \times \not{2}^1 \times \not{2}^1 \times 2}{\not{2}_1 \times \not{2}_1 \times \not{2}_1 \times \not{2}_1 \times 3} = \frac{2}{3}$$

Cancellation by 0 is not permitted when expressing a fraction in its simplest form, for this will make nonsense of the fraction. It could possibly be reasoned that 0 is an integer and to simplify $\frac{9}{15}$, merely multiply numerator and denominator by 0. Then $\frac{9}{15} \times \frac{0}{0} = \frac{9 \times 0}{15 \times 0} = \frac{0}{0}$, which is meaningless. Thus multiplication or division by zero will usually lead to a meaningless expression.

Addition and subtraction of fractions

To add or subtract two fractions the lowest common multiple of the denominators is found and each fraction expressed with the L.C.M. as a denominator. This is done by multiplying both numerator and denominator by the number required to give the L.C.M. as the denominator. Multiplying both numerator and denominator by the same number does not alter the value of the fraction at all. To simplify $\frac{1}{4}+\frac{1}{18}$ the procedure is as shown:

$4 = 2 \times 2$

$18 = 2 \times 3 \times 3$

giving the L.C.M. as $2 \times 2 \times 3 \times 3 = 36$ and

$$\frac{1}{4} \times \frac{9}{9} = \frac{9}{36}, \frac{1}{18} \times \frac{2}{2} = \frac{2}{36}$$

Thus

$$\frac{1}{4} + \frac{1}{18} = \frac{9}{36} + \frac{2}{36} = \frac{11}{36}$$

For subtraction

$$\frac{1}{4} - \frac{1}{18} = \frac{9}{36} - \frac{2}{36} = \frac{7}{36}$$

For mixtures of integers and fractions either improper fractions can be formed or the integers can be worked out independently of the fractions. To find the value of $3\frac{1}{4} + 5\frac{1}{2} - \frac{6}{11}$ using the improper fraction method, the terms are first turned into improper fractions.

$$3\frac{1}{4} = \frac{3 \times 4}{4} + \frac{1}{4} = \frac{12}{4} + \frac{1}{4} = \frac{13}{4}$$

$$5\frac{1}{2} = \frac{5 \times 2}{2} + \frac{1}{2} = \frac{10}{2} + \frac{1}{2} = \frac{11}{2}$$

Then

$$3\frac{1}{4} + 5\frac{1}{2} - \frac{6}{11} = \frac{13}{4} + \frac{11}{2} - \frac{6}{11}$$

The L.C.M. is thus 44. Expressing each term with the L.C.M. as its denominator gives:

$$\frac{13}{4} \times \frac{11}{11} + \frac{11}{2} \times \frac{22}{22} - \frac{6}{11} \times \frac{4}{4}$$

$$= \frac{143}{44} + \frac{242}{44} - \frac{24}{44}$$

or, since each term has the same denominator, this can be written as

$$\frac{143 + 242 - 24}{44} = \frac{361}{44}$$

Now 361 ÷ 44 is 8 remainder 9. Hence

$$\frac{361}{44} = 8\frac{9}{44}$$

The same problem could have been solved by treating the integers separately from the fractions.

$$3\frac{1}{4} + 5\frac{1}{2} - \frac{6}{11} = 3 + \frac{1}{4} + 5 + \frac{1}{2} - \frac{6}{11}$$

$$3 + 5 = 8$$

$$\frac{1}{4} + \frac{1}{2} - \frac{6}{11} = \frac{11 + 22 - 24}{44} = \frac{9}{44}$$

Then

$$3\frac{1}{4} + 5\frac{1}{2} - \frac{6}{11} = 8 + \frac{9}{44} = 8\frac{9}{44} \text{ as before.}$$

Multiplication of fractions

To multiply two or more fractions together the numerators are first multiplied to give a single number, and this becomes the numerator of the combined fraction. The denominators are then multiplied to give the denominator of the combined fraction. Then

$$\frac{3}{10} \times \frac{5}{7} = \frac{3 \times 5}{10 \times 7} = \frac{15}{70}$$

and, dividing both numerator and denominator by the H.C.F., 5 gives:

$$\frac{15}{70} = \frac{3}{14}$$

Questions involving the multiplication of fractions sometimes use the word 'of' instead of multiply. Then $\frac{3}{5}$ of 12 becomes

$$\frac{3}{5} \times \frac{12}{1} = \frac{3 \times 12}{5 \times 1} = \frac{36}{5} = 7\frac{1}{5}$$

Positive and negative integers can always be expressed as a fraction by dividing by 1, if this assists the calculation.
Mixed numbers must be expressed as improper fractions before multiplication.

Division of fractions

The procedure for division is to change the division sign into a multiplication sign and invert the second fraction. That is, the numerator is written for the denominator and vice versa. Then

$$\frac{3}{4} \div \frac{4}{5} = \frac{3}{4} \times \frac{5}{4} = \frac{15}{16}$$

Order of precedence

The order of performing operations is the same as for integers, namely **B**racket, **O**f, **D**ivision, **M**ultiplication, **A**ddition and **S**ubtraction: that is, as given by BODMAS. If any of the operations appears more than once, say there were two multiplications, then the order in which these two multiplications are done is immaterial, provided the overall order dictated by BODMAS is maintained. The operation 'Of' merely means change the word 'Of' to a multiplication sign and then continue with the order of precedence.

Worked problems on fractions

Problem 1. Add $\frac{7}{12}$ and $\frac{2}{5}$

The L.C.M. of 12 and 5 is 60, thus

$$\frac{7}{12}+\frac{2}{5}=\frac{7\times 5}{12\times 5}+\frac{2\times 12}{5\times 12}$$

$$=\frac{35}{60}+\frac{24}{60}=\frac{59}{60}$$

Problem 2. Evaluate $3\frac{3}{5}+2\frac{1}{10}$

Either

$$3\frac{3}{5}+2\frac{1}{10}=3+\frac{3}{5}+2+\frac{1}{10}$$

$$=5+\frac{3}{5}+\frac{1}{10}$$

Now the L.C.M. of 5 and 10 is 10

and

$$\frac{3}{5}+\frac{1}{10}=\frac{3\times 2+1}{10}=\frac{7}{10}$$

Thus

$$3\frac{3}{5}+2\frac{1}{10}=5+\frac{7}{10}\text{ or }5\frac{7}{10}$$

Or

$$3\frac{3}{5}+2\frac{1}{10}=\frac{3\times 5}{5}+\frac{3}{5}+\frac{2\times 10}{10}+\frac{1}{10}$$

$$=\frac{18}{5}+\frac{21}{10}$$

$$= \frac{18 \times 2 + 21}{10}$$

$$= \frac{57}{10} = 5\frac{7}{10}$$

Problem 3. Find the value of $\frac{5}{8} - \frac{3}{40}$

$$\frac{5}{8} - \frac{3}{40} = \frac{5 \times 5 - 3}{40} = \frac{25 - 3}{40}$$

$$= \frac{22}{40} = \frac{2 \times 11}{2 \times 20} = \frac{11}{20}$$

Problem 4. Evaluate $3\frac{4}{19} - 2\frac{7}{57}$

$$3\frac{4}{19} - 2\frac{7}{57} = (3 - 2) + \left(\frac{4}{19} - \frac{7}{57}\right)$$

$$= 1 + \frac{3 \times 4 - 7}{57} = 1 + \frac{5}{57}$$

$$= 1\frac{5}{57}$$

Problem 5. Find the value of $3\frac{3}{5} - 2\frac{5}{8}$

$$3\frac{3}{5} - 2\frac{5}{8} = (3 - 2) + \left(\frac{3}{5} - \frac{5}{8}\right)$$

$$= 1 + \frac{3 \times 8 - 5 \times 5}{40}$$

$$= 1 + \frac{(-1)}{40}$$

The $+\frac{(-1)}{40}$ gives an overall minus due to unlike signs and the problem becomes $1 - \frac{1}{40}$

$$= \frac{40}{40} - \frac{1}{40} = \frac{39}{40}$$

Problem 6. Simplify $2\frac{7}{16} - 3\frac{1}{5} + 1\frac{5}{8}$

$$2\frac{7}{16} - 3\frac{1}{5} + 1\frac{5}{8} = (2 - 3 + 1) + \left(\frac{7}{16} - \frac{1}{5} + \frac{5}{8}\right)$$

$$= 0 + \frac{7 \times 5 - 1 \times 16 + 5 \times 10}{80}$$

$$= \frac{35 - 16 + 50}{80} = \frac{69}{80}$$

Problem 7. Evaluate $\frac{3}{8} \times \frac{4}{15}$

Multiplying the numerators together and denominators together

$$\frac{3}{8} \times \frac{4}{15} = \frac{3 \times 4}{8 \times 15} = \frac{12}{120} = \frac{1}{10}$$

In this example cancelling can be done before multiplication. Thus

$$\frac{3}{8} \times \frac{4}{15} = \frac{3}{\cancel{8}_2} \times \frac{\cancel{4}^1}{15} = \frac{\cancel{3}^1}{2 \times \cancel{15}_5} = \frac{1}{10}$$

The cancelling was firstly by 4, that is, dividing both numerator and denominator by 4 and then by 3.

Problem 8. Simplify $3\frac{2}{5} \times 1\frac{3}{13} \times 2\frac{3}{8}$

The fractions must be expressed as improper fractions before multiplication can be done.

$$3\frac{2}{5} \times 1\frac{3}{13} \times 2\frac{3}{8} = \frac{17}{5} \times \frac{16}{13} \times \frac{19}{8}$$

since $3\frac{2}{5} = \frac{3 \times 5}{5} + \frac{2}{5} = \frac{15 + 2}{5} = \frac{17}{5}$

$$1\frac{3}{13} = \frac{1 \times 13}{13} + \frac{3}{13} = \frac{16}{13}$$

and $2\frac{3}{8} = \frac{2 \times 8}{8} + \frac{3}{8} = \frac{16 + 3}{8} = \frac{19}{8}$

$$\frac{17}{5} \times \frac{\cancel{16}^2}{13} \times \frac{19}{\cancel{8}_1} = \frac{17 \times 2 \times 19}{5 \times 13}$$

When all the numbers in either the numerator or the denominator are prime numbers, no further cancelling is possible. Thus

$$\frac{17 \times 2 \times 19}{5 \times 13} = \frac{646}{65} = 9\frac{61}{65}$$

Problem 9. Evaluate $\frac{3}{5} \div \frac{27}{25}$

The rule is to invert the dividing fraction and change the division to a mutliplication sign

Then $\frac{3}{5} \div \frac{27}{25} = \frac{\not{3}^{1}}{\not{5}_{1}} \times \frac{\not{25}^{5}}{\not{27}_{9}} = \frac{5}{9}$

Problem 10. Find the value of $5\frac{1}{6} \div \frac{62}{16}$

Division can only be performed by using improper fractions.

$$\text{Then } 5\frac{1}{6} \div \frac{62}{16} = \frac{31}{6} \div \frac{62}{16}$$

$$= \frac{\not{31}^{1}}{\not{6}_{3}} \times \frac{\not{16}^{\not{8}\,4}}{\not{62}_{\not{2}\,1}} = \frac{4}{3} = 1\frac{1}{3}$$

Problem 11. Simplify $\frac{3}{4} + \frac{11}{5} \times \frac{3}{8} - \frac{3}{10}$

Several operations are present therefore apply BODMAS to obtain the order of precedence.

$$\frac{3}{4} + \frac{11}{5} \times \frac{3}{8} - \frac{3}{10} = \frac{3}{4} + \frac{33}{40} - \frac{3}{10} \qquad \text{(M)}$$

$$= \frac{30 + 33 - 12}{40} \qquad \text{(A)}$$

$$= \frac{63 - 12}{40} = \frac{51}{40} = 1\frac{11}{40} \qquad \text{(S)}$$

Problem 12. Simplify $(\frac{5}{2} - \frac{1}{4}) \div \frac{3}{(\frac{4}{3} + \frac{2}{5})} \times \frac{7}{12}$

Applying BODMAS, the brackets must be worked out before doing the divisions and finally the multiplication. The brackets are sometimes omitted when writing fractions as the numerator or denominator of a term. If this is done then it must always be assumed that they exist. For example, the second term $\frac{3}{\left(\frac{4}{3} + \frac{2}{5}\right)}$ could also be written as $\frac{3}{\frac{4}{3} + \frac{2}{5}}$, but it must then be assumed that $\frac{4}{3} + \frac{2}{5}$ is a bracketed term.

$$\text{Then } \left(\frac{5}{2} - \frac{1}{4}\right) \div \frac{3}{\left(\frac{4}{3} + \frac{2}{5}\right)} \times \frac{7}{12} = \frac{9}{4} \div \frac{3}{\frac{26}{15}} \times \frac{7}{12}$$

$$= \frac{9}{4} \div \frac{45}{26} \times \frac{7}{12}$$

$$= \frac{\not{9}^{1}}{4} \times \frac{\not{26}^{13}}{\not{45}_{5}} \times \frac{7}{\not{12}_{6}}$$

$$= \frac{91}{120}$$

Further problems on fractions may be found in Section 1.8 (Problems 41–69), page 32.

1.5 Ratio and proportion

All fractions are ratios. Ratio can be defined as the number of times which one quantity is contained in another quantity **of the same kind.** Ratios are usually used to compare the relative size, quantity or cost of two things. For example, to find the ratio of 25 days to 1 year, expressing each of these quantities in the same units, gives

$$\frac{25 \text{ days}}{365 \text{ days}} = \frac{5}{73}$$

which can be written as 5:73. The : means 'is to' and the ratio of 25 days to 1 year is as 5 is to 73. Ratios can apply to more than two quantities. When mixing the materials for a concrete floor, the ratio of sand to gravel to cement is 2:5:1, that is 2 parts of sand to 5 parts of gravel to 1 part of cement. The volume of sand in 1 cubic metre of the mix is given by

$$\frac{2 \text{ parts of sand}}{(2+5+1) \text{ parts of the whole}} = \frac{2}{8} = \frac{1}{4} \text{ of the whole.}$$

Then since the whole is 1 cubic metre, the amount of sand is $\frac{1}{4}$ cubic metre.

A proportion is formed when the ratio of two quantities is equal to the ratio of a second pair of quantities. There are three ways of expressing a proportion, for example, since $\frac{10}{4} = \frac{5}{2}$

(i) 10 is to 4 as 5 is to 2
(ii) 10:4::5:2 (where :: means 'as')
and (iii) $\frac{10}{4} = \frac{5}{2}$

Worked problems on ratio and proportion

Problem 1. Divide £192 in the ratio of 9 to 7

There are a total of (9 + 7) = 16 parts.

One part will be $\frac{192}{16} = £12$

Nine parts will be £12 × 9 = **£108**

and seven parts will be £12 × 7 = **£84**

Having divided a quantity in a certain ratio, a check should be made to make sure that the parts add up to the original amount. In this case 108 + 84 = 192 showing that it is unlikely that an error has been made.

Problem 2. Divide 2 601 bricks in the ratio 3:5:9

There are a total of (3 + 5 + 9) = 17 parts

One part will be $\frac{2\ 601}{17} = 153$ bricks

Hence

3 parts will be $3 \times 153 = 459$ bricks
5 parts will be $5 \times 153 = 765$ bricks

and

9 parts will be $9 \times 153 = 1\ 377$ bricks

(Check: 459 + 765 + 1 377 = 2 601)

Then for 2 601 bricks **3:5:9::459:765:1 377.**

Problem 3. A piece of metal bar 120 cm long is cut in the ratio of 7:3. Find the length of the two pieces.

There are a total of (7 + 3) = 10 parts

Length of 1 part $= \frac{120}{10} = 12$ cm

Length of 7 parts $= 12 \times 7 =$ **84 cm**

and length of 3 parts $= 12 \times 3 =$ **36 cm**

(Check: 84 + 36 = 120)

Problem 4. Three men paint a house in 5 days. Find how long it will take two men to paint the house.

Three men paint a house in 5 days. Since 1 man will take three times as long, then 1 man will paint the house in $5 \times 3 = 15$ days. Two men will do the work in half the time it takes 1 man to do it, hence

2 men will paint the house in $\frac{15}{2}$ or $7\frac{1}{2}$ days.

Problem 5. Express 75p as a ratio of £2.50.

Since £2.50 = 250p, and as ratio problems can only be done when working in quantities of the same kind, the problem becomes $\frac{75}{250} = \frac{3}{10}$.

Thus 75p:**250p::3:10**

Problem 6. A bonus is paid to 4 men in the ratio 9:11:13:15. If the total bonus is £240, find how much each man receives.

There are a total of (9 + 11 + 13 + 15) = 48 parts to the bonus. Then one part will be $\frac{240}{48}$ = £5, and the required amounts are 9 × 5 = £45; 11 × 5 = £55; 13 × 5 = £65; and 15 × 5 = £75. Thus for £240 9:11:13:15::**£45:£55:£65:£75**

(**Check:** 45 + 55 + 65 + 75 = 240)

Further problems on ratio and proportion may be found in Section 1.8 (Problems 70–75), page 33.

1.6 Decimals

The decimal system of numbers is the one in everyday use based on the numbers 0 to 9. The number 23.61 is called a **decimal fraction** and means $(2 \times 10) + (3 \times 1) + (6 \times \frac{1}{10}) + (1 \times \frac{1}{100})$.

To add 17.4 and 5.8 becomes

$$\left[(1 \times 10) + (7 \times 1) + \left(4 \times \frac{1}{10}\right)\right] + \left[(5 \times 1) + \left(8 \times \frac{1}{10}\right)\right]$$

and collecting like terms this becomes

$$(1 \times 10) + (7 + 5) \times 1 + (4 + 8) \times \frac{1}{10}$$

$$\text{or } 10 + 12 + \frac{12}{10}$$

$$\text{i.e. } 10 + 12 + 1\frac{2}{10}$$

$$\text{or } 23\frac{2}{10} \text{ or } 23.2$$

The importance of the **decimal point** for the exact placing of numbers can be seen by adding these numbers as shown:

```
17.4
 5.8
----
23.2
----
```

The decimal point (a dot) is used to distinguish between multiplication by 1 (units), and multiplication by $\frac{1}{10}$ (tenths of a unit).

Addition

Adding two or more decimal fractions is a similar process to adding integers, the only difference being the positioning of the number before addition. Decimal fractions are arranged for addition in a column with each of the decimal points beneath one another. The numbers 23.4, 0.07 and 128.714 can be added as shown.

$$\begin{array}{lr} & 23.4 \\ & 0.07 \\ & 128.714 \\ \hline \text{Adding} & \underline{152.184} \end{array}$$

Subtraction

To take 73.71 from 186.317 the decimal points are arranged beneath one another and subtraction carried out, as for integers. Then

$$\begin{array}{lr} & 186.317 \\ & 73.71 \\ \hline \text{Subtracting} & \underline{112.607} \end{array}$$

Multiplication

Again, the same procedure is used as when multiplying integers but in this case care must be taken in placing the decimal point in the correct position in the answer. Two decimal fractions are multiplied together ignoring the decimal points and when an answer has been obtained, a count is taken of the total number of figures to the right of the decimal point in both the numbers being multiplied. These are added together and the answer will contain this number of figures to the right of the decimal point. For example, to multiply 36.81 by 2.4, ignore the decimal points and treat these numbers as integers.

$$\begin{array}{r} 3\,681 \\ 24 \\ \hline 14\,724 \\ 73\,62 \\ \hline \underline{88\,344} \end{array}$$

The numbers $36.\underline{81}$ and $2.\underline{4}$ contain 3 figures altogether on the right-hand side of their decimal points. Hence the answer must contain 3 figures altogether to the right of its decimal point. Then

$$36.\underline{81} \times 2.\underline{4} = 88.\underline{344}$$

Division

To divide one decimal fraction by another the denominator of the decimal fraction is made into an integer by multiplying the denominator by 10 or by a multiple of 10. The numerator must be multiplied by the same number. For example, $0.06 \div 0.12$ can be written as $\frac{0.06}{0.12}$. Now

$$\frac{0.06}{0.12} \times \frac{100}{100} = \frac{6}{12} = \frac{1}{2}$$

Since both numerator and denominator have been multiplied by the same number, the answer obtained by dividing 0.06 by 0.12 will be the same as dividing 6 by 12.

To convert a proper fraction to a decimal fraction, the numerator is divided by the denominator. Noughts can be added after the decimal point without affecting the value of the number. Thus $\frac{1}{2}$ becomes

$$2\,\overline{)\,1.000}\quad 0.5$$

that is $\frac{1}{2} = 0.5$

To convert a decimal fraction to a proper fraction, the part of the decimal fraction to the right of the decimal point is divided by 10 or by a multiple of 10. The number 23.64 has 0.64 to the right of the decimal point. This is made up of $6 \times \frac{1}{10} + 4 \times \frac{1}{100}$ i.e. $\frac{6}{10} + \frac{4}{100}$. Now

$$\frac{6}{10} + \frac{4}{100} = \frac{60+4}{100} = \frac{64}{100}$$

and by cancellation

$$\frac{64}{100} = \frac{16}{25}$$

Then

$$23.64 = 23\frac{16}{25}$$

Significant figures and decimal places

A number which can be expressed exactly as a decimal fraction is called a **terminating decimal.** In the previous example $23\frac{16}{25} = 23.64$ exactly and is a terminating decimal. Many proper fractions when expressed as decimal fractions do not terminate and these are called **non-terminating decimals.** To convert $\frac{7}{3}$ to a decimal fraction, 7 is divided by 3.

$$3\,\overline{)\,7.000\ 000\ 00}\quad 2.333\ 33\ldots$$

The result of this division would continue indefinitely and mathematically, the answer would be written $2.\dot{3}$, called 'two point three recurring'. The dot over the three indicates that 3 recurs or repeats indefinitely. Since many decimal fractions are non-terminating, errors are introduced in calculations when these are used, but the extent of the error can be stated.

If the number $2\frac{1}{3}$ is required to an accuracy of 4 figures only, it would be written as 2.333 and the words 'correct to 3 decimal places' would be added to the answer. These words indicate that 3 figures are written after the decimal point. To express the number $\frac{2}{7}$ correct to 4 decimal places, 2 would be divided by 7 giving 0.285 71 . . . and the answer would be terminated after the digit 7. Then $\frac{2}{7} = 0.285\ 7$ correct to 4 decimal places. An alternative way of expressing accuracy is to specify the number of figures to be written. If the accuracy of $\frac{7}{3}$ is required to 4 figures, then the words used are 'correct to 4 significant figures'. Thus $\frac{7}{3}$ = 2.333 correct to 4 significant figures.

When writing a decimal fraction to a certain degree of accuracy, the last digit in the answer is corrected depending on the value of the digit following it. If this digit is 0, 1, 2, 3 or 4, the last digit in the answer is left as it is. However, if it is 5, 6, 7, 8 or 9, the last digit in the answer is increased by one. Then an answer of 0.285 714 3 . . . would be written 0.286 correct to 3 decimal places, since the digit 7 following the 5 is in the group 5, 6, 7, 8 or 9. Similarly the answer would be 0.285 71 correct to 5 significant figures, since the digit 4 following the 1 is in the group 0, 1, 2, 3 or 4, indicating no increase in the last digit of the answer.

When determining the number of significant figures, the first digit on the left which signifies something starts the count. Noughts written before a number starts do not signify anything. Thus 0.003 174 = 0.003 17 correct to 3 significant figures, since the noughts in the $\frac{1}{10}$th and $\frac{1}{100}$th places do not signify anything. However, noughts within a number are significant and 20.007 17 = 20.007 2 correct to 6 significant figures, or 20.0 correct to 3 significant figures.

Worked problems on decimals

Problem 1. Add 3.27, 821.7, 9.042 7, 7 263 and 0.000 46

Each number is written so that the decimal points are in a column down the page and they are then added.

```
          3.27
        821.7
          9.042 7
      7 263.0
          0.000 46
Adding 8 097.013 16
```

Problem 2. Take 73.71 from 286.714

The numbers are written so that the decimal points are beneath one another and then subtracted. Care is taken to see what is being taken from what.

```
             286.714
              73.710
Subtracting  213.004
```

Problem 3. Evaluate 683 − 21.347

Noughts can be added after the decimal point as required: this does not alter the value of the number.

$$\begin{array}{lr} & 683.000 \\ & 21.347 \\ \hline \text{Subtracting} & 661.653 \\ \hline \end{array}$$

Problem 4. Subtract 0.005 from 0.000 513

As with integers, 0.000 513 − 0.005 = −1 (0.005 − 0.000 513). Thus subtracting the smaller number from the larger and multiplying by (−1) gives

$$\begin{array}{lr} & 0.005\ 000 \\ & 0.000\ 513 \\ \hline \text{Subtracting} & 0.004\ 487 \\ \hline \end{array}$$

Remembering to multiply by (−1) gives 0.000 513 − 0.005 = **−0.004 487**

Problem 5. Find the value of 0.006 × 26.53

Multiplying the numbers as if integers (6 × 2 653)

$$\begin{array}{lr} & 2\ 653 \\ & 6 \\ \hline \text{Multiplying} & 15\ 918 \\ \hline \end{array}$$

To determine the position of the decimal point in the answer, count the total number of figures to the right of the decimal point and add together. Then $0.\underline{006} \times 26.\underline{53}$ gives (3 + 2) = 5 figures to the right of the decimal point and so the answer must have 5 figures to the right of its decimal point. Thus 0.006 × 26.53 = **0.159 18.**

Problem 6. Evaluate 270 × 0.006 3

As for problem 5, multiply 270 by 63

$$\begin{array}{r} 270 \\ 63 \\ \hline 810 \\ 16\ 20 \\ \hline 17\ 010 \\ \hline \end{array}$$

$270 \times 0.\underline{006\ \ 3}$ gives (0 + 4) figures after the decimal point, hence 270 × 0.006 3 = **1.7010**

Problem 7. Divide 46.3 by 7 and give the answer correct to 4 decimal places.

Since 4 decimal place accuracy is required in the answer, an answer having 5 figures after the decimal point is necessary for the last figure correction.

The decimal point in the answer is written above the decimal point in the number being divided.

$$7\,\overline{)46.300\ 00}\quad = 6.614\ 28$$

Since the 5th figure after the decimal point is in the group 5, 6, 7, 8 or 9, the fourth figure after the decimal point is increased by one. Thus $43.6 \div 7 =$ **6.614 3** correct to 4 decimal places.

Problem 8. Divide 27.46 by 0.07 expressing the answer correct to 5 significant figures.

The answer must be worked to 6 figures to give 5-figure correction. Multiplying each quantity by 100 gives

$$\frac{27.46}{0.07} \times \frac{100}{100} = \frac{2\ 746}{7}$$

$$7\,\overline{)2\ 746.000}\quad = 392.285$$

Since the 6th figure in the answer is in the group 5, 6, 7, 8 or 9, the 5th figure is increased by one.
Hence $27.46 \div 0.07 =$ **392.29**

Problem 9. Evaluate $\dfrac{0.281}{0.03}$

Multiplying numerator and denominator by 100 gives $\dfrac{28.1}{3}$

$$3\,\overline{)28.100\ 0}\quad = 9.366\ 6\ldots$$

and 6 will recur in the answer as long as division is continued.

Thus $\dfrac{0.281}{0.03} = \mathbf{9.3\dot{6}}$

Problem 10. Convert 0.0176 to a proper fraction.

The number 0.0176 comprises

$$0 \times \frac{1}{10} + 1 \times \frac{1}{100} + 7 \times \frac{1}{1\ 000} + 6 \times \frac{1}{10\ 000}$$

Converting to $\dfrac{1}{10\ 000}$ ths gives

$$0 + \frac{100}{10\ 000} + \frac{70}{10\ 000} + \frac{6}{10\ 000}$$

or $0.017\ 6 = \dfrac{176}{10\ 000} = \mathbf{\dfrac{11}{625}}$ by cancellation.

As a general rule, one figure after the decimal point is expressed in $\frac{1}{10}$ths, 2 figures as $\frac{1}{100}$ths, 3 figures as $\frac{1}{1\,000}$ths and so on.

Problem 11. Convert 23.82 to an improper fraction.

23.82 = 23 + $\frac{82}{100}$ since there are 2 figures after the decimal point. Then, by cancellation,

$$23.82 = 23 + \frac{41}{50}$$

$$= \frac{23 \times 50 + 41}{50} = \frac{1\,150 + 41}{50}$$

$$= \frac{\mathbf{1\,191}}{\mathbf{50}}$$

Problem 12. Express $\frac{9}{32}$ as a decimal fraction.

$\frac{9}{32}$ means 9 divided by 32. Hence

```
     0.281 25
32 ) 9.000 00
     64
     260
     256
       40
       32
        8 0
        6 4
        1 60
        1 60
```

Thus $\frac{9}{32}$ = **0.281 25**

Further problems on decimals may be found in Section 1.8 (Problems 76–92), page 33.

1.7 Percentages

The word percentage implies a common standard of 100, and per cent, written %, means 'for every 100' or per 100. Thus in all percentages, a fraction is formed whose denominator is 100. Thus 62% becomes $\frac{62}{100}$ or $\frac{31}{50}$ and 35% becomes $\frac{35}{100}$ or $\frac{7}{20}$.

To convert a proper fraction into a percentage, the numerator and denominator must be multiplied by a suitable number to make the denominator 100. Thus $\frac{3}{5}$ becomes $\frac{3 \times 20}{5 \times 20} = \frac{60}{100}$ or 60%.

Decimal fractions can readily be converted to percentages. The decimal fraction 0.37 is

$$\frac{3}{10} + \frac{7}{100} = \frac{3 \times 10 + 7}{100} = \frac{37}{100}$$

That is, 0.37 is 37%. Thus to express a decimal fraction as a percentage the procedure is to multiply by 100 and this means moving the decimal point 2 places to the right. Then 0.125 is 12.5% and 0.017 is 1.7%. To express a percentage as a decimal fraction can also be readily done,

$$\text{for} \quad 6\% = \frac{6}{100} = 0.06 \text{ and}$$

$$27\% = \frac{27}{100} = 0.27$$

When converting proper fractions to a percentage, conversion to a decimal fraction, as a first step, is often the easiest method. For example, to convert $\frac{9}{17}$ to a percentage, 9 is divided by 17.

$$\begin{array}{r} 0.529\,41 \\ 17\overline{)9.000\,00} \\ 85 \\ \hline 50 \\ 34 \\ \hline 160 \\ 153 \\ \hline 70 \\ 68 \\ \hline 20 \\ 17 \\ \hline 3 \end{array}$$

That is $\frac{9}{17}$ is 0.529 4 correct to 4 decimal places. Hence $\frac{9}{17}$ = 52.94% correct to 2 decimal places.

Worked problems on percentages

Problem 1. Find 5% of 140 metres.

1% of 140 m is $\frac{1}{100}$th of 140 m or $\frac{140}{100}$ = 1.4 m

Therefore 5% of 140 m is **1.4 × 5 = 7 metres.**

Problem 2. Express $\frac{3}{11}$ as a percentage, correct to 3 significant figures.

Expressing $\frac{3}{11}$ as a decimal fraction

$$\begin{array}{r} 0.272\,7 \\ 11\overline{)3.000\,0} \end{array}$$

Correcting the 3rd figure gives $\frac{3}{11}$ = 0.273 correct to 3 significant figures.

Multiplying by 100 to turn the decimal fraction into a percentage gives $\frac{3}{11} = 27.3\%$ correct to 3 significant figures.

Problem 3. Express 18 seconds as a percentage of 1 minute.

One minute is 60 s and 18 s is $\frac{18}{60}$ of 60 s. $\frac{18}{60} = \frac{3}{10}$ by cancellation and $\frac{3}{10} = 0.3$. Multiplying by 100 to find the percentage gives 18 seconds as 30% of 1 minute.

Problem 4. An alloy consists of 65% by weight of zinc, the remainder being copper. Find the weight of copper in 300 grams of alloy.

One gram of alloy contains $\frac{65}{100}$ g of zinc and $\frac{100-65}{100} = \frac{35}{100}$ g of copper.

Then 300 grams of alloy contains $\frac{35}{100} \times 300$ g of copper

or **105 grams** of copper.

Further problems on percentages may be found in the following section (1.8) (Problems 93–100), page 33.

1.8 Further problems

Arithmetic operations

1. Add 2 961, 743, 82 615 and 27 [86 346]
2. Take 73 from 262 [189]
3. From 8 249 take 93 714 [−85 465]
4. Subtract 3 116 from 7 271 [4 155]
5. Add 264, −78, 3 142, 683 and −2 117 [1 894]
6. Find the value of −2 198 + 1 371 + 823 − 317 − 4 142 [−4 463]
7. Multiply 746 by 11 [8 206]
8. Multiply 236 by −13 [−3 068]
9. Multiply −4 312 by 9 [−38 808]
10. Find the value of −143 × −15 [2 145]
11. Evaluate 4 × −53 × 6 [−1 272]
12. Divide 736 by 16 [46]
13. Find the value of 29 040 divided by 120 [242]
14. Evaluate $\frac{4\,450}{13}$ [$342\frac{4}{13}$]
15. Evaluate 1 626/11 [$146\frac{9}{11}$]

H.C.F. and L.C.M.

16. Find the H.C.F. of the numbers 42, 154 and 231 [7]
17. Find the H.C.F. of the numbers 30, 105 and 165 [15]
18. Find the H.C.F. of 37, 74, 148 and 296 [37]

19. Find the H.C.F. of 196, 350, 770 and −910 [14]
20. Find the H.C.F. of 616, −308, 3 696 and −13 860 [308]
21. Find the L.C.M. of the numbers 36 and 210 [1 260]
22. Find the L.C.M. of 30, 105 and 165 [2 310]
23. Find the L.C.M. of 42, 154 and 231 [462]
24. Find the L.C.M. of 196, 350, 770 and −910 [700 700]
25. Find the L.C.M. of −308, 616, 3 696 and −13 860 [55 440]

Brackets and basic laws

26. Find the value of $53 + (61 + 70)$ [184]
27. Find the value of $53 - (61 + 70)$ [−78]
28. Evaluate $(53 - 61) + 70$ [62]
29. Evaluate $53 - (61 - 70)$ [62]
30. Evaluate $(53 - 61) - 70$ [−78]
31. Find the value of $(6 + 9)(15 - 7)$ [120]
32. Evaluate $(6 - 9)(15 + 7)$ [−66]
33. Simplify $(17 - 5) \div 4$ [3]
34. Simplify $(173 - 52) \div 11$ [11]
35. Evaluate $(8 + 7)\,4 \div 2\,(9 + 6)$ [450]
36. Find the value of $12 \times 60 \div 3\,(26 - 21)$ [1 200]
37. Find the value of $(3 \times 4)(12 \times 4) \div (16 + 3 - 7)$ [48]
38. Evaluate $(7 \times 6) \div 14 + 23 - 2 \times 39 \div 3$ [0]
39. Evaluate $17 \times 3 - (-16 + 19)(86 - 22) \div 8 + 7$ [34]
40. Find the value of
$4 \times 6 \div (3 - 7) + (12 - 3) \div 3 - (16 \div 2) \div (3 - 7)(7 - 5)$ [1]

Fractions

Simplify the following:

41. $\frac{1}{4} + \frac{1}{5}$ $[\frac{9}{20}]$
42. $\frac{3}{20} + \frac{7}{12}$ $[\frac{11}{15}]$
43. $2\frac{3}{4} + \frac{3}{10}$ $[3\frac{1}{20}]$
44. $3\frac{5}{8} + 2\frac{1}{9} + \frac{7}{16}$ $[6\frac{25}{144}]$
45. $\frac{1}{7} - \frac{1}{9}$ $[\frac{2}{63}]$
46. $\frac{4}{9} - \frac{4}{15}$ $[\frac{8}{45}]$
47. $6\frac{5}{7} - 4\frac{7}{8}$ $[1\frac{47}{56}]$
48. $\frac{1}{4} - \frac{1}{7} + \frac{3}{16}$ $[\frac{33}{112}]$
49. $\frac{7}{16} + \frac{3}{5} - \frac{19}{20}$ $[\frac{7}{80}]$
50. $3\frac{1}{5} - 2\frac{5}{6} + 6\frac{3}{15}$ $[6\frac{17}{30}]$
51. $26\frac{1}{3} + 33\frac{3}{5} - 61\frac{7}{15}$ $[-1\frac{8}{15}]$
52. $6\frac{3}{22} - 3\frac{1}{55} - 2\frac{4}{5}$ $[\frac{7}{22}]$
53. $\frac{1}{3} \times \frac{2}{5}$ $[\frac{2}{15}]$
54. $\frac{4}{5} \times \frac{3}{8} \times \frac{15}{16}$ $[\frac{9}{32}]$
55. $\frac{3}{13} \times \frac{39}{11} \times \frac{55}{19}$ $[2\frac{7}{19}]$
56. $3\frac{1}{4} \times \frac{19}{26}$ $[2\frac{3}{8}]$
57. $4\frac{2}{9} \times 1\frac{4}{5} \times 1\frac{10}{13}$ $[13\frac{29}{65}]$
58. $\frac{3}{5} \div \frac{4}{7}$ $[1\frac{1}{20}]$
59. $\dfrac{\frac{3}{4}}{\frac{5}{16}}$ $[2\frac{2}{5}]$
60. $1\frac{7}{8} \div 1\frac{15}{16}$ $[\frac{30}{31}]$
61. $\dfrac{1\frac{3}{4}}{2\frac{3}{5}} \div \frac{10}{13}$ $[\frac{7}{8}]$
62. $3\frac{1}{2} \div \frac{3}{7} \div 3\frac{1}{3}$ $[2\frac{9}{20}]$
63. $3\frac{1}{4} + \frac{3}{5} \times 1\frac{1}{6}$ $[3\frac{19}{20}]$
64. $2\frac{1}{2} - \frac{3}{4} \times 1\frac{1}{7}$ $[1\frac{9}{14}]$
65. $(3\frac{1}{4} + 2\frac{3}{8}) \times 4\frac{4}{7}$ $[25\frac{5}{7}]$
66. $2\frac{2}{15}\;(2\frac{1}{2} - 3\frac{1}{8})$ $[-1\frac{1}{3}]$
67. $\dfrac{2\frac{1}{3} + \frac{3}{4}}{4\frac{2}{5} - 3\frac{17}{30}}$ $[3\frac{7}{10}]$
68. $3 - \dfrac{(\frac{1}{6} \times \frac{3}{4})}{(\frac{1}{2} \div \frac{3}{5})} + \frac{7}{8}$ $[3\frac{29}{40}]$

69. $\frac{1}{3}$ of $\frac{1}{(\frac{1}{3} \div \frac{5}{8})} \times \frac{(4\frac{1}{4} - 3\frac{2}{5})}{(2\frac{1}{4} - 1\frac{3}{16})}$ [$\frac{1}{2}$]

Ratio and proportion

70. Divide 221 in the ratio 4:9. [68:153]
71. Express 16 months as a ratio of 2 years. [2:3]
72. A piece of wood 8 metres long is cut in the ratio 9:7. Determine the length of the two pieces of wood. [$4\frac{1}{2}$m; $3\frac{1}{2}$m]
73. A plot of land is divided between 3 men in the ratio 13:9:2. If the plot of land is 1 440 square metres, determine how much land each man gets. [780 m^2; 540 m^2; 120 m^2]
74. Two ploughmen can plough a field in 10 days. Find how long it would take 5 ploughmen to plough the same field working at the same rate. [4 days]
75. To supply a factory with energy costs £570 per week when it is open for 5 days a week, for 8 hours a day. Find how much it will cost to supply the factory with energy if the working day is extended to 9½ hours. [£676.87½p]

Decimals

76. Add together 137.917 7, 1 271.73, 4.027 and 0.014 5 [1 413. 689 2]
77. Evaluate 0.74 + 137 + 8.371 + 0.007 4 + 387.7 [533.818 4]
78. Find the value of 147 − 33.28 [113.72]
79. Evaluate 0.006 7 − 0.000 324 [0.006 376]
80. Find the value of 27.281 − 149.112 4 [−121.831 4]
81. Evaluate 427.8 × 0.006 4 [2.737 92]
82. Evaluate 0.018 2 × 0.674 [0.012 266 8]
83. Find the value of 75.814 × 27.63 [2094.740 82]
84. Find the value of 0.000 746 × 3 281.4 [2.447 924 4]
85. Evaluate 25.71 ÷ 0.6 [42.85]
86. Find the value of 67.418 ÷ 36.71 correct to 5 significant figures. [1.836 5]
87. Evaluate 0.255 ÷ 0.001 5 [170]
88. Determine the value of 237.41 ÷ 0.74 correct to 3 decimal places. [320.824]
89. Express $\frac{37}{63}$ as a decimal fraction correct to 4 significant figures. [0.587 3]
90. Express $3\frac{26}{35}$ as a decimal fraction correct to 4 decimal places. [3.742 9]
91. Express 0.343 75 as a proper fraction. [$\frac{11}{32}$]
92. Express 7.562 5 as an improper fraction. [$\frac{121}{16}$]

Percentages

93. Find 17% of 3 kilograms. [0.51 kg]
94. Determine 38% of £47. [£17.86]

95. Express $\frac{5}{13}$ as a percentage correct to three significant figures. [38.5%]

96. A pay rise of 4½% is paid to a man earning £63 a week. Determine how much he will then earn. [£65.83½p]

97. A deduction of 7½% is given for paying a bill of £26.80 promptly. Find out how much will have to be paid if the bill is settled promptly. [£24.79]

98. A pile of coal weighing 2 000 kilograms is divided between three people, so that one receives 15%, another 45% and the third person the remainder. Find the weight of coal each person receives. [300 kg; 900 kg; 800 kg]

99. A piece of elastic 0.7 metres long stretches by 73% when a weight is attached to it. Determine the new length. [1.211 m]

100. A company employing 800 people increases its staff to 830. Determine the percentage increase in staff. [3.75%]

Chapter 2

Indices and standard form

2.1 Definitions

The system of numbers which uses the numbers 0 to 9 is called the **decimal or denary system.** There are ten figures or digits in this system and it is said to have a **radix** ten. If a counting system used only the numbers 0, 1, 2, 3 and 4 it would have radix five. Comparatively recently, and associated with the growth of the computer industry, the **binary system** of numbers, having radix two and using the two digits 0 and 1, has come more into general use.

When a number is multiplied by itself, it is said to be **squared.** Thus for the number 3 the square is obtained by multiplying 3 by itself. Three squared is 3×3 and is written 3^2. This process is also called **raising to a power** and three squared is also called three raised to the power two. Numbers can be multiplied by themselves more than once and three raised to the power four is written 3^4 where

$$3^4 = 3 \times 3 \times 3 \times 3 = 81$$

and seven raised to the power five is 7^5 where

$$7^5 = 7 \times 7 \times 7 \times 7 \times 7 = 16\,807$$

When raising to a power of three, the number is said to be **cubed,** but special names are not normally used for powers greater than three. The number which has to be multiplied by itself is called the **base** and the number of times the number is repeated is called the **index**. Thus 5^6 has a base of 5 and an index of 6.

The **reciprocal** of an integer or decimal fraction is found by writing the number as the denominator of a fraction and giving the numerator the value 1. The reciprocal of 5 is 1/5 and the reciprocal of 3.128 is $\frac{1}{3.128}$. The reciprocal of a proper or improper fraction is found by interchanging the positions of the numbers in the numerator and denominator: hence the reciprocal of 3/4 is 4/3, and of 81/5 is 5/81.

The **square root** of a positive number can be found by looking for a base which when multiplied by itself gives the positive number. For the number 4, for example, a base must be selected which when multiplied by itself is equal to 4. Since $2 \times 2 = 4$, the base is 2. Thus the square root of 4 is 2 and is written $\sqrt{4} = 2$.

There is also another base of -2, since $(-2) \times (-2) = 4$: hence the square root of 4 is also -2. There will always be two square roots of a number and so $\sqrt{4} = \pm 2$, called 'plus or minus two'.

Since $3 \times 3 = (-3) \times (-3) = 9$, then
$\sqrt{9} = \pm 3$, that is $+3$ or -3.

2.2 Indices

To manipulate numbers which are raised to a power and to simplify calculations there are some fundamental rules or laws which can be applied. These are known as the **laws of indices** and those relating to arithmetic expressions are given below.

Law 1

When multiplying a number raised to a power by **the same number** raised to a power, the indices (plural of index) must be added. For example:

$5^3 \times 5^4 = (5 \times 5 \times 5) \times (5 \times 5 \times 5 \times 5) = 5^7$, and

$5^{(3+4)} = 5^7$, that is, $5^3 \times 5^4 = 5^{(3+4)} = 5^7$

Applying the same law, $17^2 \times 17^4 = 17^{2+4} = 17^6$.

When a number is written without an index, the index is 1 and so

$5 \times 5^3 = 5^1 \times 5^3 = 5^{1+3} = 5^4$, and

$13^2 \times 13^5 \times 13 = 13^{(2+5+1)} = 13^8$.

Law 2

When dividing a number raised to a power by the same number raised to a power, the indices must be subtracted. For example, by the cancellation method

$$\frac{5^6}{5^4} = \frac{5 \times 5 \times 5 \times 5 \times 5 \times 5}{5 \times 5 \times 5 \times 5} = 5^2$$

$5^{6-4} = 5^2$, that is $\dfrac{5^6}{5^4} = 5^{6-4} = 5^2$

Law 3

When a number raised to a power is raised to a further power, the indices must be multiplied. For example:

$(5^2)^3 = 5^2 \times 5^2 \times 5^2 = 5^{2+2+2} = 5^6$ by law 1. And by law 3

$(5^2)^3 = 5^{2\times 3} = 5^6$

Law 4

Any number raised to the power nought is equal to one. This rule can be verified as follows:

$\frac{5^3}{5^3} = 5^{3-3} = 5^0$ by law 2, but by cancellation

$\frac{5^3}{5^3} = \frac{5 \times 5 \times 5}{5 \times 5 \times 5} = 1$. Hence $\frac{5^3}{5^3} = 5^0 = 1$

Similarly it can be shown that any other number raised to the power nought is equal to 1.

Law 5

A number raised to a negative power is the reciprocal of that number raised to the same positive power. Since $\frac{5^2}{5^4} = 5^{2-4} = 5^{-2}$ by law 2, and

$\frac{5^2}{5^4} = \frac{5 \times 5}{5 \times 5 \times 5 \times 5} = \frac{1}{5 \times 5} = \frac{1}{5^2}$ then $5^{-2} = \frac{1}{5^2}$.

Similarly $13^{-3} = \frac{1}{13^3}$ and $7^{-1} = \frac{1}{7}$

Law 6

When a number is raised to a fractional power the denominator of the fraction is the root of the number and the numerator is the power.

Thus $8^{\frac{2}{3}} = \sqrt[3]{8^2} = 2^2 = 4$.

(Note that it does not matter whether the third root of 8 is found first, or whether 8 squared is found first – the same answer will result.)

Similarly, $9^{\frac{1}{2}} = \sqrt{9^1} = \pm 3$.

Thus when $\frac{1}{2}$ is the fractional power a square root is indicated and is written as $\sqrt{}$ (and not $\sqrt[2]{}$).

Worked problems on indices

Problem 1. Simplify $3^3 \times 3^5$ and express the answer in power form.

Applying law 1, when multiplying the same quantities the indices are added, which gives

$3^3 \times 3^5 = 3^{3+5} = 3^8$

Problem 2. Simplify $6^2 \times 6^9 \times 6$ and express the answer in power form.

Any base with no index indicates an index of 1.

By law 1, $6^2 \times 6^9 \times 6 = 6^{2+9+1} = \mathbf{6^{12}}$

Problem 3. Evaluate $2^3 \times 2^4 \times 2$

By law 1, $2^3 \times 2^4 \times 2 = 2^{3+4+1} = 2^8$

and $2^8 = 2 \times 2 \times 2 \ldots$ to 8 terms = **256**

Problem 4. Find the value of $\frac{3^5}{3^2}$

By law 2, $\frac{3^5}{3^2} = 3^{5-2} = 3^3$

and $3^3 = 3 \times 3 \times 3 = \mathbf{27}$

Problem 5. Simplify $\frac{13^{17}}{13^9}$ and express the answer in power form.

By law 2, $\frac{13^{17}}{13^9} = 13^{17-9} = \mathbf{13^8}$

Problem 6. Evaluate $3^3 \times 3^4 \div 3^6$

This can be written as $\frac{3^3 \times 3^4}{3^6}$

$= \frac{3^{3+4}}{3^6}$ by law 1

$= \frac{3^7}{3^6} = 3^{7-6}$ by law 2

$= \mathbf{3}$

Problem 7. Evaluate $(7^2 \times 7^4) \div (7 \times 7^3)$

Applying law 1 for multiplication and then writing in fraction form

$(7^2 \times 7^4) \div (7 \times 7^3) = \frac{7^{2+4}}{7^{1+3}} = \frac{7^6}{7^4}$

Applying law 2, $\frac{7^6}{7^4} = 7^{6-4} = 7^2 = \mathbf{49}$

Problem 8. Evaluate $(3^2)^2$

By law 3 when raising to a power the indices are multiplied, so

$(3^2)^2 = 3^{2 \times 2} = 3^4 = \mathbf{81}$

Problem 9. Evaluate $\dfrac{4^3 \times 4^{10}}{4^8 \times 4^7}$

By laws 1 and 2

$$\frac{4^3 \times 4^{10}}{4^8 \times 4^7} = \frac{4^{13}}{4^{15}} = 4^{13-15} = 4^{-2}$$

And by law 5

$$4^{-2} = \frac{1}{4^2} = \mathbf{\frac{1}{16}}$$

Problem 10. Evaluate $\dfrac{17^7 \times (17^2)^4}{(17^3)^5}$

By law 3, $\dfrac{17^7 \times (17^2)^4}{(17^3)^5} = \dfrac{17^7 \times 17^8}{17^{15}}$

By law 1 $= \dfrac{17^{15}}{17^{15}}$

By law 2 $= 17^{15-15} = 17^0$

and by law 4, 17^0 $= \mathbf{1}$

Problem 11. Find the value of $\dfrac{3^2 \times 2^5}{2^3 \times 3^3}$

The five laws of indices apply to the same base. As this problem can be written as

$$\frac{3^2}{3^3} \times \frac{2^5}{2^3}$$

the laws can be applied to each of these two bases independently. Then

$$\frac{3^2}{3^3} \times \frac{2^5}{2^3} = 3^{2-3} \times 2^{5-3}$$

$$= 3^{-1} \times 2^2 = \frac{2^2}{3} = \mathbf{\frac{4}{3}}$$

Problem 12. Evaluate (*a*) $4^{\frac{1}{2}}$ (*b*) $27^{\frac{2}{3}}$ (*c*) $16^{\frac{3}{4}}$

From law 6:

(*a*) $4^{\frac{1}{2}} = \sqrt{4} = \pm 2$

(*b*) $27^{\frac{2}{3}} = \sqrt[3]{27^2} = 3^2 = 9$

(*c*) $16^{\frac{3}{4}} = \sqrt[4]{16^3} = \pm 2^3 = \pm 8$

Problem 13. Evaluate (*a*) $16^{-\frac{1}{4}}$ (*b*) $81^{-0.25}$ (*c*) $\left(\frac{1}{2^2}\right)^{-1}$

From laws 5 and 6:

(*a*) $16^{-\frac{1}{4}} = \dfrac{1}{16^{\frac{1}{4}}} = \dfrac{1}{\sqrt[4]{16}} = \pm\dfrac{1}{2}$

(*b*) $81^{-0.25} = 81^{-\frac{1}{4}} = \dfrac{1}{81^{\frac{1}{4}}} = \dfrac{1}{\sqrt[4]{81}} = \pm\dfrac{1}{3}$

(*c*) $\left(\dfrac{1}{2^2}\right)^{-1} = \left(\dfrac{1}{4}\right)^{-1} = \left(\dfrac{4}{1}\right)^{+1} = 4$

Problem 14. Evaluate $\dfrac{(3^2)^{\frac{3}{2}} \times (8^{\frac{1}{3}})^2}{(4^3)^{\frac{1}{2}} \times 9^{-\frac{1}{2}} \times 3^3}$, taking positive square roots only.

Using laws 3, 5 and 6, $\dfrac{(3^2)^{\frac{3}{2}} \times (8^{\frac{1}{3}})^2}{(4^3)^{\frac{1}{2}} \times 9^{-\frac{1}{2}} \times 3^3} = \dfrac{3^{(2\times 3/2)} \times 8^{(\frac{1}{3}\times 2)}}{4^{(3\times\frac{1}{2})} \times \dfrac{1}{9^{\frac{1}{2}}} \times 3^3} = \dfrac{3^3 \times 8^{\frac{2}{3}}}{4^{\frac{3}{2}} \times \dfrac{1}{9^{\frac{1}{2}}} \times 3^3}$

$$= \frac{3^3 \times \sqrt[3]{8^2}}{\sqrt{4^3} \times \dfrac{1}{\sqrt{9}} \times 3^3} = \frac{2^2}{2^3 \times \dfrac{1}{3}} = \frac{1}{2 \times \dfrac{1}{3}}$$

$$= \frac{3}{2} = 1\tfrac{1}{2}$$

Problem 15. Simplify $\dfrac{9^4 \times 8^2 \times 7^3}{7^2 \times 8^4 \times 9^5}$

This expression can be re-written as

$$\frac{9^4}{9^5} \times \frac{8^2}{8^4} \times \frac{7^3}{7^2} = 9^{4-5} \times 8^{2-4} \times 7^{3-2}$$

$$= 9^{-1} \times 8^{-2} \times 7^1$$

$$= \frac{7}{8^2 \times 9} = \frac{7}{576}$$

Problem 16. Evaluate $\dfrac{2^2 \times 3^3 + 3^2 \times 2^3}{3^4 \times 2^5}$

(Remember that the fraction $\dfrac{4+7}{11}$ can be split into two fractions, $\dfrac{4}{11} + \dfrac{7}{11}$.)

Applying this principle to the problem

$$\frac{2^2 \times 3^3 + 3^2 \times 2^3}{3^4 \times 2^5} = \frac{2^2 \times 3^3}{3^4 \times 2^5} + \frac{3^2 \times 2^3}{3^4 \times 2^5}$$

$$= 2^{2-5} \times 3^{3-4} + 3^{2-4} \times 2^{3-5}$$

$$= 2^{-3} \times 3^{-1} + 3^{-2} \times 2^{-2}$$

$$= \frac{1}{2^3 \times 3} + \frac{1}{3^2 \times 2^2}$$

$$= \frac{1}{24} + \frac{1}{36} = \frac{3+2}{72}$$

$$= \frac{5}{72}$$

Problem 17. Simplify $\dfrac{5^{-6} \times 13^7}{7 \times 5^{-4} \times 13^3}$ and express the answer in index form.

Negative powers can either be treated directly by the laws of indices, in which case

$$\frac{5^{-6} \times 13^7}{7 \times 5^{-4} \times 13^3} = \frac{5^{-6}}{5^{-4}} \times \frac{13^7}{13^3} \times \frac{1}{7}$$

$$= \frac{5^{-6+4} \times 13^{7-3}}{7}$$

$$= \frac{5^{-2} \times 13^4}{7} = \frac{13^4}{5^2 \times 7}$$

or they can be converted into positive indices by law 5 and

$$\frac{5^{-6} \times 13^7}{7 \times 5^{-4} \times 13^3} = \frac{5^4 \times 13^7}{7 \times 5^6 \times 13^3}$$

by taking the reciprocal of the negative indices. Then as previously shown

$$\frac{5^4 \times 13^7}{7 \times 5^6 \times 13^3} = \frac{13^{7-3}}{5^{6-4} \times 7}$$

$$= \frac{13^4}{5^2 \times 7}$$

Problem 18. Evaluate $\dfrac{2^5 \times 3^4}{2^2 \times 3^3 + 2^4 \times 3^2}$

To simplify the calculations, each term may be divided by the highest common factor of the three terms. By doing this, each term is reduced to its smallest integer number without altering the value of the expression. The H.C.F. is $2^2 \times 3^2$ and gives:

$$\frac{2^5 \times 3^4}{2^2 \times 3^3 + 2^4 \times 3^2} = \frac{\dfrac{2^5 \times 3^4}{2^2 \times 3^2}}{\dfrac{2^2 \times 3^3}{2^2 \times 3^2} + \dfrac{2^4 \times 3^2}{2^2 \times 3^2}} = \frac{2^{5-2} \times 3^{4-2}}{2^{2-2} \times 3^{3-2} + 2^{4-2} \times 3^{2-2}}$$

$$= \frac{2^3 \times 3^2}{2^0 \times 3^1 + 2^2 \times 3^0} = \frac{8 \times 9}{1 \times 3 + 4 \times 1} = \frac{72}{7} = \mathbf{10\tfrac{2}{7}}$$

Problem 19. Find the value of $\dfrac{3^3 \times 4^4}{2^3 \times 9^2 + 4^2 \times 27}$

This expression can be written in terms of bases 2 and 3 only, since 4 can be written as 2^2, 9 as 3^2 and 27 as 3^3. Thus

$$\frac{3^3 \times 4^4}{2^3 \times 9^2 + 4^2 \times 27} = \frac{3^3 \times (2^2)^4}{2^3 \times (3^2)^2 + (2^2)^2 \times 3^3}$$

$$= \frac{3^3 \times 2^8}{2^3 \times 3^4 + 2^4 \times 3^3}$$

The H.C.F. of all the terms is $2^3 \times 3^3$ and dividing each term by this gives

$$\frac{\dfrac{3^3 \times 2^8}{2^3 \times 3^3}}{\dfrac{2^3 \times 3^4}{2^3 \times 3^3} + \dfrac{2^4 \times 3^3}{2^3 \times 3^3}} = \frac{2^5}{3+2} = \frac{32}{5} = 6\frac{2}{5}$$

Problem 20. Find the value of $\dfrac{\left(\dfrac{3}{8}\right)^5 - \left(\dfrac{16}{3}\right)^{-3}}{\left(\dfrac{4}{3}\right)^{-4}}$

When proper or improper fractions are raised to a power, both numerator and denominator are raised to the power shown. Thus $\left(\dfrac{3}{8}\right)^5$ means $\dfrac{3^5}{8^5}$. To change a negative index to a positive index, the position of the numbers in the numerator and denominator are interchanged. $\left(\dfrac{16}{3}\right)^{-3}$ becomes $\dfrac{3^3}{16^3}$ and

$\left(\frac{4}{3}\right)^{-4}$ becomes $\left(\frac{3}{4}\right)^{4}$ or $\frac{3^4}{4^4}$. The problem can be rewritten with positive indices as

$$\frac{\left(\frac{3}{8}\right)^5 - \left(\frac{3}{16}\right)^3}{\left(\frac{3}{4}\right)^4} = \frac{\frac{3^5}{8^5} - \frac{3^3}{16^3}}{\frac{3^4}{4^4}}$$

And, in terms of bases 2 and 3 only, as

$$\frac{\frac{3^5}{2^{15}} - \frac{3^3}{2^{12}}}{\frac{3^4}{2^8}}$$

The L.C.M. of 2^{15} and 2^{12} is 2^{15} and replacing the division by multiplication gives

$$\frac{3^5 - 2^3 \times 3^3}{3^4} \times \frac{2^8}{2^{15}}$$

and, by dividing each term in the first expression by 3^3, this becomes

$$\frac{3^2 - 2^3}{3} \times \frac{2^8}{2^{15}} = \frac{3^2 - 2^3}{3} \times \frac{1}{2^7}$$

$$= \frac{9 - 8}{3 \times 128} = \frac{1}{384}$$

Further problems on indices may be found in Section 2.4. (Problems 1–23), page 48.

2.3 Standard form

To multiply a decimal fraction by 10 the decimal point is moved one place to the right, by a 100 two places to the right and so on. To divide a decimal fraction by 10, the decimal point is moved one place to the left and to divide by 100, two places to the left. The value of a number is unaltered if the number is both multiplied and divided by the same number. For example, the number 3 is not altered if multiplied by 1 000 and divided by 1 000, for $3 \times 10^3 \div 10^3 = \frac{3 \times 10^3}{10^3} = 3$. When solving problems containing decimal or other fractions, the fractions can be expressed in decimal fraction form with one figure only in front of the decimal point by multiplying or dividing the number by 10 raised to some power. When this way of writing a number is used it is said to be written in **standard form**. Thus a number written in standard form is a number between 1 and $9.\dot{9}$ multiplied by 10 raised to a power.

To write 43.7 in standard form, for example, it is first divided by 10 by moving the decimal point one place to the left to give 4.37. But it must now be multiplied by 10 to retain the value of the original number. So, $43.7 = 4.37 \times 10$ when written in standard form.

Again, to write 0.043 7 in standard form, it is multiplied by 100 or 10^2 by moving the decimal point two places to the right and then divided by 100 (or multiplied by 10^{-2}) to retain its original value. Thus $0.043\,7 = 4.37 \times 10^{-2}$ when written in standard form.

Writing a number in standard form enables a quick check to be made on the approximate value of a calculation to make sure an error in the position of the decimal point has not occurred. Also a similar principle is used to denote the size of certain physical quantities. The SI system of units has adopted the metre as its basic unit of linear measure (length or distance). To measure the distance between two towns, thousands or tens of thousands of metres would be required, whereas the length of a small insect such as an ant would be expressed in thousandths of a metre. Since length and distance can vary so much, large distances are measured in kilometres or metres $\times$ 10^3. Very small distances or lengths are measured in millimetres or metres $\times$ 10^{-3}. Table 2.1 gives some of the powers of 10 used to express numbers as a reasonable size, together with the abbreviations used for these powers of 10 and the name given to them.

Table 2.1 Powers of ten in common use to keep numbers to a reasonable size, their names and abbreviations.

When multiplying a number by	The prefix used is	The abbreviation used is
10^6	mega	M
10^3	kilo	k
10^{-1}	deci	d
10^{-2}	centi	c
10^{-3}	milli	m
10^{-6}	micro	μ
10^{-12}	pico	p

To measure the power output from a large modern alternator in a power station, megawatts (MW) are used, but the power to drive a small transistor radio would be measured in milliwatts (mW). The distance between London and Birmingham would be stated in kilometres (km) but the distance between the ends of a pencil would be measured in centimetres (cm). These units are selected to keep numbers to a reasonable size. Other units used in this text such as velocity, whose SI unit is metres per second, will be written as m s^{-1} and acceleration, having an SI unit of metres per second squared, will be written as m s^{-2}.

When a number is written in standard form, the number is called the **mantissa** and the factor by which it is multiplied the **exponent.** Thus 4.3×10^5 has a mantissa of 4.3 and an exponent of 10^5. Addition and subtraction of numbers in standard form can be achieved by adding the mantissae provided the exponent is the same for each of the numbers being added. For example:

$4 \times 10^2 + 5.6 \times 10^2 = 9.6 \times 10^2$

This can be verified by writing the numbers as integers, for

$4 \times 10^2 + 5.6 \times 10^2 = 400 + 560 = 960$

also $9.6 \times 10^2 = 960$

hence $4 \times 10^2 + 5.6 \times 10^2 = 9.6 \times 10^2$

When the exponents are not the same it is usually better to write the numbers in decimal fraction form before adding or subtracting.

The laws of indices are used when multiplying or dividing numbers given in standard form. For example:

$(3 \times 10^3) \times (5 \times 10^2) = (3 \times 5) \times (10^{3+2}) = 15 \times 10^5 = 1.5 \times 10^6$

Similarly, $\dfrac{8 \times 10^5}{2 \times 10^3} = (\dfrac{8}{2}) \times (10^{5-3}) = 4 \times 10^2$

Worked problems on standard form

Problem 1. Change the following numbers to standard form:
(*a*) 7·3.82 (*b*) 526 (*c*) 83 470

(*a*) To express 73.82 in standard form (that is, with one figure to the left of the decimal point), it is divided by 10. Then 73.82 ÷ 10 = 7.382. To compensate for dividing by 10 and to give the number its original value it must now be multiplied by 10. Thus
$73.82 = \mathbf{7.382 \times 10}$

(*b*) To obtain standard form, 526 must be divided by 100 and multiplied by 100 to retain its original value. Then

$526 = \dfrac{526}{100} \times 100 = \mathbf{5.26 \times 10^2}$

(*c*) 83 470 has to be divided by 10^4 to obtain a mantissa having only one figure to the left of the decimal point and multiplied by 10^4 to retain its original value. Thus

$83\,470 = \dfrac{83\,470}{10^4} \times 10^4 = \mathbf{8.347 \times 10^4}$

Problem 2. Express the following numbers in standard form:
(*a*) 0.3716 (*b*) 0.002 (*c*) 0.000 071 3

(*a*) To express 0.371 6 in standard form, 0.371 6 is multiplied by 10 to give 3.716 but also divided by 10 to retain its original value. Thus

$0.371\,6 = \dfrac{3.716}{10} = \mathbf{3.716 \times 10^{-1}}$

since $\frac{1}{10} = 10^{-1}$ by law 5 of indices.

(*b*) To express 0.002 in standard form, multiply by 10^3 to obtain the mantissa, divide by 10^3 (or multiply by 10^{-3}) to retain the original value. Thus

$$0.002 = \frac{0.002 \times 10^3}{10^3} = \frac{2.0}{10^3} = \mathbf{2.0 \times 10^{-3}}$$

(*c*) 0.000 071 3 must be multiplied by 10^5 and divided by 10^5. Thus

$$0.000\,071\,3 = \frac{0.000\,071\,3 \times 10^5}{10^5}$$

$$= \frac{7.13}{10^5} = \mathbf{7.13 \times 10^{-5}}$$

Problem 3. Write the following numbers, which are in standard form, as integers or decimal fractions:
(*a*) $1.371\,2 \times 10^2$ (*b*) 6.67×10^{-3} (*c*) 9.871×10^4 (*d*) 5.57×10^0

(*a*) $1.371\,2 \times 10^2 = 1.371\,2 \times 100 = \mathbf{137.12}$

(*b*) $6.67 \times 10^{-3} = \frac{6.67}{10^3} = \frac{6.67}{1\,000} = \mathbf{0.006\,67}$

(*c*) $9.871 \times 10^4 = 9.871 \times 10\,000 = \mathbf{98\,710}$

(*d*) $5.57 \times 10^0 = 5.57 \times 1 = \mathbf{5.57}$ (law 4 of indices states that any number raised to a power of 0 is 1).

Problem 4. Rewrite the following statements without using powers of ten:
(*a*) the speed of light is 2.998×10^8 metres per second;
(*b*) the Crab Nebula is 6×10^3 light years from earth; and
(*c*) the reciprocal of 20.116 8 is 4.971×10^{-2}

(*a*) The exponent 10^8 indicates moving the decimal place eight places to the right, thus $2.998 \times 10^8 = 299\,800\,000$ metres per second, and inserting this value in the statement gives:
the speed of light is 299 800 000 metres per second or 299 800 km per second

(*b*) $6 \times 10^3 = 6\,000$, the 10^3 indicating moving the decimal point three places to the right. Thus
the Crab Nebula is 6 000 light years from earth

(*c*) Since the exponent 10^{-2} means divide by 100 or move the decimal point two places to the left, the statement becomes:
the reciprocal of 20.116 8 is 0.049 71.

Problem 5. Express the following in standard form, correct to 4 significant figures:
(*a*) $\frac{9}{57}$ (*b*) $7\frac{7}{32}$ (*c*) $76\frac{3}{4}$

(*a*) Since $\frac{9}{57} = 0.157\,89\ldots$, this can be written in standard form as

$1.578\,9\ldots \times 10^{-1}$ and correcting to 4 significant figures gives:
$\frac{9}{57} = \mathbf{1.579 \times 10^{-1}}$ correct to 4 significant figures.

(*b*) $\frac{7}{32} = 0.218\,75$, then $7\frac{7}{32} = 7.218\,75$ or 7.219 correct to 4 significant figures. By the definition of standard form, this should be written as a mantissa multiplied by an exponent, that is 7.219×10^0, but the 10^0 which is equal to 1 is usually omitted. Thus
$7\frac{7}{32} = \mathbf{7.219}$ correct to 4 significant figures.

(*c*) $76\frac{3}{4} = 76.75$ and writing this in standard form gives:
$76\frac{3}{4} = \mathbf{7.675 \times 10}$

Problem 6. Express the following as proper or improper fractions:
(*a*) 7.5×10^{-1} (*b*) 5.625×10^{-2} (*c*) 3.375×10^2

(*a*) $7.5 \times 10^{-1} = 0.75 = \frac{75}{100}$. Dividing numerator and denominator by 25 gives:

$$7.5 \times 10^{-1} = \frac{3}{4}$$

(*b*) $5.625 \times 10^{-2} = 0.056\,25$ or $\frac{5\,625}{100\,000}$. By cancellation

$$5.625 \times 10^{-2} = \mathbf{\frac{9}{160}}$$

(*c*) $3.375 \times 10^2 = 337.5$ or $337\frac{1}{2}$. Expressing this as an improper fraction gives:

$$3.375 \times 10^2 = \mathbf{\frac{675}{2}}$$

Problem 7. Evaluate:
(*a*) $6.71 \times 10^4 + 7.76 \times 10^4$, (*b*) $7.32 \times 10^{-3} - 6.77 \times 10^{-3}$;
(*c*) $5.47 \times 10^{-2} - 6.16 \times 10^{-2}$; and (*d*) $3.67 \times 10^2 + 4.21 \times 10^3$; and express answers in standard form.

(*a*) Since the exponents are the same

$$6.71 \times 10^4 + 7.76 \times 10^4 = (6.71 + 7.76)10^4 = \mathbf{1.447 \times 10^5}$$

(*b*)

$$7.32 \times 10^{-3} - 6.77 \times 10^{-3} = (7.32 - 6.77)10^{-3} = 0.55 \times 10^{-3}$$

and expressing the answer in standard form

$$0.55 \times 10^{-3} = \mathbf{5.5 \times 10^{-4}}$$

(*c*)

$$5.47 \times 10^{-2} - 6.16 \times 10^{-2} = (5.47 - 6.16)10^{-2} = -0.69 \times 10^{-2}$$

and written in standard form this becomes $\mathbf{-6.9 \times 10^{-3}}$

(*d*) To evaluate $3.67 \times 10^2 + 4.21 \times 10^3$, since the exponents are different, straight addition of the mantissae cannot be done. However, 3.67×10^2 is the same value as 0.367×10^3 and having made the exponents the same, the mantissae can now be added. Thus,

$0.367 \times 10^3 + 4.21 \times 10^3 = 4.577 \times 10^3$

An alternative approach would be to write the expression in non-standard form, when

$$3.67 \times 10^2 + 4.21 \times 10^3 = 367 + 421\,0$$
$$= 457\,7$$
$$= \mathbf{4.577 \times 10^3} \text{ in standard form.}$$

Problem 8. Evaluate (*a*) $(5.75 \times 10^3) \times (6 \times 10^2)$

(*b*) $\dfrac{2.25 \times 10^5}{9 \times 10^2}$, expressing answers in standard form.

(*a*) $(5.75 \times 10^3) \times (6 \times 10^2) = (5.75 \times 6) \times (10^{3+2})$

$$= 34.5 \times 10^5 = \mathbf{3.45 \times 10^6}$$

(*b*) $\dfrac{2.25 \times 10^5}{9 \times 10^2} = (\dfrac{2.25}{9}) \times (10^{5-2}) = 0.25 \times 10^3 = \mathbf{2.5 \times 10^2}$

Further problems on standard form may be found in the following Section 2.4 (Problems 24–37), page 50.

2.4 Further problems

Indices

1. Simplify and express the answer in index form
   (*a*) $5^3 \times 5^2$ (*b*) 13×13^4 (*c*) $2^2 \times 2^3 \times 2^4$ (*d*) $7^2 \times 7$
   (*e*) $8^5 \times 8^2 \times 8 \times 8^4$
   (*a*) [5^5] (*b*) [13^5] (*c*) [2^9] (*d*) [7^3] (*e*) [8^{12}]
2. Find the value of
   (*a*) $6^2 \times 6^3$ (*b*) $2^2 \times 2^7 \times 2^3$ (*c*) $9^3 \times 9^3$
   (*a*) [7 776] (*b*) [4 096] (*c*) [531 441]
3. Simplify and express the answer in index form
   (*a*) $\dfrac{5^8}{5^3}$ (*b*) $\dfrac{9^{10}}{9^7}$ (*c*) $\dfrac{8^6}{8}$ (*d*) $13^6 \div 13^2$ (*e*) $17^7 \div 17^6$
   (*a*) [5^5] (*b*) [9^3] (*c*) [8^5] (*d*) [13^4] (*c*) [17 or 17^1]
4. Find the value of
   (*a*) $13^{15} \div 13^{13}$ (*b*) $\dfrac{2^{10}}{2^6}$ (*c*) $\dfrac{5^6}{5^3}$
   (*a*) [169] (*b*) [16] (*c*) [125]
5. Simplify and express the answer in index form
   (*a*) $(2^3)^2$ (*b*) $(8^4)^3$ (*c*) $(8^3)^4$ (*d*) $(31^6)^3$ (*e*) $(14^7)^5$
   (*a*) [2^6] (*b*) [8^{12}] (*c*) [8^{12}] (*d*) [31^{18}] (*e*) [14^{35}]

6. Find the value of
$(a)\ (2^2)^3 \quad (b)\ (3^2)^2 \quad (c)\ (3^1)^4$
(a) [64] (b) [81] (c) [81]

7. Evaluate (a) $25^{\frac{1}{2}}$ (b) $8^{\frac{-1}{3}}$ (c) $(\frac{4}{9})^{\frac{1}{2}}$ $\left[(a)\ \pm 5\quad (b)\ \frac{1}{2}\quad (c)\ \pm\frac{2}{3}\right]$

8. Evaluate (a) $(\frac{1}{3^2})^{-1}$ (b) $81^{\frac{3}{4}}$ (c) $32^{-0.4}$ $\left[(a)\ 9\quad (b)\ 27\quad (c)\ \frac{1}{4}\right]$

9. Evaluate
(a) $\dfrac{9^2 \times 9 \times 9^4}{9 \times 9^5}$ (b) $(13^3 \times 13^5) \div (13^7 \times 13)$ (c) $\dfrac{2^3 \times 2^{12}}{2^{17}}$
(a) [9] (b) [1] (c) [$\frac{1}{4}$]

10. Simplify and express in index form
(a) $(3^2 \times (3^2)^3) \div (3 \times 3^4)$ (b) $(17^2)^5 \times (17^5)^2$ (c) $\dfrac{(6^2)^7 \times (6^3)^6}{(6^8)^4}$
(a) [3^3] (b) [17^{20}] (c) [1^1 or 1]

11. Simplify and express in index form
(a) $\dfrac{4^3 \times 5^2}{4^2 \times 5^3}$ (b) $\dfrac{3^7 \times 8^4 \times 13^6}{8^5 \times 13^7 \times 3^9}$
(a) [$\frac{4}{5}$] (b) $\left[\dfrac{1}{3^2 \times 8 \times 13}\right]$

12. Evaluate (a) $\dfrac{3 \times 8^2 + 5 \times 8^4}{4 \times 8^3}$ (b) $\dfrac{4^6 \times 6^4 + 6^7 \times 4^3}{4^5 \times 6^6}$ $\left[(a)\ 10\frac{3}{32}\quad (b)\ \frac{35}{72}\right]$

13. Evaluate $\dfrac{8^{\frac{2}{3}} \times 16^{-0.25}}{2^3 \times (\frac{1}{4})^{-1}}$ $\left[\dfrac{1}{16}\right]$

14. Evaluate $\dfrac{27^{\frac{2}{3}} \times 4^{-\frac{1}{2}}}{2^{-5} \times (\frac{1}{3})^{-2}}$, taking positive square roots only [16]

15. Find the value of
(a) $\dfrac{2^{-4} \times 3 \times 4^3}{3^{-3} \times 4^2 \times 2^{-6}}$ (b) $\dfrac{3 \times 7^8 \times 9^{-6} \times 7^{-6}}{5 \times 9^{-5} \times 7}$ (c) $\dfrac{6^{-3} \times 8^3}{7 \times 6^{-4} \times 8^4}$

$\left[(a)\ 1\,296\quad (b)\ \frac{7}{15}\quad (c)\ \frac{3}{28}\right]$

16. Find the value of the following expressions in decimal fraction form correct to 4 significant figures.
(a) $\dfrac{4^2 \times 5^3 + 5^2 \times 4^3}{4^4 \times 5^4}$ (b) $\dfrac{3 \times 6^2 + 6 \times 8^3}{3^2 \times 8^2}$ (c) $\dfrac{4 \times 7^{-4} + 7^{-3}}{6 \times 7^{-2}}$

[(a) 0.022 50 (b) 5.521 (c) 0.037 41]

17. Find the value of
(a) $\dfrac{4^6 \times 5^7}{5^6 \times 4^7 + 4^5 \times 5^8}$ (b) $\dfrac{3 \times 4^3}{4 \times 5^2 + 3^2 \times 4^2}$ (c) $\dfrac{2^3 \times 3^2 \times 4}{2^2 \times 3^3 + 3 \times 4^2}$

$$\left[(a)\frac{20}{41}\quad(b)\frac{48}{61}\quad(c)\frac{24}{13}\text{ or }1\frac{11}{13}\right]$$

18. Evaluate (*a*) $\dfrac{5^3 \times 7^4}{25^2 \times 3 \times 49^2}$ (*b*) $\dfrac{6^3 \times 4^2 \times 13}{2^4 \times 3^3 \times 13^2 + 2^3 \times 3^2 \times 13}$

$$\left[(a)\frac{1}{15}\quad(b)\frac{48}{79}\right]$$

Simplify the following and express in index form with positive indices.

19. $\dfrac{3^{-4} \times 2^3 \times 5^{-4}}{2^{-2} \times 3^{-3} \times 5}$ $\left[\frac{1}{3}(\frac{2}{5})^5\right]$

20. $\dfrac{2^4 \times 3^{-3} \times 4^2 + 2^{-2} \times 3^2 \times 4^{-3}}{9^2 \times 8^3}$ $\left[\dfrac{2^{16} + 3^5}{2^{17} \times 3^7}\right]$

21. $\dfrac{(2^3)^4 \times (3^2)^2}{16^2 \times 9^3}$ $\left[\frac{16}{9}\text{ or }1\frac{7}{9}\right]$

22. $\dfrac{3^{-2} \times 5^4 \times 6^{-7}}{(\frac{2}{5})^3 \times (\frac{5}{3})^{-2}}$ $\left[\dfrac{5^9}{2^{10} \times 3^{11}}\right]$

23. $\dfrac{(\frac{7}{8})^{-2} - (\frac{1}{4})^4}{(\frac{1}{16})^2 \times (\frac{1}{32})^{-3}}$ $\left[\dfrac{2^{14} - 7^2}{2^{15} \times 7^2}\right]$

Standard form

In problems 24 to 28 express the numbers in standard form:

24. (*a*) 47.44 (*b*) 83.6 (*c*) 91.274 (*d*) 387.7 (*e*) 671.772
(*a*) [4.744×10] (*b*) [8.36×10] (*c*) [$9.127\,4 \times 10$]
(*d*) [3.877×10^2] (*e*) [$6.717\,72 \times 10^2$]

25. (*a*) 563 (*b*) 7 210 (*c*) 630 000 000 (*d*) 76 271.85
(*a*) [5.63×10^2] (*b*) [7.21×10^3] (*c*) [6.3×10^8]
(*d*) [$7.627\,185 \times 10^4$]

26. (*a*) 0.375 (*b*) 0.14 (*c*) 0.6 (*d*) 0.002 6 (*e*) 0.003 02
(*a*) [3.75×10^{-1}] (*b*) [1.4×10^{-1}] (*c*) [6×10^{-1}]
(*d*) [2.6×10^{-3}] (*e*) [3.02×10^{-3}]

27. (*a*) 0.000 001 7 (*b*) 0.000 101 5 (*c*) 0.100 02 (*d*) 0.070 73
(*a*) [1.7×10^{-6}] (*b*) [1.015×10^{-4}] (*c*) [$1.000\,2 \times 10^{-1}$]
(*d*) [7.073×10^{-2}]

28. (*a*) $63\frac{7}{8}$ (*b*) $\frac{3}{20}$ (*c*) $468\frac{4}{5}$ (*d*) $\frac{1}{500}$
(*a*) [$6.387\,5 \times 10$] (*b*) [1.5×10^{-1}] (*c*) [4.688×10^2]
(*d*) [2×10^{-3}]

In problems 29–31 change the numbers from standard form to integers or decimal fractions:

29. (*a*) 3.72×10^2 (*b*) $6.217\,4 \times 10^{-2}$ (*c*) $1.100\,4 \times 10^3$

(*d*) 3.27×10^4 (*e*) 8.27×10^{-1}
(*a*) [372] (*b*) [0.062 174] (*c*) [1 100.4] (*d*) [32 700]
(*e*) [0.827]

30. (*a*) 5.21×10^0 (*b*) 3×10^{-6} (*c*) $1.477\,1 \times 10^{-3}$ (*d*) 5.87×10
(*a*) [5.21] (*b*) [0.000 003] (*c*) [0.001 477 1] (*d*) [58.7]

31. (*a*) 7.176×10^6 (*b*) 9.98×10^{-4} (*c*) 4×10^{-5} (*d*) $4.000\,1 \times 10^5$
(*a*) [717 600 0] (*b*) [0.000 998] (*c*) [0.000 04] (*d*) [400 010]

32. Change the following numbers from standard form to proper or improper fractions:
(*a*) 9.375×10^{-2} (*b*) $1.873\,5 \times 10^2$ (*c*) 5.625×10^{-1}
(*d*) 3.2475×10^2
(*a*) [$\frac{3}{32}$] (*b*) [$\frac{3\,747}{20}$] (*c*) [$\frac{9}{16}$] (*d*) [$\frac{1\,299}{4}$]

33. Evaluate and express the answer in standard form:
(*a*) $3.774 \times 10^{-2} + 7.28 \times 10^{-2}$ (*b*) $6.3 \times 10^3 + 5.381 \times 10^3$
(*c*) $1.476 \times 10^{-6} - 1.471 \times 10^{-6}$ (*d*) $3.576 \times 10^4 - 4.211 \times 10^4$
(*a*) [$1.105\,4 \times 10^{-1}$] (*b*) [$1.168\,1 \times 10^4$] (*c*) [5×10^{-9}]
(*d*) [-6.35×10^3]

34. Find the value of the following, giving the answer in standard form:
(*a*) $1.874 \times 10^{-2} + 2.227 \times 10^{-3}$ (*b*) $5.27 \times 10^{-10} + 8.371\,42 \times 10^{-10}$
(*c*) $3.877\,1 \times 10^{-4} - 7.287\,3 \times 10^{-4}$ (*d*) $9.71 \times 10^2 - 9.998 \times 10^3$
(*a*) [$2.096\,7 \times 10^{-2}$] (*b*) [$1.364\,142 \times 10^{-9}$] (*c*) [$-3.410\,2 \times 10^{-4}$]
(*d*) [-9.027×10^3]

35. Rewrite the following statements without using powers of 10:
(*a*) the freezing temperature of copper is $1.357\,6 \times 10^3$ Kelvin
[1 357.6 K]
(*b*) one kilowatt hour has the same energy as 3.6×10^6 joules
[3 600 000 joules]
(*c*) the reciprocal of 1.609×10^3 is 6.214×10^{-4} [1609; 0.000 621 4]
(*d*) the volume of one fluid ounce is 2.841×10^{-5} cubic metres
[0.000 028 41 m^3]
(*e*) the square root of 4×10^{-4} is $\pm 2 \times 10^{-2}$ [0.000 4, ± 0.02]

In problems 36 and 37, evaluate, giving the answers in standard form:

36. (*a*) $4.75 \times 10^2)(8 \times 10^3)$ (*b*) $3 \times (4.4 \times 10^3)$
[(*a*) 3.8×10^6 (*b*) 1.32×10^4]

37. (*a*) $\dfrac{8 \times 10^{-3}}{5 \times 10^{-5}}$ (*b*) $\dfrac{(4.5 \times 10^3)(3 \times 10^{-2})}{2.7 \times 10^4}$ [(*a*) 1.6×10^2 (*b*) 5×10^{-3}]

Chapter 3

Calculations

3.1 Errors, accuracy and approximate values

In order to find out how long it takes a person to walk one kilometre at constant speed, several methods may be adopted. A reasonable walking speed is 5 or 6 kilometres per hour: using this data, a result of 10 to 12 minutes can be calculated. So in integer values, it will take, say, 11 minutes to walk one kilometre. For a more accurate result, a distance of one kilometre can be marked, using the mileage recording meter on a car as a measuring device. A person can be timed by a wrist watch walking this distance and the result obtained may be accurate correct to one decimal place. By using a stop watch and carefully measuring the distance using surveying equipment a result which is correct to two decimal places may be achieved. Ultimately, using a device calibrated at the National Physics Laboratory and an atomic clock which only varies from the correct time by about 5 seconds in 700 years, a result which is correct to several decimal places can be achieved.

In all these cases the result is correct to a stated accuracy only and is not the exact time taken by a person to walk 1 kilometre. In all problems in which the measurement of distance, time, weight, or other quantities occurs, an exact answer cannot be given, only an answer which is correct to a stated degree of accuracy. To take account of this an **error due to measurement** is said to exist.

When numbers are expressed as decimal fractions, an error frequently exists. The proper fraction $\frac{9}{17}$ is equal to 0.529 411 7 . . . but will be expressed as, say, 0.529 correct to 3 significant figures or 0.529 4 correct to 4 decimal places. The error introduced in this case is called a **rounding off error**.

Errors can occur in calculations. When evaluating an expression errors can be made when performing arithmetic operations such as multiplication, division, addition and subtraction. For example, if when multiplying 27.63 by 62.7 the answer obtained is 1 731.401 then an error would have been made since $27.63 \times 62.7 = 1\,732.401$. This type of error is known as a **blunder.**

Another kind of error may occur due to the incorrect positioning of the decimal point after a calculation has been completed, and this is called an **order of magnitude error.** If, when determining the value of $\frac{274.3}{0.067\,1}$ the result is found to be 408.79 instead of 4 087.9, correct to 5 significant figures, then an order of magnitude error will have been made.

To take account of measurement errors it is usual to limit answers so that the result given **is not more than one significant figure greater than the least accurate number given in the data.** For example, it is required to find the resistance of a resistor in an electrical circuit when a voltage of 1.34 volts results in a current flow of 7.3 amperes. The relationship is $R = \frac{V}{I}$. Thus

$$R = \frac{1.34}{7.3} = 0.183\,561\,6 \ldots \text{ ohms.}$$

The least accurate number in the data is 7.3 A, since its value is only given correct to two significant figures. Thus the answer should not be expressed to more than a three significant figure accuracy. Hence

$$R = \frac{1.34}{7.3} = 0.184 \text{ ohms}$$

would be the normal way of writing this answer.

Blunders and order of magnitude errors can be reduced by determining approximate values of calculations. There is no hard and fast rule for doing this, but generally the problem is reduced to a mental arithmetic type problem by expressing numbers correct to one or two significant figures only and reducing the calculation to standard form if necessary. Then the approximate value of

$$\frac{289.63 \times 0.047\,1}{73.824} \text{ could be } \frac{3 \times 10^2 \times 5 \times 10^{-2}}{7 \times 10}$$

Using the laws of indices this becomes $\frac{15}{7} \times 10^{-1} \simeq 0.2$. That is, an answer of about 0.2 will be expected. On performing the calculation, if an answer of 0.184 8 is obtained, it will indicate that the calculation has probably been done correctly. An answer of 0.482 7 will suggest that a blunder has occurred or an answer of 0.018 48 that an order of magnitude error has occurred. If either of these last two answers are obtained, the calculation should be checked and repeated if necessary.

Answers which do not seem feasible must be checked and repeated as necessary. If as a result of a calculation the area of a triangle is found to be thousands of square kilometres, or the height of a man tens of metres, then it is likely that an error has occurred. For example, if 5 kg of sugar cost 125p and it is required to find out how much 2 kg cost, then an answer of $125 \times \frac{5}{2}$

or 312.5p is obviously incorrect since it must cost less to buy 2 kg than 5 kg.

When determining an answer to a calculation the points listed below should be considered to ascertain whether the result obtained is likely to be correct and expressed to a reasonable degree of accuracy:

(*a*) the answer should not be more accurate than the data on which it is based;
(*b*) errors do exist in all problems based on measurements;
(*c*) errors do exist in all problems where rounding off has occurred;
(*d*) blunders and order of magnitude errors should be eliminated as far as possible by determining the approximate value of a calculation; and
(*e*) a check should be made to see whether the answer obtained seems feasible.

Worked problems on errors

Problem 1. The area of a rectangle $A = bh$. The base, b, of the rectangle when measured was found to be 4.63 centimetres and the height, h, 9.4 centimetres. Determine the area of the rectangle.

Area $= bh = 9.4 \times 4.63 = 43.522\ \text{cm}^2$.

The approximate value is $9 \times 5 = 45\ \text{cm}^2$, so there are no obvious blunder or order of magnitude errors. However it is not usual in a measurement type question to state the answer to an accuracy greater than one significant figure more than the least accurate number in the data; this is 9.4 cm, so the result should not have more than 3 significant figures.

Thus

Area = **43.5 cm^2**

would be the normal way of expressing the result.

Problem 2. State which type of error or errors have been made in the following problems:

(*a*) $\dfrac{37.4}{16.8 \times 0.071} = 1.335$ correct to 3 decimal places;

(*b*) $16.814 \times 0.003\,8 = 0.036\,89$ correct to 4 significant figures;

(*c*) In the formula $C = 2\pi r$, r was measured as 1.72 cm, giving a value for C of 10.808 5 correct to 4 decimal places when π was taken to be 3.142.; and

(*d*) $127.8 \times 1.632 = 208.6$

(*a*) $\dfrac{37.4}{16.8 \times 0.071}$ is approximately equal to $\dfrac{4 \times 10}{2 \times 10 \times 7 \times 10^{-2}}$ or $\frac{4}{14} \times 10^2$, that is about 30. **Thus two errors have occurred, both a blunder and an order of magnitude error.**

(*b*) $16.814 \times 0.003\,8$ is approximately equal to $1.7 \times 10 \times 4 \times 10^{-3}$, that is about 7×10^{-2} or 0.07. **Thus a blunder has been made in the calculation.**

(*c*) $C = 2\pi r = 2 \times 3.142 \times 1.72$. The answer should be expressed to 4 significant figures only (one more than the least accurate data figure, 1.72). **Hence the error is that of introducing a greater accuracy than exists in the data.**

(*d*) 127.8 × 1.632 = 208.569 6, **hence a rounding off error has occurred.** This answer should have stated 127.8 × 1.632 = 208.6 correct to 4 significant figures.

Further problems on errors may be found in Section 3.6 (Problems 1–10), page 88.

3.2 The use of four-figure tables to find squares, square roots and reciprocals

To do calculations by long division and multiplication and other basic arithmetic means can be time consuming and tedious and blunders and order of magnitude errors tend to occur. Several aids to calculations are available. One of these is the use of four-figure tables.

Squares

Most books of four-figure tables give the values of squares of numbers from 1.000 to 9.999. The results stated in the tables are correct to three significant figures, the fourth significant figure being correct to ±1. Such a set of tables is shown in Table 3.3 (on pp. 76 and 77).

To determine the square of a number between 1.000 and 9.999, the first two significant figures of the number are found in the column on the left of the table, this giving the **row**. The third significant figure of the number is found at the top centre of the page, this giving the **column**. A number corresponding to both the row and column called the **matrix number** can now be located. Finally, the fourth significant figure is found at the top of the table but on the right-hand side under 'mean differences' and the number corresponding to this column and the appropriate row is ascertained. The mean difference number is added to the matrix number, both numbers being treated as integers for this purpose.

To find the square of 2.596, the row containing 2.5 is selected. Often the easiest way of reading tables is to put a ruler immediately beneath the row required. The third significant figure is 9 and the appropriate column is found at the top centre of the table, giving a matrix number of 6.708. The fourth significant figure is 6 and the column under mean differences headed 6, for row 2.5 gives a mean difference of 31. Treating the numbers as integers, the matrix mean difference numbers are added, that is, 6 708 + 31 = 6 739. Then $2.596^2 = 6.739$. When using tables, it is usual to state this fact at the start of a caculation and the degree of accuracy is then known.

The squares of some other numbers between 1 and 9.999, obtained by using Table 3.3, are given below:

$3.271^2 = 10.70$
$7.42^2 = 55.06$ (the mean difference column is not required for a number having 3 significant figures)
$9.377^2 = 87.93$

To find the squares of numbers outside of the range 1.000 to 9.999 using tables, the number should be expressed in standard form. Thus

$$\begin{aligned}
27.41^2 &= (2.741 \times 10)^2 \\
&= (2.741 \times 10) \times (2.741 \times 10) \\
&= 2.741^2 \times 10^2 \\
&= 7.513 \times 10^2 \\
&= 751.3
\end{aligned}$$

$$\begin{aligned}
\text{and } 0.069\,47^2 &= (6.947 \times 10^{-2})^2 \\
&= 6.947^2 \times 10^{-4} \\
&= 48.26 \times 10^{-4} \\
0.069\,47^2 &= 0.004\,826
\end{aligned}$$

Square roots

These are usually listed in two sets of tables, one for numbers 1.000 to 9.999 and a second table for numbers 10.00 to 99.99. These are shown in Tables 3.4 and 3.5 (on pp. 78 to 81).

As for squares, the matrix number is found for the row containing the first two significant figures and the column containing the third significant number of the square root required. The mean difference is determined using the appropriate row and the mean difference column which has the fourth significant figure at the top. The mean difference number is added to the matrix number, treating both numbers as integers. Then $\sqrt{6.275}$ has a matrix number of 2.504, and a mean difference number of 1 and so $\sqrt{6.275} = \pm 2.505$, since a square root always has two answers. To find $\sqrt{62.75}$, the 10.00 to 99.99 table must be used and this gives $\sqrt{62.75} = \pm 7.921$.

By using Tables 3.4 and 3.5 the square roots of some other numbers between 1 and 99.99 are shown below:

$\sqrt{5.27} = \pm 2.296$ (the mean difference column is not required for a number having 3 significant figures)
$\sqrt{7.415} = \pm 2.723$
$\sqrt{33.47} = \pm 5.785$
$\sqrt{67.41} = \pm 8.211$

For the square root of a number outside of the range 1.000 to 99.99, that number should be expressed as a number between 1.000 and 99.99 multiplied by **10 raised to an even power.** Thus:

$$\begin{aligned}
\sqrt{372} &= (372)^{\frac{1}{2}} \\
&= (3.72 \times 10^2)^{\frac{1}{2}} \\
&= 3.72^{\frac{1}{2}} \times (10^2)^{\frac{1}{2}} \\
&= \sqrt{3.72} \times 10 \\
&= \pm 1.929 \times 10 = \pm 19.29 \\
\sqrt{174\,300} &= (17.43 \times 10^4)^{\frac{1}{2}} \\
&= \sqrt{17.43} \times 10^2
\end{aligned}$$

$$= \pm 4.175 \times 10^2 = \pm 417.5$$

$$\sqrt{0.028\,4} = (2.84 \times 10^{-2})^{\frac{1}{2}}$$
$$= 2.84^{\frac{1}{2}} \times 10^{-1}$$
$$= \pm 1.685 \times 10^{-1} = \pm 0.168\,5$$

$$\sqrt{0.009\,26} = (92.6 \times 10^{-4})^{\frac{1}{2}}$$
$$= \sqrt{92.6} \times 10^{-2}$$
$$= \pm 9.623 \times 10^{-2} = \pm 0.096\,23$$

Reciprocals

A similar procedure to that used to determine squares is adopted. To find the reciprocal of 1.735 using Table 3.6 (on pp. 82 and 83), the row containing 1.7 and the column containing 3 are located, giving a matrix number of 0.578 0. The mean difference number is located using the 1.7 row and the mean difference column headed 5, giving 16. This is **subtracted** from the last two digits of the matrix number, because the matrix numbers are decreasing as the numbers increase. A note that 'Numbers in difference columns to be subtracted, not added' will be found at the top of the tables, to act as a reminder. Thus the reciprocal of 1.735 is 0.576 4 (5 780 − 16 or 5 764 treating the matrix and mean difference numbers as integers). By using table 3.6 the reciprocals of some other numbers between 1.000 and 9.999 are shown below:

$$\frac{1}{2.743} = 0.364\,6$$

$$\frac{1}{4.62} = 0.216\,5$$ (the mean difference column is not required for a number having 3 significant figures)

$$\frac{1}{7.987} = 0.125\,2$$

There are no numbers listed in the mean difference columns for matrix numbers corresponding to the range 1.000 to 1.399. This is because the mean differences in this region cease to be sufficiently accurate. For this range numbers should be rounded off to three significant figures before using the tables. Outside of the range 1.000 to 9.999, numbers should be expressed in standard form. Thus

$$\frac{1}{39.62} = \frac{1}{3.962 \times 10} = \frac{1}{3.962} \times 10^{-1}$$
$$= 0.252\,4 \times 10^{-1} = 0.025\,24$$

and

$$\frac{1}{0.082\,7} = \frac{1}{8.27 \times 10^{-2}} = \frac{1}{8.27} \times 10^{2}$$
$$= 0.120\,9 \times 10^{2} = 12.09$$

Problem 1. Use four-figure tables to find the value of
(*a*) 7.214^2 (*b*) 827.4^2 (*c*) $0.037\ 1^2$

(*a*) To find 7.214^2, using Table 3.3 (p. 77), the row containing 7.2 on the left of the table is located. Then
$7.2^2 = 51.84$
The column at the centre of the page containing 1 at its head is located and the matrix number corresponding to column 1 and row 7.2 is found, giving
$7.21^2 = 51.98$
The mean difference column containing 4 at its head is found and the mean difference number corresponding to this column and the 7.2 row is 6. Adding 51.98 and 6, treating both as integers gives 5 204. Thus
$7.214^2 = \mathbf{52.04}$

(*b*) To find 827.4^2 the number is firstly expressed in standard form, thus
$827.4^2 = (8.274 \times 10^2)^2$
$= 8.274^2 \times 10^4$
Using the tables to find 8.274^2 as for problem 1 (*a*) gives a matrix number of 68.39 and a mean difference number of 7. Then
$8.274^2 = 68.46$
and
$827.4^2 = 68.46 \times 10^4 = \mathbf{684\ 600}$

(*c*) Expressing $0.037\ 1^2$ in standard form gives
$0.037\ 1^2 = (3.71 \times 10^{-2})^2$
$= 3.71^2 \times 10^{-4}$
The matrix number is 13.76. The mean difference number is not required here, since there is not a fourth significant figure. Hence
$0.037\ 1^2 = 13.76 \times 10^{-4} = \mathbf{0.001\ 376}$

Problem 2. Use four-figure tables to find the value of
(*a*) $\sqrt{3.947}$ (*b*) $\sqrt{92.84}$ (*c*) $\sqrt{3\ 927}$ (*d*) $\sqrt{0.012\ 8}$

(*a*) Since 3.947 lies between 1.000 and 9.999, Table 3.4 (p. 78) is used. The row containing 3.9 on the left of the table is located. Then
$\sqrt{3.9} = \pm 1.975$
The column at the centre of the page containing 4 at its head is located and the matrix number corresponding to column 4 and row 3.9 is found, giving
$\sqrt{3.94} = \pm 1.985$

The mean difference column headed 7 and the mean difference number corresponding with this column and row 3.9 is found to be 2. Adding 1.985 and 2, treating both as integers give the required result that
$\sqrt{3.947} = \mathbf{\pm 1.987}$

(*b*) Using Table 3.5 (p. 81) shows that $\sqrt{92} = \pm 9.592$ and that the matrix number corresponding to $\sqrt{92.8} = \pm 9.633$. The mean difference for the 4 column and 92 row is 2, hence
$\sqrt{92.84} = \mathbf{\pm 9.635}$

(*c*) To find $\sqrt{3\ 927}$, expressing 3 927 as a number between 1.000 and

99.99 multiplied by 10 raised to an even power gives

$$3\,927 = 39.27 \times 10^2$$

and $\sqrt{3\,927} = (39.27 \times 10^2)^{\frac{1}{2}}$

$$= 39.27^{\frac{1}{2}} \times (10^2)^{\frac{1}{2}}$$

$$= \sqrt{39.27} \times 10$$

Using the tables as for problem 2 (*b*)

$\sqrt{39.27} \times 10 = \pm 6.267 \times 10$

Then

$$\sqrt{3\,927} = \pm \mathbf{62.67}$$

(*d*) As for problem 2 (*c*)

$$\sqrt{0.012\,8} = (1.28 \times 10^{-2})^{\frac{1}{2}}$$

$$= \sqrt{1.28} \times 10^{-1}$$

$$= \pm 1.131 \times 10^{-1}$$

$$= \pm \mathbf{0.113\,1}$$

Problem 3. Use four-figure tables to find the value of

(*a*) $\dfrac{1}{7.384}$ (*b*) $\dfrac{1}{1.248}$ (*c*) $\dfrac{1}{764.5}$ (*d*) $\dfrac{1}{0.004\,281}$

(*a*) Using Table 3.6 (pp. 82 and 83)

$\dfrac{1}{7.3} = 0.137\,0$ from the appropriate row and the 0 column.

$\dfrac{1}{7.38} = 0.135\,5$, the matrix number from the appropriate row and column.

The mean difference is 1 for the column headed 4 and this is **subtracted** (see note at the top of the tables). Thus

$\dfrac{1}{7.384} = \mathbf{0.135\,4}$

(*b*) The value of $\dfrac{1}{1.248}$ cannot be found to the required degree of accuracy and the denominator must be rounded off to 1.25, because no mean differences are given in this region. Thus $\dfrac{1}{1.2} = 0.833\,3$

and, using the matrix number $\dfrac{1}{1.25} = \mathbf{0.800\,0}$

(*c*) Now $\dfrac{1}{764.5} = \dfrac{1}{7.645 \times 10^2} = \dfrac{1}{7.645} \times 10^{-2}$

Using the same method as for problem 3 (*a*)

$\dfrac{1}{7.645} = 0.130\,8$

and $\dfrac{1}{764.5} = 0.130\,8 \times 10^{-2} = \mathbf{0.001\,308}$

(*d*) $\dfrac{1}{0.004\,281} = \dfrac{1}{4.281 \times 10^{-3}} = \dfrac{1}{4.281} \times 10^3$

As for problem 3 (*a*)

$$\frac{1}{4.281} \times 10^3 = 0.233\,5 \times 10^3 = 233.5$$

Further problems on squares, square roots and reciprocals may be found in Section 3.6 (Problems 11–16), page 88.

3.3 Logarithms

Generally speaking, when dealing with numbers which are not integers or outside of the range 1 to 12, addition and subtraction are easier processes than multiplication and division. By using logarithms the process of multiplication is converted to addition and that of division is converted to subtraction, thus simplifying calculations. The table of logarithms shown in Table 3.7 (on pp. 84 and 85) uses a base of 10. The logarithm of a number shown in the table is the power to which 10 has to be raised to be equal to the number. If the number is N and x is the power to which 10 has to be raised to be equal to N, then x is called the **logarithm** of N to base 10. It follows that since $100 = 10^2$, then 2 is the logarithm of 100 to base 10 and is written

$2 = \lg 100$

Similarly, since $10\,000 = 10^4$, then 4 is the logarithm of 10 000 to base 10 and is written

$4 = \lg 10\,000$

In general, the definition of a logarithm is:

if $y = a^x$ then $x = \log_a y$.

To find the logarithm of a number using four-figure tables

To find the logarithm of any number between 1.000 and 9.999 the table of logarithms shown in Table 3.7 (on pp. 84 and 85) may be used. The structure of these tables is similar to those of squares, square roots and reciprocals. The column on the left is associated with the first two significant figures of the number, the columns headed 0 to 9 adjacent to this column are associated with the third significant figure and the columns headed 0 to 9 on the right are associated with the fourth significant figure.

To find the logarithm of 4.758 the 47 row is located and lg 4.7 is shown as 0.672 1. Although the decimal points are not shown in the table, their position will become apparent as this section develops. The matrix number corresponding to row 47 and column 5 is 0.676 7, hence lg 4.75 is 0.676 7. The fourth significant figure is 8, and the number associated with the mean difference column 8 and row 47 is 7. This number is added to the matrix number, treating both numbers as integers for this purpose. Then 6 767 + 7 is 6 774 and lg 4.758 = 0.677 4. By the definition of a logarithm, the mathematical statement lg 4.758 = 0.677 4 means that $4.758 = 10^{0.6774}$.

The logarithms of some other numbers between 1.000 and 9.999 found by using Table 3.7 are given below:

$\lg 2.714 = 0.433\ 6$ or $2.714 = 10^{0.433\ 6}$
$\lg 5.789 = 0.762\ 6$ or $5.789 = 10^{0.762\ 6}$
$\lg 8.27 = 0.917\ 5$ or $8.27 = 10^{0.917\ 5}$

(as there is no fourth significant figure in 8.27, the columns on the right were not required).

To find the logarithm of a number larger than 9.999 it must be expressed in standard form. To find the logarithm of, say, 230 it is written as

$230 = 2.3 \times 10^2$

By using Table 3.7

$\lg 2.3 = 0.361\ 7$ or $2.3 = 10^{0.361\ 7}$

Then $230 = 10^{0.361\ 7} \times 10^2$
$= 10^{0.361\ 7 + 2}$

by the laws of indices and $\lg 230 = 2.361\ 7$ by the definition of a logarithm.

The integer in front of the decimal point is called the **characteristic.** The figures after the decimal point are called the **mantissa**, and are always positive. Rather than go through the procedure shown above each time the logarithm of a number has to be found, the following procedure is usually adopted. When expressing a number in standard form, it will be mutliplied by 10 raised to some power. This power is always the characteristic of the logarithm. Then since

$2.345 = 2.345 \times 10^0$, $\lg 2.345 = 0.370\ 1$

and

$23.45 = 2.345 \times 10^1$, $\lg 23.45 = 1.370\ 1$

Similarly

$234.5 = 2.345 \times 10^2$, $\lg 234.5 = 2.370\ 1$

and

$2\ 345 = 2.345 \times 10^3$, $\lg 234\ 5 = 3.370\ 1$

Numbers in the range 0 to 0.999 should again be expressed in standard form, and since

$0.234\ 5 = 2.345 \times 10^{-1}$, $\lg 0.234\ 5 = -1 + 0.370\ 1$

since the mantissa is always positive. Similarly 0.023 45 is 2.345×10^{-2} and $\lg 0.023\ 45 = -2 + 0.370\ 1$.

It is necessary for the characteristic and mantissa to keep their separate identities, particularly when this is only part of a larger calculation, and also to enable conversion from logarithmic to ordinary number form. To simplify the writing, $-2 + 0.370\ 1$ is written as $\bar{2}.370\ 1$ called 'bar two point 3 701'.

If $y = 10^x$, y will always be positive whether x is a positive or negative number. Thus it is impossible to obtain the logarithm of a negative number.

To enable speedy determination of the characteristic of a logarithm, the rules shown below may be used:

1. The characteristic for a number greater than one is positive and is one less than the number of digits to the left of the decimal point. For example, the characteristic for 567.2 is 3 (the number of digits before the decimal point) $-1 = 2$.
2. The characteristic of a number less than one is negative and is numerically one more than the number of noughts to the right of the decimal point before figures other than nought begin.

For example, the characteristic for 0.004 28 is 2 (the number of noughts after the decimal point) + 1 = 3. Hence the characteristic is $\bar{3}$.

To find the antilogarithm of a number

The reverse process of finding the logarithm of a number is to find the antilogarithm of a number. Tables of antilogarithms for numbers 0 to 0.999 are shown in Table 3.8 (on pp. 86 and 87).

These tables are similar in their layout to tables met previously. If lg N = 0.117 3, the row containing the first two significant figures is located, and the antilogarithm of 0.11 is 1 288.

The matrix number for this row and the column headed 7 is 1 309, thus the antilogarithm of 0.117 is 1 309. Using the columns on the right, the antilogarithm of 0.117 3 is 1 309 + 1 or 1 310. The characteristic number '0' indicates a number between 1 and 9.999. Thus if

lg N = 0.117 3
N = 1.310

The antilogarithms of some other numbers between 0 and 0.999 9 found by using the table of antilogarithms are given below:

when lg N = 0.721 4, N = 5.265
when lg N = 0.577, N = 3.776
and when lg N = 0.924 7, N = 8.409

In each case, it is the characteristic (0.) which fixes the position of the decimal point and the mantissa (.*XXXX*) which fixes the value, neglecting the decimal point.

For numbers outside of the range 0 to 0.999 9 the position of the decimal point is fixed by the value of the characteristic and the value by the mantissa. To find N when lg N is 2.217 8 the value is found by determining the antilogarithm of 0.217 8, that is 1 651; and the characteristic 2 then shows that N = (a number in standard form) $\times 10^2$. Thus $N = 1.651 \times 10^2$ or 165.1. The same procedure is carried out for bar numbers. To find N when lg $N = \bar{2}.671\ 8$ tables are first used to find the antilogarithm of 0.671 8, giving a result of 4 697. The characteristic $\bar{2}$ then shows that N is (a number of standard form) $\times 10^{-2}$. Hence $N = 4.697 \times 10^{-2}$ or 0.046 97.

Multiplication

To find the value of, say, 15×47 by using logarithms:
the logarithm of 15 is 1.176 1 and the logarithm of 47 is 1.672 1. Thus $15 = 10^{1.176\,1}$ and $47 = 10^{1.672\,1}$ by the definition of a logarithm. Also

$$\begin{aligned} 15 \times 47 &= 10^{1.176\,1} \times 10^{1.672\,1} = 10^{1.176\,1 + 1.672\,1} \\ &= 10^{2.848\,2} \text{ by the laws of indices.} \end{aligned}$$

Using tables of antilogarithms $10^{2.848\,2} = 705.0$. This procedure shows that to multiply two numbers together using logarithms the logarithm of each of the numbers is first found; these values are then added; and the antilogarithm of the sum is finally determined. In mathematical terms, if one number is represented by A and the other by B

$$\lg (A \times B) = \lg A + \lg B \qquad [I]$$

Applying this rule of logarithm to multiplying 15 by 47:

$$\begin{aligned} \lg (15 \times 47) &= \lg 15 + \lg 47 \\ &= 1.171\,6 + 1.672\,1 \\ &= 2.848\,2 \end{aligned}$$

The process of finding an antilogarithm reverses the logarithmic process so that

$$\begin{aligned} 15 \times 47 &= \text{the antilogarithm of } 2.848\,2 \\ &= 705.0 \end{aligned}$$

The same procedure is used for multiplying more than two numbers; the logarithms of each of the numbers are determined, added together and the antilogarithm of the sum found to obtain the answer. For example, to find the value of $23.7 \times 0.037\,14 \times 720$ the calculation can be laid out as shown below:

	Number	Logarithm	
	23.7	1.374 7	
	0.037 14	$\bar{2}.569\,9$	
	720	2.857 3	
Multiplication	$23.7 \times 0.037\,14 \times 720$	2.801 9	Adding

Finding the antilogarithm of 2.801 9 shows that
$23.7 \times 0.037\,14 \times 720 = 633.7$

To find the value of $\frac{15}{47}$ by logarithms:

Since $15 = 10^{1.176\,1}$ and $47 = 10^{1.672\,1}$, then $\frac{15}{47} = \dfrac{10^{1.176\,1}}{10^{1.672\,1}}$.

By the laws of indices this is $10^{1.176\,1 - 1.672\,1}$

To evaluate 1.176 1 − 1.672 1, lay out as a normal subtraction problem:

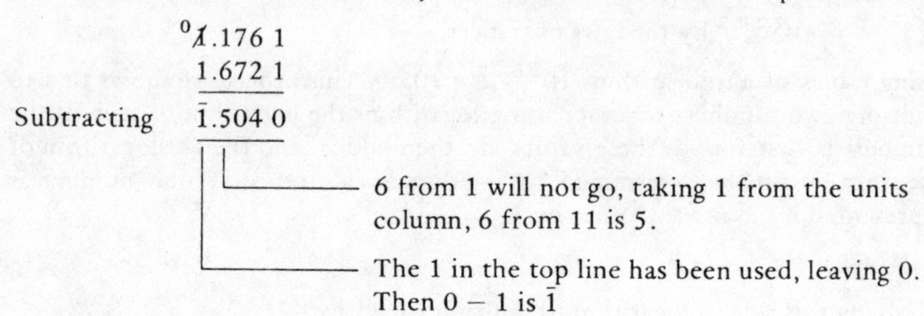

Thus $10^{1.176\,1 - 1.672\,1} = 10^{\bar{1}.504\,0}$

And using tables of antilogarithms, $10^{\bar{1}.504\,0} = 0.319\,2$.

This procedure shows that to divide two numbers using logarithms the logarithm of the denominator is taken from the logarithm of the numerator and the antilogarithm of the result is determined. In mathematical terms, if the numerator is number A and the denominator number B.

$$\lg\left(\frac{A}{B}\right) = \lg A - \lg B \qquad \text{[II]}$$

Applying this rule of logarithms to dividing 15 by 47:

$$\begin{aligned} \lg\left(\tfrac{15}{47}\right) &= \lg 15 - \lg 47 \\ &= 1.176\,1 - 1.672\,1 \\ &= \bar{1}.504\,0 \end{aligned}$$

and using the table of antilogarithms, $\frac{15}{47} = 0.319\,2$

When finding the reciprocal of a number, the numerator is always 1 and lg 1.000 0 is 0.000 0. Then to find the reciprocal of, say, 15 using rule [II]

$\lg \frac{1}{15} = 0.000\,0 - 1.176\,1$

To evaluate 0.000 0 − 1.176 1 lay out as a normal subtraction problem.

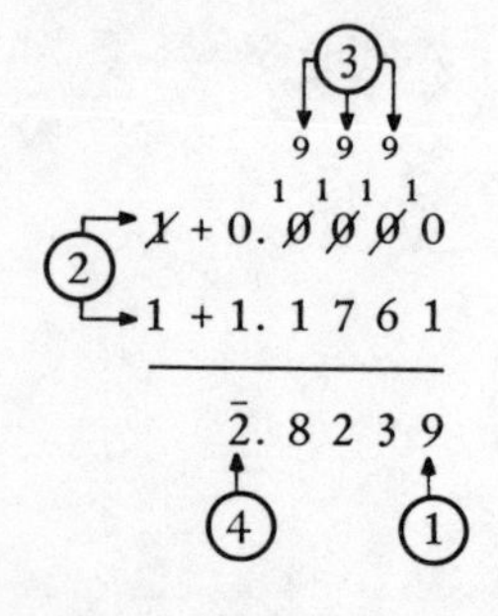

(1) 1 from 0 cannot be done.

(2) Add 1 to the characteristic of both top and bottom lines: this will not alter the value of the result.

(3) Borrow the 1 introduced in the top line and subtract normally.

(4) Since the top line 1 has been used $0 - 2 = -2 = \bar{2}$.

Then $\lg \frac{1}{15} = \bar{2}.823\,9$

Using table of antilogarithms, $\frac{1}{15} = 0.066\,67$

Powers

To find the value of 15^3 using logarithms:
$15^3 = 15 \times 15 \times 15$ and $\lg 15 = 1.176\,1$ or $15 = 10^{1.176\,1}$ then

$$\begin{aligned} 15^3 &= 10^{1.176\,1} \times 10^{1.176\,1} \times 10^{1.176\,1} \\ &= 10^{1.176\,1 + 1.176\,1 + 1.176\,1} \\ &= 10^{3 \times 1.176\,1} \\ &= 10^{3.528\,3} \end{aligned}$$

By using Table 3.8, $10^{3.528\,3} = 3\,375$

This shows that to find the power of a number using logarithms, the logarithm of the number is found and multiplied by the power and the antilogarithm of the result is determined. In mathematical terms if the number is A and the power n

$$\lg A^n = n \times \lg A \qquad \text{[III]}$$

Applying this rule of logarithms to finding the value of 15^3

$$\begin{aligned} \lg 15^3 &= 3 \lg 15 \\ &= 3 \times 1.176\,1 \\ &= 3.528\,3 \end{aligned}$$

Using Table 3.8, $15^3 = 337\,5$.

The same rule applies when n is a fraction. Thus to find $\sqrt{15}$ or $15^{\frac{1}{2}}$ rule [III] is applied:

$$\begin{aligned} \lg 15^{\frac{1}{2}} &= \tfrac{1}{2} \lg 15 \\ &= \frac{1.176\,1}{2} \\ &= 0.588\,1 \end{aligned}$$

The antilogarithm of 0.588 1 is 3.874. Hence

$$\sqrt{15} = \pm 3.874$$

When using 'bar numbers' a complication can arise and can be overcome as shown. To find $\sqrt{0.476\,3}$:

$$\begin{aligned} \lg (0.476\,3)^{\frac{1}{2}} &= \tfrac{1}{2} \lg 0.476\,3 \\ &= \frac{\bar{1}.677\,9}{2} \end{aligned}$$

But $\bar{1}.677\,9$ means $-1 + 0.677\,9$. By adding -1 to the characteristic and $+1$ to the mantissa, the overall value is not changed. Thus

$$\begin{aligned} -1 + 0.677\,9 &= (-1 - 1) + (1 + 0.677\,9) \\ &= \bar{2} + 1.677\,9 \end{aligned}$$

which can now be divided by 2. Then

$$\frac{\bar{1}.677\,9}{2} = \frac{\bar{2} + 1.677\,9}{2} = \frac{\bar{2}}{2} + \frac{1.677\,9}{2}$$
$$= \bar{1} + 0.839\,0 \text{ or } \bar{1}.839\,0$$

and finding the antilogarithm gives

$\sqrt{0.476\,3} = \pm 0.690\,2$

Summary of the rules of logarithms

(*a*) To multiply two or more numbers, the rule is $\lg (A \times B) = \lg A + \lg B$ (rule [I]).
Then (*i*) add the logarithms of each of the numbers and
(*ii*) find the antilogarithms of the result of (*i*).

(*b*) To divide two numbers, the rule is $\lg \frac{A}{B} = \lg A - \lg B$ (rule [II]).
Then (*i*) take the logarithm of the denominator number from the logarithm of the numerator number and
(*ii*) find the antilogarithm of the result of (*i*).

(*c*) To raise a number to a power, the rule is $\lg A^n = n \times \lg A$ (rule [III]).
Then (*i*) multiply the logarithm of the number by the power and
(*ii*) find the antilogarithm of the result of (*i*).

(*d*) To find the square root of a number, the rule is $\lg A^n = n \times \lg A$ (rule [III]), for n being $\frac{1}{2}$.
Then (*i*) divide the logarithm of the number by 2 and
(*ii*) find the antilogarithm of the result of (*i*).

Some of the worked problems on this section show how to apply these rules of logarithms governing the multiplication, division and powers of numbers to the solution of problems containing more than one operation.

Worked problems on logarithms

Problem 1. Find the value of:
(*a*) lg 4.281 (*b*) lg 284.7 (*c*) lg 0.067 1

(*a*) The number 4.281 will have a characteristic of 0 since in standard form $4.281 = 4.281 \times 10^0$. Using Table 3.7 to find the mantissa of 4 281, the row containing 42 gives lg 42 = 623 2. The matrix number for the centre column headed 8 and row 42 is 6 314. The column on the right headed 1 and row 42 gives 1. Then
lg 428 = 6 314
and
lg 4 281 = 6 314 + 1 = 6 315
Applying both the characteristic and value gives
lg 4.281 = **0.631 5**

(*b*) 284.7 will have a characteristic of 2 since in standard form, 284.7 is equal to 2.847×10^2. Using Table 3.7, the mantissa of 2 847 is 4 544. Thus
lg 284.7 = **2.454 4**

(*c*) $0.067\,1 = 6.71 \times 10^{-2}$ hence the characteristic is −2. From Table 3.7

the mantissa of 671 is 8 267. Thus lg 0.067 1 = −2 + 0.826 7, written as $\bar{2}$**.826 7.**

Problem 2. Find the antilogarithm of:
(*a*) 0.147 4   (*b*) 2.387   (*c*) $\bar{1}$.928 3

(*a*) The '0' shows that there will be one digit in front of the decimal point, since N = (a standard number) × 10^0. Using Table 3.8 to determine the value of 0.147 4, the row containing .14 is selected and the adjacent column headed 7, gives a matrix number of 1 403. The column on the right headed 4 and row .14 indicates 1. Hence the value is 1 403 + 1 or 1 404. Then the antilogarithm of 0.147 4 is **1.404.**

(*b*) For 2.387, the value is obtained by finding the antilogarithm of 0.387 from Table 3.8. Thus
antilog 0.387 = 2 438
The characteristic, 2, shows that the required antilogarithm of the number is **2.438 × 10^2 or 243.8**

(*c*) For $\bar{1}$.928 3, the value is 8 478, obtained from the table of antilogarithms and the characteristic, $\bar{1}$, shows that the antilogarithm of $\bar{1}$.928 3 is **8.478 × 10^{-1} or 0.847 8.**

Problem 3. Use logarithms to find the value of:
(*a*) 21.73 × 0.017 62   (*b*) 38.3 × 82.97 × 0.371 4

One of the methods commonly used for doing calculations by logarithms is shown below. It has the advantages of showing the stage reached during a calculation and making it easier to check for blunders.

(*a*) 21.73 × 0.017 62. For multiplication the logarithms are added.

	Number	Logarithm	
	21.73	1.337 1	
	0.017 62	$\bar{2}$.246 0	
Multiplying	0.382 9	$\bar{1}$.583 1	Adding

To get from a logarithm to a number, the antilogarithm is taken. The antilogarithm of $\bar{1}$.583 1 shown in the logarithm column is equal to 0.382 9, shown under the number column. Thus
21.73 × 0.017 62 = **0.382 9**

(*b*) 38.3 × 82.97 × 0.371 4. For multiplication add the logarithms.

	Number	Logarithm	
	38.3	1.583 2	
	82.97	1.919 0	
	0.371 4	$\bar{1}$.569 9	
Multiplying	1 180	3.072 1	Adding

Thus 38.3 × 82.97 × 0.371 4 = **1 180.**

Problem 4. Use logarithms to find the value of:
(*a*) $\frac{2.714}{19.83}$ (*b*) $\frac{1}{271.4}$

(*a*) $\frac{2.714}{19.83}$. For division the logarithm of the denominator is subtracted from the logarithm of the numerator.

	Number	Logarithm	
	2.714	0.433 6	
	19.83	1.297 3	
Dividing	0.136 9	$\bar{1}.136\ 3$	Subtracting

Thus $\frac{2.714}{19.83} = \mathbf{0.136\ 9}$

(*b*) $\frac{1}{271.4}$. To find a reciprocal, the logarithm of the denominator is subtracted from 0.000 0 (the logarithm of 1.000 0).

	Number	Logarithm	
	1	1 + 0.000 0	
	271.4	1 + 2.433 6	
Dividing	0.003 684	$\bar{3}.566\ 4$	Subtracting

One was added to both top and bottom lines of the logarithm to keep the mantissa positive and to make the subtraction feasible. Thus the reciprocal of 271.4 is $\mathbf{3.684 \times 10^{-3}}$.

Problem 5. Use logarithms to find the value of:
(*a*) 23.49^3 (*b*) $0.734\ 6^4$

(*a*) to find the nth power of a number, the logarithm of the number is multiplied by n. Thus to find 23.49^3

	Number	Logarithm	
	23.49	1.370 9	
		3	The power
Raising to a power	12 960	4.112 7	Multiplying

Thus $23.49^3 = \mathbf{12\ 960}$

(b) Multiplying the logarithm of 0.734 6 by 4 gives $0.734\,6^4$

	Number	Logarithm	
	0.734 6	$\bar{1}.866\,1$	
		4	The power
Raising to power	0.291 4	$\bar{1}.464\,4$	Multiplying

4 × 8 = 32, + (carry 2) =34 write 4, carry 3.
4 × (−1) = −4 + (carry 3) = −1 or $\bar{1}$

Thus $0.734\,6^4$ = **0.291 4**

The accuracy of logarithms is correct to 3 significant figures with the 4th significant figure ±1. However, when raising to a power or doing repeated calculations using logarithms, these errors are magnified and it is likely that the accuracy of the fourth significant figure of this calculation is only correct within the range ±4.

Problem 6. Find the value of: (a) $\sqrt{465.7}$ (b) $\sqrt{0.009\,283}$

(a) Since $\sqrt{465.7}$ is $465.7^{\frac{1}{2}}$, then the procedure is to divide the logarithm of 465.7 by 2.

	Number	Logarithm	
	465.7	2) 2.668 2	
Square rooting	21.58	1.334 1	Dividing by 2

Thus $\sqrt{465.7}$ = **±21.58**

(b) To find $\sqrt{0.009\,283}$, divide lg 0.009 283 by 2

Number	Logarithm
0.009 283	$\bar{3}.967\,6$

The characteristic is negative and the mantissa is positive, and these must keep their separate identities. To make $\bar{3}$ divisible by 2, −1 is added to $\bar{3}$ to make it $\bar{4}$. Then 1 must be added to the mantissa to keep the overall value of the logarithm the same. Thus lg 0.009 283 = $\bar{3} + \bar{1} + 1 + .967\,6$ or $\bar{4} + 1.967\,6$

	Number	Logarithm	
	0.009 283	2) $\bar{4} + 1.967\,6$	
Square rooting	0.096 34	$\bar{2}.983\,8$	Dividing by 2

Thus $\sqrt{0.009\,283}$ = **±0.096 34**

Problem 7. Use logarithms to find the value of the energy, E, stored in a capacitor in joules given $E = \frac{1}{2} CV^2$ when the capacitance $C = 3 \times 10^{-6}$ F and the voltage $V = 240$ V.

The approach to this problem will be to find lg V and multiply by 2 (giving V^2) and then add lg C. By finding the antilogarithms the value of CV^2 will be obtained which can be halved to give the required result.

	Letter	Number	Logarithm	
	V	240	2.380 2	
			2	
Squaring	V^2		4.760 4	Multiplying by 2
	C	3×10^{-6}	$\bar{6}.477\ 1$	
Multiplying	CV^2	0.172 8	$\bar{1}.237\ 5$	Adding

Thus $\frac{1}{2} CV^2 = \frac{0.172\ 8}{2}$ = **0.086 4 joules**

Problem 8. Use logarithms to find the value of the kinetic energy in joules, E, given that $E = \frac{1}{2} m(v^2 - u^2)$ and that the mass $m = 250$ kg, the final velocity, $v = 64.2$ m s^{-1} and the initial velocity $u = 38.9$ m s^{-1}.

Addition and subtraction cannot be done by logarithms and it will be necessary to determine the values of v^2 and u^2 by using logarithms, but the value of $(v^2 - u^2)$ must be determined by subtraction.

Number	Logarithm	Number	Logarithm
$v = 64.2$	1.807 5	$u = 38.9$	1.589 9
	2		2
$v^2 = 4\ 121$	3.615 0	$u^2 = 1\ 513$	3.179 8

Thus $v^2 - u^2 = 4\ 121 - 1\ 513 = 2\ 608$

	Number	Logarithm	
	$v^2 - u^2 = 2\ 608$	3.416 3	
	$m = 250$	2.397 9	
Multiplying	$m(v^2 - u^2) = 651\ 900$	5.814 2	Adding

Thus $\frac{1}{2} m (v^2 - u^2) = \frac{651\ 900}{2}$ = **325 950 joules**

That is, **326 000 joules** since the values given refer to measured data and the least accurate data was correct to two significant figures.

Problem 9. Use logarithms to evaluate $\left(\frac{23.6 \times 0.079}{2.847}\right)^3$

Adding lg 23.6 to lg 0.079 will give the logarithm of the numerator, and by subtracting lg 2.847 from the logarithm of the numerator the logarithm of the bracketed term will be given. Multiplying this logarithm by 3 will give the logarithm of the bracketed term cubed and the result will be obtained by finding the antilogarithm of this figure.

Multiplying 23.6 × 0.079

	Number	Logarithm	
	23.6	1.372 9	
	0.079	$\bar{2}.897\ 6$	
Multiplying	23.6 × 0.079	0.270 5	Adding

	Number	Logarithm	
	23.6 × 0.079	0.270 5	
	2.847	0.454 4	
Dividing	$\frac{23.6 \times 0.079}{2.847}$	$\bar{1}.816\ 1$ 3	
Power	$\left(\frac{23.6 \times 0.079}{2.847}\right)^3 = 0.280\ 7$	$\bar{1}.448\ 3$	Multiplying

Thus $\left(\frac{23.6 \times 0.079}{2.847}\right)^3 = \mathbf{0.280\ 7}$

Problem 10. Use logarithms to find the value of $\frac{53.47^2 \times \sqrt{28.71}}{0.071\ 4^3}$

The square, square root and cube should be determined in logarithmic form before evaluating the fraction.

Number	Logarithm
53.47	1.728 1 2
53.47^2	3.456 2

Number	Logarithm
28.71	2)1.458 1
$\sqrt{28.71}$	0.729 1

Number	Logarithm
0.071 4	$\bar{2}.853\,7$
	3
$0.071\,4^3$	$\bar{4}.561\,1$

Number	Logarithm
53.47^2	3.456 2
$\sqrt{28.71}$	0.729 1
$53.47^2 \times \sqrt{28.71}$	4.185 3
$0.071\,4^3$	$\bar{4}.561\,1$
4.209×10^7	7.624 2

Thus $\dfrac{53.47^2 \times \sqrt{28.71}}{(0.071\,4)^3} = \mathbf{4.209 \times 10^7}$

Problem 11. Use logarithms to find the value of $\dfrac{83^2 - 9.241^3}{15.74}$

Since addition and subtraction cannot be done by logarithms, the values of 83^2 and 9.241^3 must be determined by logarithms and the antilogarithms of each found before subtracting in the normal way.

Number	Logarithm
83	1.919 1
	2
$83^2 = 6\,890$	3.838 2

Number	Logarithm
9.241	0.965 7
	3
$9.241^3 = 789.1$	2.897 1

Then $83^2 - 9.241^3 = 6\,890 - 789.1 = 6\,100.9$ or $6\,101$

Number	Logarithm
6 101	3.785 4
15.74	1.197 0
387.7	2.588 4

Thus $\dfrac{83^2 - 9.241^3}{15.74} = \mathbf{387.7}$

Further problems on logarithms may be found in Section 3.6 (Problems (17–60), page 89.

3.4 Calculators

The most modern aid to calculations is the pocket-sized electronic calculator. With one of these, calculations can be quickly and accurately performed correct to about nine significant figures. For about the cost of a good quality slide rule, a small arithmetic calculator, which can perform addition, subtraction, multiplication and division, can be purchased. However, for about twice this price, a calculator having scientific notation with the principal mathematical functions can be purchased, and will prove to be a valuable asset to students. Since detailed instructions on the use of electronic calculators are given in leaflets supplied with the machine, and since there are many types which are all different in their operation, no attempt will be made to describe their use here.

The scientific type of electronic calculator has made the use of tables, logarithms and the slide rule for performing calculations largely redundant and of little use except towards gaining dexterity in mathematics; however, books of tables are still by far the cheapest method of obtaining an aid to calculation.

3.5 Use of conversion tables and charts

It is often necessary to make calculations from various conversion tables and charts. Examples include imperial to metric unit conversions, currency exchange rates, bus or train timetables and production schedules. Some of these are demonstrated in the following worked problems.

Worked problems on conversion tables and charts

Problem 1. Table 3.1 shows some approximate metric to imperial conversions. Use the table to determine:

(*a*) the number of millimetres in 18 inches,
(*b*) a speed of 30 miles per hour in kilometres per hour,
(*c*) the number of miles in 500 km,
(*d*) the number of kilograms in 25 pounds weight,
(*e*) the number of pounds and ounces in 36 kilograms (correct to the nearest ounce),
(*f*) the number of litres in 12 gallons, and
(*g*) the number of gallons in 50 litres.

Table 3.1

Length:	2.54 cm = 1 inch	
	1.61 km = 1 mile	
Weight:	1 kg	= 2.2 lb (1 lb = 16 ozs)
Capacity:	1 litre	= 1.76 pints (8 pints = 1 gallon)

(*a*) 18 inches = 18 × 2.54 cm = **45.72 cm**
45.72 cm = 45.72 × 10 mm = **457.2 mm**
(*b*) 30 m.p.h. = 30 × 1.61 km h^{-1} = **48.3 km h^{-1}**

(*c*) 500 km $= \dfrac{500}{1.61}$ miles = **310.6 miles**

(*d*) 25 lb $= \dfrac{25}{2.2}$ kg = **11.36 kg**

(*e*) 36 kg = 36 × 2.2 lb = 79.2 lb

0.2 lb = 0.2 × 16 oz = 3.2 oz = [illegible] oz, correct to the nearest ounce.

Thus, 36 kg = **79 lb 3 oz**, correct to the nearest ounce.

(*f*) 12 gallons = 12 × 8 pints = 96 pints

96 pints $= \dfrac{96}{1.76}$ litres = **54.55 litres**

(*g*) 50 litres = 50 × 1.76 pints = 88 pints

88 pints $= \dfrac{88}{8}$ gallons = **11 gallons.**

Problem 2.

Table 3.2

France	£1 = 11.0 francs (f)
West Germany	£1 = 4.76 Deutschmarks (Dm)
Italy	£1 = 2300 lira (l)
U.S.A.	£1 = 1.90 dollars ($)
Spain	£1 = 190 pesetas (pes)

Currency exchange rates for five countries are shown in Table 3.2. Calculate (*a*) how many French francs £14.50 will buy, (*b*) the number of West German Deutschmarks which can be bought for £62, (*c*) the pounds sterling which can be exchanged for 55 200 lira, (*d*) the number of American dollars which can be purchased for £83.50, and (*e*) the pounds sterling which can be exchanged for 1 805 pesetas.

(*a*) £1 = 11.0 francs, hence £14.50 = 14.50 × 11.0 francs = **159.5 f**

(*b*) £1 = 4.76 Deutschmarks, hence £62 = 62 × 4.76 Dm = **295.12 Dm**

(*c*) £1 = 2 300 lira, hence 55 200 lira = £$\dfrac{55\,200}{2\,300}$ = **£24**

(*d*) £1 = 1.90 dollars, hence £83.50 = 83.50 × 1.90 dollars = **$158.65**

(*e*) £1 = 190 pesetas, hence 1 805 pesetas = £$\dfrac{1\,805}{190}$ = **£9.50**

Problem 3. A production schedule used in the manufacture of a machine tool is as shown on p. 75. Each process requires the work of one man.

(*a*) How many man-hours are required to produce the machine tool?
(*b*) How long does it take to make the tool?
(*c*) The men employed in processes A, B and C are paid £2 per hour whereas those employed in processes D, E and F are paid £2.50 per hour. Materials for the production of one tool cost £25. If each tool is sold for

£120 determine the percentage profit of each tool based on the sale price.

Process	Start hour	End hour
A	1	5
B	4	12
C	2	7
D	7	9
E	8	11
F	12	14

(*a*) Since process A starts in the first hour and ends in the fifth hour then 5 hours are required altogether for the process. This is obtained by (end hour − start hour + 1). Similarly, process C takes (7 − 2 + 1), i.e. 6 hours to complete. Hence:

Length of time for process A = 5 hours
B = 9 hours
C = 6 hours
D = 3 hours
E = 4 hours
F = 3 hours

Total time for production: = 30 hours

Since each process requires the work of one man, then **30 man-hours** are required for the production of the tool.

(*b*) The machine tool is completed after process F (i.e. the highest 'end-hour' value), i.e. after **14 hours.**

(*c*) For each machine tool produced:

Men on process A earn 5 × 2 = £10.00
B earn 9 × 2 = £18.00
C earn 6 × 2 = £12.00
D earn 3 × 2.5 = £ 7.50
E earn 4 × 2.5 = £10.00
F earn 3 × 2.5 = £ 7.50

Total labour costs = £65.00

Cost of labour + materials = £65 + £25 = £90.
If each machine tool is sold for £120 then £(120 − 90). i.e. £30 profit is made on each. Percentage profit = $\frac{30}{120} \times 100\%$ = **25%.**

Further problems on conversion tables and charts may be found in Section 3.6 (Problems 61 to 65), page 92.

 Table 3.3 Squares. From 1.00 - 5.49

	0	1	2	3	4	5	6	7	8	9	Mean Differences								
											1	2	3	4	5	6	7	8	9
1·0	1·000	1·020	1·040	1·061	1·082	1·103	1·124	1·145	1·166	1·188	2	4	6	8	10	13	15	17	19
1·1	1·210	1·232	1·254	1·277	1·300	1·323	1·346	1·369	1·392	1·416	2	5	7	9	11	14	16	18	21
1·2	1·440	1·464	1·488	1·513	1·538	1·563	1·588	1·613	1·638	1·664	2	5	7	10	12	15	17	20	22
1·3	1·690	1·716	1·742	1·769	1·796	1·823	1·850	1·877	1·904	1·932	3	5	8	11	13	16	19	22	24
1·4	1·960	1·988	2·016	2·045	2·074	2·103	2·132	2·161	2·190	2·220	3	6	9	12	14	17	20	23	26
1·5	2·250	2·280	2·310	2·341	2·372	2·403	2·434	2·465	2·496	2·528	3	6	9	12	15	19	22	25	28
1·6	2·560	2·592	2·624	2·657	2·690	2·723	2·756	2·789	2·822	2·856	3	7	10	13	16	20	23	26	30
1·7	2·890	2·924	2·958	2·993	3·028	3·063	3·098	3·133	3·168	3·204	3	7	10	14	17	21	24	28	31
1·8	3·240	3·276	3·312	3·349	3·386	3·423	3·460	3·497	3·534	3·572	4	7	11	15	18	22	26	30	33
1·9	3·610	3·648	3·686	3·725	3·764	3·803	3·842	3·881	3·920	3·960	4	8	12	16	19	23	27	31	35
2·0	4·000	4·040	4·080	4·121	4·162	4·203	4·244	4·285	4·326	4·368	4	8	12	16	20	25	29	33	37
2·1	4·410	4·452	4·494	4·537	4·580	4·623	4·666	4·709	4·752	4·796	4	9	13	17	21	26	30	34	39
2·2	4·840	4·884	4·928	4·973	5·018	5·063	5·108	5·153	5·198	5·244	4	9	13	18	22	27	31	36	40
2·3	5·290	5·336	5·382	5·429	5·476	5·523	5·570	5·617	5·664	5·712	5	9	14	19	23	28	33	38	42
2·4	5·760	5·808	5·856	5·905	5·954	6·003	6·052	6·101	6·150	6·200	5	10	15	20	24	29	34	39	44
2·5	6·250	6·300	6·350	6·401	6·452	6·503	6·554	6·605	6·656	6·708	5	10	15	20	25	31	36	41	46
2·6	6·760	6·812	6·864	6·917	6·970	7·023	7·076	7·129	7·182	7·236	5	11	16	21	26	32	37	42	48
2·7	7·290	7·344	7·398	7·453	7·508	7·563	7·618	7·673	7·728	7·784	5	11	16	22	27	33	38	44	49
2·8	7·840	7·896	7·952	8·009	8·066	8·123	8·180	8·237	8·294	8·352	6	11	17	23	28	34	40	46	51
2·9	8·410	8·468	8·526	8·585	8·644	8·703	8·762	8·821	8·880	8·940	6	12	18	24	29	35	41	47	53
3·0	9·000	9·060	9·120	9·181	9·242	9·303	9·364	9·425	9·486	9·548	6	12	18	24	30	37	43	49	55
3·1	9·610	9·672	9·734	9·797	9·860	9·923	9·986				6	13	19	25	31	38	44	50	57
3·1								10·05	10·11	10·18	1	1	2	3	3	4	4	5	6
3·2	10·24	10·30	10·37	10·43	10·50	10·56	10·63	10·69	10·76	10·82	1	1	2	3	3	4	5	5	6
3·3	10·89	10·96	11·02	11·09	11·16	11·22	11·29	11·36	11·42	11·49	1	1	2	3	3	4	5	5	6
3·4	11·56	11·63	11·70	11·76	11·83	11·90	11·97	12·04	12·11	12·18	1	1	2	3	3	4	5	6	6
3·5	12·25	12·32	12·39	12·46	12·53	12·60	12·67	12·74	12·82	12·89	1	1	2	3	4	4	5	6	6
3·6	12·96	13·03	13·10	13·18	13·25	13·32	13·40	13·47	13·54	13·62	1	1	2	3	4	4	5	6	7
3·7	13·69	13·76	13·84	13·91	13·99	14·06	14·14	14·21	14·29	14·36	1	2	2	3	4	5	5	6	7
3·8	14·44	14·52	14·59	14·67	14·75	14·82	14·90	14·98	15·05	15·13	1	2	2	3	4	5	5	6	7
3·9	15·21	15·29	15·37	15·44	15·52	15·60	15·68	15·76	15·84	15·92	1	2	2	3	4	5	6	6	7
4·0	16·00	16·08	16·16	16·24	16·32	16·40	16·48	16·56	16·65	16·73	1	2	2	3	4	5	6	6	7
4·1	16·81	16·89	16·97	17·06	17·14	17·22	17·31	17·39	17·47	17·56	1	2	2	3	4	5	6	7	7
4·2	17·64	17·72	17·81	17·89	17·98	18·06	18·15	18·23	18·32	18·40	1	2	3	3	4	5	6	7	8
4·3	18·49	18·58	18·66	18·75	18·84	18·92	19·01	19·10	19·18	19·27	1	2	3	3	4	5	6	7	8
4·4	19·36	19·45	19·54	19·62	19·71	19·80	19·89	19·98	20·07	20·16	1	2	3	4	5	5	6	7	8
4·5	20·25	20·34	20·43	20·52	20·61	20·70	20·79	20·88	20·98	21·07	1	2	3	4	5	5	6	7	8
4·6	21·16	21·25	21·34	21·44	21·53	21·62	21·72	21·81	21·90	22·00	1	2	3	4	5	6	7	7	8
4·7	22·09	22·18	22·28	22·37	22·47	22·56	22·66	22·75	22·85	22·94	1	2	3	4	5	6	7	8	9
4·8	23·04	23·14	23·23	23·33	23·43	23·52	23·62	23·72	23·81	23·91	1	2	3	4	5	6	7	8	9
4·9	24·01	24·11	24·21	24·30	24·40	24·50	24·60	24·70	24·80	24·90	1	2	3	4	5	6	7	8	9
5·0	25·00	25·10	25·20	25·30	25·40	25·50	25·60	25·70	25·81	25·91	1	2	3	4	5	6	7	8	9
5·1	26·01	26·11	26·21	26·32	26·42	26·52	26·63	26·73	26·83	26·94	1	2	3	4	5	6	7	8	9
5·2	27·04	27·14	27·25	27·35	27·46	27·56	27·67	27·77	27·88	27·98	1	2	3	4	5	6	7	8	9
5·3	28·09	28·20	28·30	28·41	28·52	28·62	28·73	28·84	28·94	29·05	1	2	3	4	5	6	7	9	10
5·4	29·16	29·27	29·38	29·48	29·59	29·70	29·81	29·92	30·03	30·14	1	2	3	4	6	7	8	9	10

Table 3.3 (cont'd) Squares. From 5.50 - 9.99

	0	1	2	3	4	5	6	7	8	9	Mean Differences								
											1	2	3	4	5	6	7	8	9
5·5	30·25	30·36	30·47	30·58	30·69	30·80	30·91	31·02	31·14	31·25	1	2	3	4	6	7	8	9	10
5·6	31·36	31·47	31·58	31·70	31·81	31·92	32·04	32·15	32·26	32·38	1	2	3	5	6	7	8	9	10
5·7	32·49	32·60	32·72	32·83	32·95	33·06	33·18	33·29	33·41	33·52	1	2	3	5	6	7	8	9	10
5·8	33·64	33·76	33·87	33·99	34·11	34·22	34·34	34·46	34·57	34·69	1	2	4	5	6	7	8	9	11
5·9	34·81	34·93	35·05	35·16	35·28	35·40	35·52	35·64	35·76	35·88	1	2	4	5	6	7	8	10	11
6·0	36·00	36·12	36·24	36·36	36·48	36·60	36·72	36·84	36·97	37·09	1	2	4	5	6	7	9	10	11
6·1	37·21	37·33	37·45	37·58	37·70	37·82	37·95	38·07	38·19	38·32	1	2	4	5	6	7	9	10	11
6·2	38·44	38·56	38·69	38·81	38·94	39·06	39·19	39·31	39·44	39·56	1	3	4	5	6	8	9	10	11
6·3	39·69	39·82	39·94	40·07	40·20	40·32	40·45	40·58	40·70	40·83	1	3	4	5	6	8	9	10	11
6·4	40·96	41·09	41·22	41·34	41·47	41·60	41·73	41·86	41·99	42·12	1	3	4	5	6	8	9	10	12
6·5	42·25	42·38	42·51	42·64	42·77	42·90	43·03	43·16	43·30	43·43	1	3	4	5	7	8	9	10	12
6·6	43·56	43·69	43·82	43·96	44·09	44·22	44·36	44·49	44·62	44·76	1	3	4	5	7	8	9	11	12
6·7	44·89	45·02	45·16	45·29	45·43	45·56	45·70	45·83	45·97	46·10	1	3	4	5	7	8	9	11	12
6·8	46·24	46·38	46·51	46·65	46·79	46·92	47·06	47·20	47·33	47·47	1	3	4	5	7	8	10	11	12
6·9	47·61	47·75	47·89	48·02	48·16	48·30	48·44	48·58	48·72	48·86	1	3	4	6	7	8	10	11	13
7·0	49·00	49·14	49·28	49·42	49·56	49·70	49·84	49·98	50·13	50·27	1	3	4	6	7	8	10	11	13
7·1	50·41	50·55	50·69	50·84	50·98	51·12	51·27	51·41	51·55	51·70	1	3	4	6	7	9	10	11	13
7·2	51·84	51·98	52·13	52·27	52·42	52·56	52·71	52·85	53·00	53·14	1	3	4	6	7	9	10	12	13
7·3	53·29	53·44	53·58	53·73	53·88	54·02	54·17	54·32	54·46	54·61	1	3	4	6	7	9	10	12	13
7·4	54·76	54·91	55·06	55·20	55·35	55·50	55·65	55·80	55·95	56·10	1	3	4	6	7	9	10	12	13
7·5	56·25	56·40	56·55	56·70	56·85	57·00	57·15	57·30	57·46	57·61	2	3	5	6	8	9	11	12	14
7·6	57·76	57·91	58·06	58·22	58·37	58·52	58·68	58·83	58·98	59·14	2	3	5	6	8	9	11	12	14
7·7	59·29	59·44	59·60	59·75	59·91	60·06	60·22	60·37	60·53	60·68	2	3	5	6	8	9	11	12	14
7·8	60·84	61·00	61·15	61·31	61·47	61·62	61·78	61·94	62·09	62·25	2	3	5	6	8	9	11	13	14
7·9	62·41	62·57	62·73	62·88	63·04	63·20	63·36	63·52	63·68	63·84	2	3	5	6	8	10	11	13	14
8·0	64·00	64·16	64·32	64·48	64·64	64·80	64·96	65·12	65·29	65·45	2	3	5	6	8	10	11	13	14
8·1	65·61	65·77	65·93	66·10	66·26	66·42	66·59	66·75	66·91	67·08	2	3	5	7	8	10	11	13	15
8·2	67·24	67·40	67·57	67·73	67·90	68·06	68·23	68·39	68·56	68·72	2	3	5	7	8	10	12	13	15
8·3	68·89	69·06	69·22	69·39	69·56	69·72	69·89	70·06	70·22	70·39	2	3	5	7	8	10	12	13	15
8·4	70·56	70·73	70·90	71·06	71·23	71·40	71·57	71·74	71·91	72·08	2	3	5	7	8	10	12	14	15
8·5	72·25	72·42	72·59	72·76	72·93	73·10	73·27	73·44	73·62	73·79	2	3	5	7	9	10	12	14	15
8·6	73·96	74·13	74·30	74·48	74·65	74·82	75·00	75·17	75·34	75·52	2	3	5	7	9	10	12	14	16
8·7	75·69	75·86	76·04	76·21	76·39	76·56	76·74	76·91	77·09	77·26	2	4	5	7	9	11	12	14	16
8·8	77·44	77·62	77·79	77·97	78·15	78·32	78·50	78·68	78·85	79·03	2	4	5	7	9	11	12	14	16
8·9	79·21	79·39	79·57	79·74	79·92	80·10	80·28	80·46	80·64	80·82	2	4	5	7	9	11	13	14	16
9·0	81·00	81·18	81·36	81·54	81·72	81·90	82·08	82·26	82·45	82·63	2	4	5	7	9	11	13	14	16
9·1	82·81	82·99	83·17	83·36	83·54	83·72	83·91	84·09	84·27	84·46	2	4	5	7	9	11	13	15	16
9·2	84·64	84·82	85·01	85·19	85·38	85·56	85·75	85·93	86·12	86·30	2	4	6	7	9	11	13	15	17
9·3	86·49	86·68	86·86	87·05	87·24	87·42	87·61	87·80	87·98	88·17	2	4	6	7	9	11	13	15	17
9·4	88·36	88·55	88·74	88·92	89·11	89·30	89·49	89·68	89·87	90·06	2	4	6	8	9	11	13	15	17
9·5	90·25	90·44	90·63	90·82	91·01	91·20	91·39	91·58	91·78	91·97	2	4	6	8	10	11	13	15	17
9·6	92·16	92·35	92·54	92·74	92·93	93·12	93·32	93·51	93·70	93·90	2	4	6	8	10	12	14	15	17
9·7	94·09	94·28	94·48	94·67	94·87	95·06	95·26	95·45	95·65	95·84	2	4	6	8	10	12	14	16	18
9·8	96·04	96·24	96·43	96·63	96·83	97·02	97·22	97·42	97·61	97·81	2	4	6	8	10	12	14	16	18
9·9	98·01	98·21	98·41	98·60	98·80	99·00	99·20	99·40	99·60	99·80	2	4	6	8	10	12	14	16	18

Table 3.4 Square roots. From 1 - 10

	0	1	2	3	4	5	6	7	8	9	Mean Differences								
											1	2	3	4	5	6	7	8	9
1·0	1·000	1·005	1·010	1·015	1·020	1·025	1·030	1·034	1·039	1·044	0	1	1	2	2	3	3	4	4
1·1	1·049	1·054	1·058	1·063	1·068	1·072	1·077	1·082	1·086	1·091	0	1	1	2	2	3	3	4	4
1·2	1·095	1·100	1·105	1·109	1·114	1·118	1·122	1·127	1·131	1·136	0	1	1	2	2	3	3	4	4
1·3	1·140	1·145	1·149	1·153	1·158	1·162	1·166	1·170	1·175	1·179	0	1	1	2	2	3	3	3	4
1·4	1·183	1·187	1·192	1·196	1·200	1·204	1·208	1·212	1·217	1·221	0	1	1	2	2	2	3	3	4
1·5	1·225	1·229	1·233	1·237	1·241	1·245	1·249	1·253	1·257	1·261	0	1	1	2	2	2	3	3	4
1·6	1·265	1·269	1·273	1·277	1·281	1·285	1·288	1·292	1·296	1·300	0	1	1	2	2	2	3	3	3
1·7	1·304	1·308	1·311	1·315	1·319	1·323	1·327	1·330	1·334	1·338	0	1	1	2	2	2	3	3	3
1·8	1·342	1·345	1·349	1·353	1·356	1·360	1·364	1·367	1·371	1·375	0	1	1	1	2	2	3	3	3
1·9	1·378	1·382	1·386	1·389	1·393	1·396	1·400	1·404	1·407	1·411	0	1	1	1	2	2	3	3	3
2·0	1·414	1·418	1·421	1·425	1·428	1·432	1·435	1·439	1·442	1·446	0	1	1	1	2	2	2	3	3
2·1	1·449	1·453	1·456	1·459	1·463	1·466	1·470	1·473	1·476	1·480	0	1	1	1	2	2	2	3	3
2·2	1·483	1·487	1·490	1·493	1·497	1·500	1·503	1·507	1·510	1·513	0	1	1	1	2	2	2	3	3
2·3	1·517	1·520	1·523	1·526	1·530	1·533	1·536	1·539	1·543	1·546	0	1	1	1	2	2	2	3	3
2·4	1·549	1·552	1·556	1·559	1·562	1·565	1·568	1·572	1·575	1·578	0	1	1	1	2	2	2	3	3
2·5	1·581	1·584	1·587	1·591	1·594	1·597	1·600	1·603	1·606	1·609	0	1	1	1	2	2	2	3	3
2·6	1·612	1·616	1·619	1·622	1·625	1·628	1·631	1·634	1·637	1·640	0	1	1	1	2	2	2	2	3
2·7	1·643	1·646	1·649	1·652	1·655	1·658	1·661	1·664	1·667	1·670	0	1	1	1	2	2	2	2	3
2·8	1·673	1·676	1·679	1·682	1·685	1·688	1·691	1·694	1·697	1·700	0	1	1	1	1	2	2	2	3
2·9	1·703	1·706	1·709	1·712	1·715	1·718	1·720	1·723	1·726	1·729	0	1	1	1	1	2	2	2	3
3·0	1·732	1·735	1·738	1·741	1·744	1·746	1·749	1·752	1·755	1·758	0	1	1	1	1	2	2	2	3
3·1	1·761	1·764	1·766	1·769	1·772	1·775	1·778	1·780	1·783	1·786	0	1	1	1	1	2	2	2	3
3·2	1·789	1·792	1·794	1·797	1·800	1·803	1·806	1·808	1·811	1·814	0	1	1	1	1	2	2	2	2
3·3	1·817	1·819	1·822	1·825	1·828	1·830	1·833	1·836	1·838	1·841	0	1	1	1	1	2	2	2	2
3·4	1·844	1·847	1·849	1·852	1·855	1·857	1·860	1·863	1·865	1·868	0	1	1	1	1	2	2	2	2
3·5	1·871	1·873	1·876	1·879	1·881	1·884	1·887	1·889	1·892	1·895	0	1	1	1	1	2	2	2	2
3·6	1·897	1·900	1·903	1·905	1·908	1·910	1·913	1·916	1·918	1·921	0	1	1	1	1	2	2	2	2
3·7	1·924	1·926	1·929	1·931	1·934	1·936	1·939	1·942	1·944	1·947	0	1	1	1	1	2	2	2	2
3·8	1·949	1·952	1·954	1·957	1·960	1·962	1·965	1·967	1·970	1·972	0	1	1	1	1	2	2	2	2
3·9	1·975	1·977	1·980	1·982	1·985	1·987	1·990	1·992	1·995	1·997	0	1	1	1	1	2	2	2	2
4·0	2·000	2·002	2·005	2·007	2·010	2·012	2·015	2·017	2·020	2·022	0	0	1	1	1	1	2	2	2
4·1	2·025	2·027	2·030	2·032	2·035	2·037	2·040	2·042	2·045	2·047	0	0	1	1	1	1	2	2	2
4·2	2·049	2·052	2·054	2·057	2·059	2·062	2·064	2·066	2·069	2·071	0	0	1	1	1	1	2	2	2
4·3	2·074	2·076	2·078	2·081	2·083	2·086	2·088	2·090	2·093	2·095	0	0	1	1	1	1	2	2	2
4·4	2·098	2·100	2·102	2·105	2·107	2·110	2·112	2·114	2·117	2·119	0	0	1	1	1	1	2	2	2
4·5	2·121	2·124	2·126	2·128	2·131	2·133	2·135	2·138	2·140	2·142	0	0	1	1	1	1	2	2	2
4·6	2·145	2·147	2·149	2·152	2·154	2·156	2·159	2·161	2·163	2·166	0	0	1	1	1	1	2	2	2
4·7	2·168	2·170	2·173	2·175	2·177	2·179	2·182	2·184	2·186	2·189	0	0	1	1	1	1	2	2	2
4·8	2·191	2·193	2·195	2·198	2·200	2·202	2·205	2·207	2·209	2·211	0	0	1	1	1	1	2	2	2
4·9	2·214	2·216	2·218	2·220	2·223	2·225	2·227	2·229	2·232	2·234	0	0	1	1	1	1	2	2	2
5·0	2·236	2·238	2·241	2·243	2·245	2·247	2·249	2·254	2·254	2·256	0	0	1	1	1	1	2	2	2
5·1	2·258	2·261	2·263	2·265	2·267	2·269	2·272	2·274	2·276	2·278	0	0	1	1	1	1	2	2	2
5·2	2·280	2·283	2·285	2·287	2·289	2·291	2·293	2·296	2·298	2·300	0	0	1	1	1	1	2	2	2
5·3	2·302	2·304	2·307	2·309	2·311	2·313	2·315	2·317	2·319	2·322	0	0	1	1	1	1	2	2	2
5·4	2·324	2·326	2·328	2·330	2·332	2·335	2·337	2·339	2·341	2·343	0	0	1	1	1	1	1	2	2

	0	1	2	3	4	5	6	7	8	9	Mean Differences								
											1	2	3	4	5	6	7	8	9
5·5	2·345	2·347	2·349	2·352	2·354	2·356	2·358	2·360	2·362	2·364	0	0	1	1	1	1	1	2	2
5·6	2·366	2·369	2·371	2·373	2·375	2·377	2·379	2·381	2·383	2·385	0	0	1	1	1	1	1	2	2
5·7	2·387	2·390	2·392	2·394	2·396	2·398	2·400	2·402	2·404	2·406	0	0	1	1	1	1	1	2	2
5·8	2·408	2·410	2·412	2·415	2·417	2·419	2·421	2·423	2·425	2·427	0	0	1	1	1	1	1	2	2
5·9	2·429	2·431	2·433	2·435	2·437	2·439	2·441	2·443	2·445	2·447	0	0	1	1	1	1	1	2	2
6·0	2·449	2·452	2·454	2·456	2·458	2·460	2·462	2·464	2·466	2·468	0	0	1	1	1	1	1	2	2
6·1	2·470	2·472	2·474	2·476	2·478	2·480	2·482	2·484	2·486	2·488	0	0	1	1	1	1	1	2	2
6·2	2·490	2·492	2·494	2·496	2·498	2·500	2·502	2·504	2·506	2·508	0	0	1	1	1	1	1	2	2
6·3	2·510	2·512	2·514	2·516	2·518	2·520	2·522	2·524	2·526	2·528	0	0	1	1	1	1	1	2	2
6·4	2·530	2·532	2·534	2·536	2·538	2·540	2·542	2·544	2·546	2·548	0	0	1	1	1	1	1	2	2
6·5	2·550	2·551	2·553	2·555	2·557	2·559	2·561	2·563	2·565	2·567	0	0	1	1	1	1	1	2	2
6·6	2·569	2·571	2·573	2·575	2·577	2·579	2·581	2·583	2·585	2·587	0	0	1	1	1	1	1	2	2
6·7	2·588	2·590	2·592	2·594	2·596	2·598	2·600	2·602	2·604	2·606	0	0	1	1	1	1	1	2	2
6·8	2·608	2·610	2·612	2·613	2·615	2·617	2·619	2·621	2·623	2·625	0	0	1	1	1	1	1	2	2
6·9	2·627	2·629	2·631	2·632	2·634	2·636	2·638	2·640	2·642	2·644	0	0	1	1	1	1	1	2	2
7·0	2·646	2·648	2·650	2·651	2·653	2·655	2·657	2·659	2·661	2·663	0	0	1	1	1	1	1	2	2
7·1	2·665	2·666	2·668	2·670	2·672	2·674	2·676	2·678	2·680	2·681	0	0	1	1	1	1	1	1	2
7·2	2·683	2·685	2·687	2·689	2·691	2·693	2·694	2·696	2·698	2·700	0	0	1	1	1	1	1	1	2
7·3	2·702	2·704	2·706	2·707	2·709	2·711	2·713	2·715	2·717	2·718	0	0	1	1	1	1	1	1	2
7·4	2·720	2·722	2·724	2·726	2·728	2·729	2·731	2·733	2·735	2·737	0	0	1	1	1	1	1	1	2
7·5	2·739	2·740	2·742	2·744	2·746	2·748	2·750	2·751	2·753	2·755	0	0	1	1	1	1	1	1	2
7·6	2·757	2·759	2·760	2·762	2·764	2·766	2·768	2·769	2·771	2·773	0	0	1	1	1	1	1	1	2
7·7	2·775	2·777	2·778	2·780	2·782	2·784	2·786	2·787	2·789	2·791	0	0	1	1	1	1	1	1	2
7·8	2·793	2·795	2·796	2·798	2·800	2·802	2·804	2·805	2·807	2·809	0	0	1	1	1	1	1	1	2
7·9	2·811	2·812	2·814	2·816	2·818	2·820	2·821	2·823	2·825	2·827	0	0	1	1	1	1	1	1	2
8·0	2·828	2·830	2·832	2·834	2·835	2·837	2·839	2·841	2·843	2·844	0	0	1	1	1	1	1	1	2
8·1	2·846	2·848	2·850	2·851	2·853	2·855	2·857	2·858	2·860	2·862	0	0	1	1	1	1	1	1	2
8·2	2·864	2·865	2·867	2·869	2·871	2·872	2·874	2·876	2·877	2·879	0	0	1	1	1	1	1	1	2
8·3	2·881	2·883	2·884	2·886	2·888	2·890	2·891	2·893	2·895	2·897	0	0	1	1	1	1	1	1	2
8·4	2·898	2·900	2·902	2·903	2·905	2·907	2·909	2·910	2·912	2·914	0	0	1	1	1	1	1	1	2
8·5	2·915	2·917	2·919	2·921	2·922	2·924	2·926	2·927	2·929	2·931	0	0	1	1	1	1	1	1	2
8·6	2·933	2·934	2·936	2·938	2·939	2·941	2·943	2·944	2·946	2·948	0	0	1	1	1	1	1	1	2
8·7	2·950	2·951	2·953	2·955	2·956	2·958	2·960	2·961	2·963	2·965	0	0	1	1	1	1	1	1	2
8·8	2·966	2·968	2·970	2·972	2·973	2·975	2·977	2·978	2·980	2·982	0	0	1	1	1	1	1	1	2
8·9	2·983	2·985	2·987	2·988	2·990	2·992	2·993	2·995	2·997	2·998	0	0	1	1	1	1	1	1	2
9·0	3·000	3·002	3·003	3·005	3·007	3·008	3·010	3·012	3·013	3·015	0	0	0	1	1	1	1	1	1
9·1	3·017	3·018	3·020	3·022	3·023	3·025	3·027	3·028	3·030	3·032	0	0	0	1	1	1	1	1	1
9·2	3·033	3·035	3·036	3·038	3·040	3·041	3·043	3·045	3·046	3·048	0	0	0	1	1	1	1	1	1
9·3	3·050	3·051	3·053	3·055	3·056	3·058	3·059	3·061	3·063	3·064	0	0	0	1	1	1	1	1	1
9·4	3·066	3·068	3·069	3·071	3·072	3·074	3·076	3·077	3·079	3·081	0	0	0	1	1	1	1	1	1
9·5	3·082	3·084	3·085	3·087	3·089	3·090	3·092	3·094	3·095	3·097	0	0	0	1	1	1	1	1	1
9·6	3·098	3·100	3·102	3·103	3·105	3·106	3·108	3·110	3·111	3·113	0	0	0	1	1	1	1	1	1
9·7	3·114	3·116	3·118	3·119	3·121	3·122	3·124	3·126	3·127	3·129	0	0	0	1	1	1	1	1	1
9·8	3·130	3·132	3·134	3·135	3·137	3·138	3·140	3·142	3·143	3·145	0	0	0	1	1	1	1	1	1
9·9	3·146	3·148	3·150	3·151	3·153	3·154	3·156	3·158	3·159	3·161	0	0	0	1	1	1	1	1	1

Table 3.5 Square roots. From 10 - 100

	0	1	2	3	4	5	6	7	8	9	Mean Differences								
											1	2	3	4	5	6	7	8	9
10	3·162	3·178	3·194	3·209	3·225	3·240	3·256	3·271	3·286	3·302	2	3	5	6	8	9	11	12	14
11	3·317	3·332	3·347	3·362	3·376	3·391	3·406	3·421	3·435	3·450	1	3	4	6	7	9	10	12	13
12	3·464	3·479	3·493	3·507	3·521	3·536	3·550	3·564	3·578	3·592	1	3	4	6	7	8	10	11	13
13	3·606	3·619	3·633	3·647	3·661	3·674	3·688	3·701	3·715	3·728	1	3	4	5	7	8	10	11	12
14	3·742	3·755	3·768	3·782	3·795	3·808	3·821	3·834	3·847	3·860	1	3	4	5	7	8	9	11	12
15	3·873	3·886	3·899	3·912	3·924	3·937	3·950	3·962	3·975	3·987	1	3	4	5	6	8	9	10	11
16	4·000	4·012	4·025	4·037	4·050	4·062	4·074	4·087	4·099	4·111	1	2	4	5	6	7	9	10	11
17	4·123	4·135	4·147	4·159	4·171	4·183	4·195	4·207	4·219	4·231	1	2	4	5	6	7	8	10	11
18	4·243	4·254	4·266	4·278	4·290	4·301	4·313	4·324	4·336	4·347	1	2	3	5	6	7	8	9	10
19	4·359	4·370	4·382	4·393	4·405	4·416	4·427	4·438	4·450	4·461	1	2	3	5	6	7	8	9	10
20	4·472	4·483	4·494	4·506	4·517	4·528	4·539	4·550	4·561	4·572	1	2	3	4	6	7	8	9	10
21	4·583	4·593	4·604	4·615	4·626	4·637	4·648	4·658	4·669	4·680	1	2	3	4	5	6	8	9	10
22	4·690	4·701	4·712	4·722	4·733	4·743	4·754	4·764	4·775	4·785	1	2	3	4	5	6	7	8	9
23	4·796	4·806	4·817	4·827	4·837	4·848	4·858	4·868	4·879	4·889	1	2	3	4	5	6	7	8	9
24	4·899	4·909	4·919	4·930	4·940	4·950	4·960	4·970	4·980	4·990	1	2	3	4	5	6	7	8	9
25	5·000	5·010	5·020	5·030	5·040	5·050	5·060	5·070	5·079	5·089	1	2	3	4	5	6	7	8	9
26	5·099	5·109	5·119	5·128	5·138	5·148	5·158	5·167	5·177	5·187	1	2	3	4	5	6	7	8	9
27	5·196	5·206	5·215	5·225	5·235	5·244	5·254	5·263	5·273	5·282	1	2	3	4	5	6	7	8	9
28	5·292	5·301	5·310	5·320	5·329	5·339	5·348	5·357	5·367	5·376	1	2	3	4	5	6	7	7	8
29	5·385	5·394	5·404	5·413	5·422	5·431	5·441	5·450	5·459	5·468	1	2	3	4	5	5	6	7	8
30	5·477	5·486	5·495	5·505	5·514	5·523	5·532	5·541	5·550	5·559	1	2	3	4	4	5	6	7	8
31	5·568	5·577	5·586	5·595	5·604	5·612	5·621	5·630	5·639	5·648	1	2	3	3	4	5	6	7	8
32	5·657	5·666	5·675	5·683	5·692	5·701	5·710	5·718	5·727	5·736	1	2	3	3	4	5	6	7	8
33	5·745	5·753	5·762	5·771	5·779	5·788	5·797	5·805	5·814	5·822	1	2	3	3	4	5	6	7	8
34	5·831	5·840	5·848	5·857	5·865	5·874	5·882	5·891	5·899	5·908	1	2	3	3	4	5	6	7	8
35	5·916	5·925	5·933	5·941	5·950	5·958	5·967	5·975	5·983	5·992	1	2	2	3	4	5	6	7	8
36	6·000	6·008	6·017	6·025	6·033	6·042	6·050	6·058	6·066	6·075	1	2	2	3	4	5	6	7	7
37	6·083	6·091	6·099	6·107	6·116	6·124	6·132	6·140	6·148	6·156	1	2	2	3	4	5	6	7	7
38	6·164	6·173	6·181	6·189	6·197	6·205	6·213	6·221	6·229	6·237	1	2	2	3	4	5	6	6	7
39	6·245	6·253	6·261	6·269	6·277	6·285	6·293	6·301	6·309	6·317	1	2	2	3	4	5	6	6	7
40	6·325	6·332	6·340	6·348	6·356	6·364	6·372	6·380	6·387	6·395	1	2	2	3	4	5	6	6	7
41	6·403	6·411	6·419	6·427	6·434	6·442	6·450	6·458	6·465	6·473	1	2	2	3	4	5	5	6	7
42	6·481	6·488	6·496	6·504	6·512	6·519	6·527	6·535	6·542	6·550	1	2	2	3	4	5	5	6	7
43	6·557	6·565	6·573	6·580	6·588	6·595	6·603	6·611	6·618	6·626	1	2	2	3	4	5	5	6	7
44	6·633	6·641	6·648	6·656	6·663	6·671	6·678	6·686	6·693	6·701	1	2	2	3	4	5	5	6	7
45	6·708	6·716	6·723	6·731	6·738	6·745	6·753	6·760	6·768	6·775	1	1	2	3	4	4	5	6	7
46	6·782	6·790	6·797	6·804	6·812	6·819	6·826	6·834	6·841	6·848	1	1	2	3	4	4	5	6	7
47	6·856	6·863	6·870	6·877	6·885	6·892	6·899	6·907	6·914	6·921	1	1	2	3	4	4	5	6	7
48	6·928	6·935	6·943	6·950	6·957	6·964	6·971	6·979	6·986	6·993	1	1	2	3	4	4	5	6	6
49	7·000	7·007	7·014	7·021	7·029	7·036	7·043	7·050	7·057	7·064	1	1	2	3	4	4	5	6	6
50	7·071	7·078	7·085	7·092	7·099	7·106	7·113	7·120	7·127	7·134	1	1	2	3	4	4	5	6	6
51	7·141	7·148	7·155	7·162	7·169	7·176	7·183	7·190	7·197	7·204	1	1	2	3	4	4	5	6	6
52	7·211	7·218	7·225	7·232	7·239	7·246	7·253	7·259	7·266	7·273	1	1	2	3	3	4	5	6	6
53	7·280	7·287	7·294	7·301	7·308	7·314	7·321	7·328	7·335	7·342	1	1	2	3	3	4	5	5	6
54	7·348	7·355	7·362	7·369	7·376	7·382	7·389	7·396	7·403	7·409	1	1	2	3	3	4	5	5	6

Table 3.5 (cont'd) Square roots. From 10 - 100

	0	1	2	3	4	5	6	7	8	9	Mean Differences								
											1	2	3	4	5	6	7	8	9
55	7·416	7·423	7·430	7·436	7·443	7·450	7·457	7·463	7·470	7·477	1	1	2	3	3	4	5	5	6
56	7·483	7·490	7·497	7·503	7·510	7·517	7·523	7·530	7·537	7·543	1	1	2	3	3	4	5	5	6
57	7·550	7·556	7·563	7·570	7·576	7·583	7·589	7·596	7·603	7·609	1	1	2	3	3	4	5	5	6
58	7·616	7·622	7·629	7·635	7·642	7·649	7·655	7·662	7·668	7·675	1	1	2	3	3	4	5	5	6
59	7·681	7·688	7·694	7·701	7·707	7·714	7·720	7·727	7·733	7·740	1	1	2	3	3	4	4	5	6
60	7·746	7·752	7·759	7·765	7·772	7·778	7·785	7·791	7·797	7·804	1	1	2	3	3	4	4	5	6
61	7·810	7·817	7·823	7·829	7·836	7·842	7·849	7·855	7·861	7·868	1	1	2	3	3	4	4	5	6
62	7·874	7·880	7·887	7·893	7·899	7·906	7·912	7·918	7·925	7·931	1	1	2	3	3	4	4	5	6
63	7·937	7·944	7·950	7·956	7·962	7·969	7·975	7·981	7·987	7·994	1	1	2	3	3	4	4	5	6
64	8·000	8·006	8·012	8·019	8·025	8·031	8·037	8·044	8·050	8·056	1	1	2	2	3	4	4	5	6
65	8·062	8·068	8·075	8·081	8·087	8·093	8·099	8·106	8·112	8·118	1	1	2	2	3	4	4	5	6
66	8·124	8·130	8·136	8·142	8·149	8·155	8·161	8·167	8·173	8·179	1	1	2	2	3	4	4	5	5
67	8·185	8·191	8·198	8·204	8·210	8·216	8·222	8·228	8·234	8·240	1	1	2	2	3	4	4	5	5
68	8·246	8·252	8·258	8·264	8·270	8·276	8·283	8·289	8·295	8·301	1	1	2	2	3	4	4	5	5
69	8·307	8·313	8·319	8·325	8·331	8·337	8·343	8·349	8·355	8·361	1	1	2	2	3	4	4	5	5
70	8·367	8·373	8·379	8·385	8·390	8·396	8·402	8·408	8·414	8·420	1	1	2	2	3	4	4	5	5
71	8·426	8·432	8·438	8·444	8·450	8·456	8·462	8·468	8·473	8·479	1	1	2	2	3	4	4	5	5
72	8·485	8·491	8·497	8·503	8·509	8·515	8·521	8·526	8·532	8·538	1	1	2	2	3	3	4	5	5
73	8·544	8·550	8·556	8·562	8·567	8·573	8·579	8·585	8·591	8·597	1	1	2	2	3	3	4	5	5
74	8·602	8·608	8·614	8·620	8·626	8·631	8·637	8·643	8·649	8·654	1	1	2	2	3	3	4	5	5
75	8·660	8·666	8·672	8·678	8·683	8·689	8·695	8·701	8·706	8·712	1	1	2	2	3	3	4	5	5
76	8·718	8·724	8·729	8·735	8·741	8·746	8·752	8·758	8·764	8·769	1	1	2	2	3	3	4	5	5
77	8·775	8·781	8·786	8·792	8·798	8·803	8·809	8·815	8·820	8·826	1	1	2	2	3	3	4	4	5
78	8·832	8·837	8·843	8·849	8·854	8·860	8·866	8·871	8·877	8·883	1	1	2	2	3	3	4	4	5
79	8·888	8·894	8·899	8·905	8·911	8·916	8·922	8·927	8·933	8·939	1	1	2	2	3	3	4	4	5
80	8·944	8·950	8·955	8·961	8·967	8·972	8·978	8·983	8·989	8·994	1	1	2	2	3	3	4	4	5
81	9·000	9·006	9·011	9·017	9·022	9·028	9·033	9·039	9·044	9·050	1	1	2	2	3	3	4	4	5
82	9·055	9·061	9·066	9·072	9·077	9·083	9·088	9·094	9·099	9·105	1	1	2	2	3	3	4	4	5
83	9·110	9·116	9·121	9·127	9·132	9·138	9·143	9·149	9·154	9·160	1	1	2	2	3	3	4	4	5
84	9·165	9·171	9·176	9·182	9·187	9·192	9·198	9·203	9·209	9·214	1	1	2	2	3	3	4	4	5
85	9·220	9·225	9·230	9· 36	9·241	9·247	9·252	9·257	9·263	9·268	1	1	2	2	3	3	4	4	5
86	9·274	9·279	9·284	9·290	9·295	9·301	9·306	9·311	9·317	9·322	1	1	2	2	3	3	4	4	5
87	9·327	9·333	9·338	9·343	9·349	9·354	9·359	9·365	9·370	9·375	1	1	2	2	3	3	4	4	5
88	9·381	9·386	9·391	9·397	9·402	9·407	9·413	9·418	9·423	9·429	1	1	2	2	3	3	4	4	5
89	9·434	9·439	9·445	9·450	9·455	9·460	9·466	9·471	9·476	9·482	1	1	2	2	3	3	4	4	5
90	9·487	9·492	9·497	9·503	9·508	9·513	9·518	9·524	9·529	9·534	1	1	2	2	3	3	4	4	5
91	9·539	9·545	9·550	9·555	9·560	9·566	9·571	9·576	9·581	9·586	1	1	2	2	3	3	4	4	5
92	9·592	9·597	9·602	9·607	9·612	9·618	9·623	9·628	9·633	9·638	1	1	2	2	3	3	4	4	5
93	9·644	9·649	9·654	9·659	9·664	9·670	9·675	9·680	9·685	9·690	1	1	2	2	3	3	4	4	5
94	9·695	9·701	9·706	9·711	9·716	9·721	9·726	9·731	9·737	9·742	1	1	2	2	3	3	4	4	5
95	9·747	9·752	9·757	9·762	9·767	9·772	9·778	9·783	9·788	9·793	1	1	2	2	3	3	4	4	5
96	9·798	9·803	9·808	9·813	9·818	9·823	9·829	9·834	9·839	9·844	1	1	2	2	3	3	4	4	5
97	9·849	9·854	9·859	9·864	9·869	9·874	9·879	9·884	9·889	9·894	1	1	1	2	3	3	4	4	5
98	9·899	9·905	9·910	9·915	9·920	9·925	9·930	9·935	9·940	9·945	0	1	1	2	2	3	3	4	4
99	9·950	9·955	9·960	9·965	9·970	9·975	9·980	9·985	9·990	9·995	0	1	1	2	2	3	3	4	4

 Table 3.6 Reciprocals of numbers. From 1 - 10

Numbers in difference columns to be subtracted, not added

	0	1	2	3	4	5	6	7	8	9	Mean Differences								
											1	2	3	4	5	6	7	8	9
1.0	1.000	9901	9804	9709	9615	9524	9434	9346	9259	9174									
1.1	.9091	9009	8929	8850	8772	8696	8621	8547	8475	8403									
1.2	.8333	8264	8197	8130	8065	8000	7937	7874	7813	7752									
1.3	.7692	7634	7576	7519	7463	7407	7353	7299	7246	7194									
1.4	.7143	7092	7042	6993	6944	6897	6849	6803	6757	6711	5	10	14	19	24	29	33	38	43
1.5	.6667	6623	6579	6536	6494	6452	6410	6369	6329	6289	4	8	13	17	21	25	29	33	38
1.6	.6250	6211	6173	6135	6098	6061	6024	5988	5952	5917	4	7	11	15	18	22	26	29	33
1.7	.5882	5848	5814	5780	5747	5714	5682	5650	5618	5587	3	6	10	13	16	20	23	26	29
1.8	.5556	5525	5495	5464	5435	5405	5376	5348	5319	5291	3	6	9	12	15	17	20	23	26
1.9	.5263	5236	5208	5181	5155	5128	5102	5076	5051	5025	3	5	8	11	13	16	18	21	24
2.0	.5000	4975	4950	4926	4902	4878	4854	4831	4808	4785	2	5	7	10	12	14	17	19	21
2.1	.4762	4739	4717	4695	4673	4651	4630	4608	4587	4566	2	4	7	9	11	13	15	17	20
2.2	.4545	4525	4505	4484	4464	4444	4425	4405	4386	4367	2	4	6	8	10	12	14	16	18
2.3	.4348	4329	4310	4292	4274	4255	4237	4219	4202	4184	2	4	5	7	9	11	13	14	16
2.4	.4167	4149	4132	4115	4098	4082	4065	4049	4032	4016	2	3	5	7	8	10	12	13	15
2.5	.4000	3984	3968	3953	3937	3922	3906	3891	3876	3861	2	3	5	6	8	9	11	12	14
2.6	.3846	3831	3817	3802	3788	3774	3759	3745	3731	3717	1	3	4	6	7	8	10	11	13
2.7	.3704	3690	3676	3663	3650	3636	3623	3610	3597	3584	1	3	4	5	7	8	9	11	12
2.8	.3571	3559	3546	3534	3521	3509	3497	3484	3472	3460	1	2	4	5	6	7	9	10	11
2.9	.3448	3436	3425	3413	3401	3390	3378	3367	3356	3344	1	2	3	5	6	7	8	9	10
3.0	.3333	3322	3311	3300	3289	3279	3268	3257	3247	3236	1	2	3	4	5	6	7	9	10
3.1	.3226	3215	3205	3195	3185	3175	3165	3155	3145	3135	1	2	3	4	5	6	7	8	9
3.2	.3125	3115	3106	3096	3086	3077	3067	3058	3049	3040	1	2	3	4	5	6	7	8	9
3.3	.3030	3021	3012	3003	2994	2985	2976	2967	2959	2950	1	2	3	4	4	5	6	7	8
3.4	.2941	2933	2924	2915	2907	2899	2890	2882	2874	2865	1	2	3	3	4	5	6	7	8
3.5	.2857	2849	2841	2833	2825	2817	2809	2801	2793	2786	1	2	2	3	4	5	6	6	7
3.6	.2778	2770	2762	2755	2747	2740	2732	2725	2717	2710	1	2	2	3	4	5	5	6	7
3.7	.2703	2695	2688	2681	2674	2667	2660	2653	2646	2639	1	1	2	3	4	4	5	6	6
3.8	.2632	2625	2618	2611	2604	2597	2591	2584	2577	2571	1	1	2	3	3	4	5	5	6
3.9	.2564	2558	2551	2545	2538	2532	2525	2519	2513	2506	1	1	2	3	3	4	4	5	6
4.0	.2500	2494	2488	2481	2475	2469	2463	2457	2451	2445	1	1	2	2	3	4	4	5	5
4.1	.2439	2433	2427	2421	2415	2410	2404	2398	2392	2387	1	1	2	2	3	3	4	5	5
4.2	.2381	2375	2370	2364	2358	2353	2347	2342	2336	2331	1	1	2	2	3	3	4	4	5
4.3	.2326	2320	2315	2309	2304	2299	2294	2288	2283	2278	1	1	2	2	3	3	4	4	5
4.4	.2273	2268	2262	2257	2252	2247	2242	2237	2232	2227	1	1	2	2	3	3	4	4	5
4.5	.2222	2217	2212	2208	2203	2198	2193	2188	2183	2179	0	1	1	2	2	3	3	4	4
4.6	.2174	2169	2165	2160	2155	2151	2146	2141	2137	2132	0	1	1	2	2	3	3	4	4
4.7	.2128	2123	2119	2114	2110	2105	2101	2096	2092	2088	0	1	1	2	2	3	3	4	4
4.8	.2083	2079	2075	2070	2066	2062	2058	2053	2049	2045	0	1	1	2	2	3	3	3	4
4.9	.2041	2037	2033	2028	2024	2020	2016	2012	2008	2004	0	1	1	2	2	2	3	3	4
5.0	.2000	1996	1992	1988	1984	1980	1976	1972	1969	1965	0	1	1	2	2	2	3	3	4
5.1	.1961	1957	1953	1949	1946	1942	1938	1934	1931	1927	0	1	1	2	2	2	3	3	3
5.2	.1923	1919	1916	1912	1908	1905	1901	1898	1894	1890	0	1	1	1	2	2	3	3	3
5.3	.1887	1883	1880	1876	1873	1869	1866	1862	1859	1855	0	1	1	1	2	2	2	3	3
5.4	.1852	1848	1845	1842	1838	1835	1832	1828	1825	1821	0	1	1	1	2	2	2	3	3

Numbers in difference columns to be subtracted, not added

	0	1	2	3	4	5	6	7	8	9	Mean Differences		
											1 2 3	**4 5 6**	**7 8 9**
5.5	.1818	1815	1812	1808	1805	1802	1799	1795	1792	1789	0 1 1	1 2 2	2 3 3
5.6	.1786	1783	1779	1776	1773	1770	1767	1764	1761	1757	0 1 1	1 2 2	2 3 3
5.7	.1754	1751	1748	1745	1742	1739	1736	1733	1730	1727	0 1 1	1 1 2	2 2 3
5.8	.1724	1721	1718	1715	1712	1709	1706	1704	1701	1698	0 1 1	1 1 2	2 2 3
5.9	.1695	1692	1689	1686	1684	1681	1678	1675	1672	1669	0 1 1	1 1 2	2 2 3
6.0	.1667	1664	1661	1658	1656	1653	1650	1647	1645	1642	0 1 1	1 1 2	2 2 3
6.1	.1639	1637	1634	1631	1629	1626	1623	1621	1618	1616	0 1 1	1 1 2	2 2 2
6.2	.1613	1610	1608	1605	1603	1600	1597	1595	1592	1590	0 1 1	1 1 2	2 2 2
6.3	.1587	1585	1582	1580	1577	1575	1572	1570	1567	1565	0 0 1	1 1 1	2 2 2
6.4	.1562	1560	1558	1555	1553	1550	1548	1546	1543	1541	0 0 1	1 1 1	2 2 2
6.5	.1538	1536	1534	1531	1529	1527	1524	1522	1520	1517	0 0 1	1 1 1	2 2 2
6.6	.1515	1513	1511	1508	1506	1504	1502	1499	1497	1495	0 0 1	1 1 1	2 2 2
6.7	.1493	1490	1488	1486	1484	1481	1479	1477	1475	1473	0 0 1	1 1 1	2 2 2
6.8	.1471	1468	1466	1464	1462	1460	1458	1456	1453	1451	0 0 1	1 1 1	2 2 2
6.9	.1449	1447	1445	1443	1441	1439	1437	1435	1433	1431	0 0 1	1 1 1	2 2 2
7.0	.1429	1427	1425	1422	1420	1418	1416	1414	1412	1410	0 0 1	1 1 1	1 2 2
7.1	.1408	1406	1404	1403	1401	1399	1397	1395	1393	1391	0 0 1	1 1 1	1 2 2
7.2	.1389	1387	1385	1383	1381	1379	1377	1376	1374	1372	0 0 1	1 1 1	1 2 2
7.3	.1370	1368	1366	1364	1362	1361	1359	1357	1355	1353	0 0 1	1 1 1	1 2 2
7.4	.1351	1350	1348	1346	1344	1342	1340	1339	1337	1335	0 0 1	1 1 1	1 1 2
7.5	.1333	1332	1330	1328	1326	1325	1323	1321	1319	1318	0 0 1	1 1 1	1 1 2
7.6	.1316	1314	1312	1311	1309	1307	1305	1304	1302	1300	0 0 1	1 1 1	1 1 2
7.7	.1299	1297	1295	1294	1292	1290	1289	1287	1285	1284	0 0 0	1 1 1	1 1 1
7.8	.1282	1280	1279	1277	1276	1274	1272	1271	1269	1267	0 0 0	1 1 1	1 1 1
7.9	.1266	1264	1263	1261	1259	1258	1256	1255	1253	1252	0 0 0	1 1 1	1 1 1
8.0	.1250	1248	1247	1245	1244	1242	1241	1239	1238	1236	0 0 0	1 1 1	1 1 1
8.1	.1235	1233	1232	1230	1229	1227	1225	1224	1222	1221	0 0 0	1 1 1	1 1 1
8.2	.1220	1218	1217	1215	1214	1212	1211	1209	1208	1206	0 0 0	1 1 1	1 1 1
8.3	.1205	1203	1202	1200	1199	1198	1196	1195	1193	1192	0 0 0	1 1 1	1 1 1
8.4	.1190	1189	1188	1186	1185	1183	1182	1181	1179	1178	0 0 0	1 1 1	1 1 1
8.5	.1176	1175	1174	1172	1171	1170	1168	1167	1166	1164	0 0 0	1 1 1	1 1 1
8.6	.1163	1161	1160	1159	1157	1156	1155	1153	1152	1151	0 0 0	1 1 1	1 1 1
8.7	.1149	1148	1147	1145	1144	1143	1142	1140	1139	1138	0 0 0	1 1 1	1 1 1
8.8	.1136	1135	1134	1133	1131	1130	1129	1127	1126	1125	0 0 0	1 1 1	1 1 1
8.9	.1124	1122	1121	1120	1119	1117	1116	1115	1114	1112	0 0 0	1 1 1	1 1 1
9.0	.1111	1110	1109	1107	1106	1105	1104	1103	1101	1100	0 0 0	1 1 1	1 1 1
9.1	.1099	1098	1096	1095	1094	1093	1092	1090	1089	1088	0 0 0	0 1 1	1 1 1
9.2	.1087	1086	1085	1083	1082	1081	1080	1079	1078	1076	0 0 0	0 1 1	1 1 1
9.3	.1075	1074	1073	1072	1071	1070	1068	1067	1066	1065	0 0 0	0 1 1	1 1 1
9.4	.1064	1063	1062	1060	1059	1058	1057	1056	1055	1054	0 0 0	0 1 1	1 1 1
9.5	.1053	1052	1050	1049	1048	1047	1046	1045	1044	1043	0 0 0	0 1 1	1 1 1
9.6	.1042	1041	1039	1038	1037	1036	1035	1034	1033	1032	0 0 0	0 1 1	1 1 1
9.7	.1031	1030	1029	1028	1027	1026	1025	1024	1022	1021	0 0 0	0 1 1	1 1 1
9.8	.1020	1019	1018	1017	1016	1015	1014	1013	1012	1011	0 0 0	0 1 1	1 1 1
9.9	.1010	1009	1008	1007	1006	1005	1004	1003	1002	1001	0 0 0	0 0 1	1 1 1

Table 3.7 Logarithms

	0	1	2	3	4	5	6	7	8	9	1	2	3	4	5	6	7	8	9
10	0000	0043	0086	0128	0170						5	9	13	17	21	26	30	34	38
						0212	0253	0294	0334	0374	4	8	12	16	20	24	28	32	36
11	0414	0453	0492	0531	0569						4	8	12	16	20	23	27	31	35
						0607	0645	0682	0719	0755	4	7	11	15	18	22	26	29	33
12	0792	0828	0864	0899	0934						3	7	11	14	18	21	25	28	32
						0969	1004	1038	1072	1106	3	7	10	14	17	20	24	27	31
13	1139	1173	1206	1239	1271						3	6	10	13	16	19	23	26	29
						1303	1335	1367	1399	1430	3	7	10	13	16	19	22	25	29
14	1461	1492	1523	1553	1584						3	6	9	12	15	19	22	25	28
						1614	1644	1673	1703	1732	3	6	9	12	14	17	20	23	26
15	1761	1790	1818	1847	1875						3	6	9	11	14	17	20	23	26
						1903	1931	1959	1987	2014	3	6	8	11	14	17	19	22	25
16	2041	2068	2095	2122	2148						3	6	8	11	14	16	19	22	24
						2175	2201	2227	2253	2279	3	5	8	10	13	16	18	21	23
17	2304	2330	2355	2380	2405						3	5	8	10	13	15	18	20	23
						2430	2455	2480	2504	2529	3	5	8	10	12	15	17	20	22
18	2553	2577	2601	2625	2648						2	5	7	9	12	14	17	19	21
						2672	2695	2718	2742	2765	2	4	7	9	11	14	16	18	21
19	2788	2810	2833	2856	2878						2	4	7	9	11	13	16	18	20
						2900	2923	2945	2967	2989	2	4	6	8	11	13	15	17	19
20	3010	3032	3054	3075	3096	3118	3139	3160	3181	3201	2	4	6	8	11	13	15	17	19
21	3222	3243	3263	3284	3304	3324	3345	3365	3385	3404	2	4	6	8	10	12	14	16	18
22	3424	3444	3464	3483	3502	3522	3541	3560	3579	3598	2	4	6	8	10	12	14	15	17
23	3617	3636	3655	3674	3692	3711	3729	3747	3766	3784	2	4	6	7	9	11	13	15	17
24	3802	3820	3838	3856	3874	3892	3909	3927	3945	3962	2	4	5	7	9	11	12	14	16
25	3979	3997	4014	4031	4048	4065	4082	4099	4116	4133	2	3	5	7	9	10	12	14	15
26	4150	4166	4183	4200	4216	4232	4249	4265	4281	4298	2	3	5	7	8	10	11	13	15
27	4314	4330	4346	4362	4378	4393	4409	4425	4440	4456	2	3	5	6	8	9	11	13	14
28	4472	4487	4502	4518	4533	4548	4564	4579	4594	4609	2	3	5	6	8	9	11	12	14
29	4624	4639	4654	4669	4683	4698	4713	4728	4742	4757	1	3	4	6	7	9	10	12	13
30	4771	4786	4800	4814	4829	4843	4857	4871	4886	4900	1	3	4	6	7	9	10	11	13
31	4914	4928	4942	4955	4969	4983	4997	5011	5024	5038	1	3	4	6	7	8	10	11	12
32	5051	5065	5079	5092	5105	5119	5132	5145	5159	5172	1	3	4	5	7	8	9	11	12
33	5185	5198	5211	5224	5237	5250	5263	5276	5289	5302	1	3	4	5	6	8	9	10	12
34	5315	5328	5340	5353	5366	5378	5391	5403	5416	5428	1	3	4	5	6	8	9	10	11
35	5441	5453	5465	5478	5490	5502	5514	5527	5539	5551	1	2	4	5	6	7	9	10	11
36	5563	5575	5587	5599	5611	5623	5635	5647	5658	5670	1	2	4	5	6	7	8	10	11
37	5682	5694	5705	5717	5729	5740	5752	5763	5775	5786	1	2	3	5	6	7	8	9	10
38	5798	5809	5821	5832	5843	5855	5866	5877	5888	5899	1	2	3	5	6	7	8	9	10
39	5911	5922	5933	5944	5955	5966	5977	5988	5999	6010	1	2	3	4	5	7	8	9	10
40	6021	6031	6042	6053	6064	6075	6085	6096	6107	6117	1	2	3	4	5	6	8	9	10
41	6128	6138	6149	6160	6170	6180	6191	6201	6212	6222	1	2	3	4	5	6	7	8	9
42	6232	6243	6253	6263	6274	6284	6294	6304	6314	6325	1	2	3	4	5	6	7	8	9
43	6335	6345	6355	6365	6375	6385	6395	6405	6415	6425	1	2	3	4	5	6	7	8	9
44	6435	6444	6454	6464	6474	6484	6493	6503	6513	6522	1	2	3	4	5	6	7	8	9
45	6532	6542	6551	6561	6571	6580	6590	6599	6609	6618	1	2	3	4	5	6	7	8	9
46	6628	6637	6646	6656	6665	6675	6684	6693	6702	6712	1	2	3	4	5	6	7	7	8
47	6721	6730	6739	6749	6758	6767	6776	6785	6794	6803	1	2	3	4	5	5	6	7	8
48	6812	6821	6830	6839	6848	6857	6866	6875	6884	6893	1	2	3	4	4	5	6	7	8
49	6902	6911	6920	6928	6937	6946	6955	6964	6972	6981	1	2	3	4	4	5	6	7	8

	0	1	2	3	4	5	6	7	8	9	1	2	3	4	5	6	7	8	9
50	6990	6998	7007	7016	7024	7033	7042	7050	7059	7067	1	2	3	3	4	5	6	7	8
51	7076	7084	7093	7101	7110	7118	7126	7135	7143	7152	1	2	3	3	4	5	6	7	8
52	7160	7168	7177	7185	7193	7202	7210	7218	7226	7235	1	2	2	3	4	5	6	7	7
53	7243	7251	7259	7267	7275	7284	7292	7300	7308	7316	1	2	2	3	4	5	6	6	7
54	7324	7332	7340	7348	7356	7364	7372	7380	7388	7396	1	2	2	3	4	5	6	6	7
55	7404	7412	7419	7427	7435	7443	7451	7459	7466	7474	1	2	2	3	4	5	5	6	7
56	7482	7490	7497	7505	7513	7520	7528	7536	7543	7551	1	2	2	3	4	5	5	6	7
57	7559	7566	7574	7582	7589	7597	7604	7612	7619	7627	1	2	2	3	4	5	5	6	7
58	7634	7642	7649	7657	7664	7672	7679	7686	7694	7701	1	1	2	3	4	4	5	6	7
59	7709	7716	7723	7731	7738	7745	7752	7760	7767	7774	1	1	2	3	4	4	5	6	7
60	7782	7789	7796	7803	7810	7818	7825	7832	7839	7846	1	1	2	3	4	4	5	6	6
61	7853	7860	7868	7875	7882	7889	7896	7903	7910	7917	1	1	2	3	4	4	5	6	6
62	7924	7931	7938	7945	7952	7959	7966	7973	7980	7987	1	1	2	3	3	4	5	6	6
63	7993	8000	8007	8014	8021	8028	8035	8041	8048	8055	1	1	2	3	3	4	5	5	6
64	8062	8069	8075	8082	8089	8096	8102	8109	8116	8122	1	1	2	3	3	4	5	5	6
65	8129	8136	8142	8149	8156	8162	8169	8176	8182	8189	1	1	2	3	3	4	5	5	6
66	8195	8202	8209	8215	8222	8228	8235	8241	8248	8254	1	1	2	3	3	4	5	5	6
67	8261	8267	8274	8280	8287	8293	8299	8306	8312	8319	1	1	2	3	3	4	5	5	6
68	8325	8331	8338	8344	8351	8357	8363	8370	8376	8382	1	1	2	3	3	4	4	5	6
69	8388	8395	8401	8407	8414	8420	8426	8432	8439	8445	1	1	2	2	3	4	4	5	6
70	8451	8457	8463	8470	8476	8482	8488	8494	8500	8506	1	1	2	2	3	4	4	5	6
71	8513	8519	8525	8531	8537	8543	8549	8555	8561	8567	1	1	2	2	3	4	4	5	5
72	8573	8579	8585	8591	8597	8603	8609	8615	8621	8627	1	1	2	2	3	4	4	5	5
73	8633	8639	8645	8651	8657	8663	8669	8675	8681	8686	1	1	2	2	3	4	4	5	5
74	8692	8698	8704	8710	8716	8722	8727	8733	8739	8745	1	1	2	2	3	4	4	5	5
75	8751	8756	8762	8768	8774	8779	8785	8791	8797	8802	1	1	2	2	3	3	4	5	5
76	8808	8814	8820	8825	8831	8837	8842	8848	8854	8859	1	1	2	2	3	3	4	5	5
77	8865	8871	8876	8882	8887	8893	8899	8904	8910	8915	1	1	2	2	3	3	4	4	5
78	8921	8927	8932	8938	8943	8949	8954	8960	8965	8971	1	1	2	2	3	3	4	4	5
79	8976	8982	8987	8993	8998	9004	9009	9015	9020	9025	1	1	2	2	3	3	4	4	5
80	9031	9036	9042	9047	9053	9058	9063	9069	9074	9079	1	1	2	2	3	3	4	4	5
81	9085	9090	9096	9101	9106	9112	9117	9122	9128	9133	1	1	2	2	3	3	4	4	5
82	9138	9143	9149	9154	9159	9165	9170	9175	9180	9186	1	1	2	2	3	3	4	4	5
83	9191	9196	9201	9206	9212	9217	9222	9227	9232	9238	1	1	2	2	3	3	4	4	5
84	9243	9248	9253	9258	9263	9269	9274	9279	9284	9289	1	1	2	2	3	3	4	4	5
85	9294	9299	9304	9309	9315	9320	9325	9330	9335	9340	1	1	2	2	3	3	4	4	5
86	9345	9350	9355	9360	9365	9370	9375	9380	9385	9390	1	1	2	2	3	3	4	4	5
87	9395	9400	9405	9410	9415	9420	9425	9430	9435	9440	0	1	1	2	2	3	3	4	4
88	9445	9450	9455	9460	9465	9469	9474	9479	9484	9489	0	1	1	2	2	3	3	4	4
89	9494	9499	9504	9509	9513	9518	9523	9528	9533	9538	0	1	1	2	2	3	3	4	4
90	9542	9547	9552	9557	9562	9566	9571	9576	9581	9586	0	1	1	2	2	3	3	4	4
91	9590	9595	9600	9605	9609	9614	9619	9624	9628	9633	0	1	1	2	2	3	3	4	4
92	9638	9643	9647	9652	9657	9661	9666	9671	9675	9680	0	1	1	2	2	3	3	4	4
93	9685	9689	9694	9699	9703	9708	9713	9717	9722	9727	0	1	1	2	2	3	3	4	4
94	9731	9736	9741	9745	9750	9754	9759	9763	9768	9773	0	1	1	2	2	3	3	4	4
95	9777	9782	9786	9791	9795	9800	9805	9809	9814	9818	0	1	1	2	2	3	3	4	4
96	9823	9827	9832	9836	9841	9845	9850	9854	9859	9863	0	1	1	2	2	3	3	4	4
97	9868	9872	9877	9881	9886	9890	9894	9899	9903	9908	0	1	1	2	2	3	3	4	4
98	9912	9917	9921	9926	9930	9934	9939	9943	9948	9952	0	1	1	2	2	3	3	4	4
99	9956	9961	9965	9969	9974	9978	9983	9987	9991	9996	0	1	1	2	2	3	3	3	4

Table 3.8 Antilogarithms

	0	1	2	3	4	5	6	7	8	9	1	2	3	4	5	6	7	8
·00	1000	1002	1005	1007	1009	1012	1014	1016	1019	1021	0	0	1	1	1	1	2	2
·01	1023	1026	1028	1030	1033	1035	1038	1040	1042	1045	0	0	1	1	1	1	2	2
·02	1047	1050	1052	1054	1057	1059	1062	1064	1067	1069	0	0	1	1	1	1	2	2
·03	1072	1074	1076	1079	1081	1084	1086	1089	1091	1094	0	0	1	1	1	1	2	2
·04	1096	1099	1102	1104	1107	1109	1112	1114	1117	1119	0	1	1	1	1	2	2	2
·05	1122	1125	1127	1130	1132	1135	1138	1140	1143	1146	0	1	1	1	1	2	2	2
·06	1148	1151	1153	1156	1159	1161	1164	1167	1169	1172	0	1	1	1	1	2	2	2
·07	1175	1178	1180	1183	1186	1189	1191	1194	1197	1199	0	1	1	1	1	2	2	2
·08	1202	1205	1208	1211	1213	1216	1219	1222	1225	1227	0	1	1	1	1	2	2	2
·09	1230	1233	1236	1239	1242	1245	1247	1250	1253	1256	0	1	1	1	1	2	2	2
·10	1259	1262	1265	1268	1271	1274	1276	1279	1282	1285	0	1	1	1	1	2	2	2
·11	1288	1291	1294	1297	1300	1303	1306	1309	1312	1315	0	1	1	1	2	2	2	2
·12	1318	1321	1324	1327	1330	1334	1337	1340	1343	1346	0	1	1	1	2	2	2	2
·13	1349	1352	1355	1358	1361	1365	1368	1371	1374	1377	0	1	1	1	2	2	2	3
·14	1380	1384	1387	1390	1393	1396	1400	1403	1406	1409	0	1	1	1	2	2	2	3
·15	1413	1416	1419	1422	1426	1429	1432	1435	1439	1442	0	1	1	1	2	2	2	3
·16	1445	1449	1452	1455	1459	1462	1466	1469	1472	1476	0	1	1	1	2	2	2	3
·17	1479	1483	1486	1489	1493	1496	1500	1503	1507	1510	0	1	1	1	2	2	2	3
·18	1514	1517	1521	1524	1528	1531	1535	1538	1542	1545	0	1	1	1	2	2	2	3
·19	1549	1552	1556	1560	1563	1567	1570	1574	1578	1581	0	1	1	1	2	2	3	3
·20	1585	1589	1592	1596	1600	1603	1607	1611	1614	1618	0	1	1	1	2	2	3	3
·21	1622	1626	1629	1633	1637	1641	1644	1648	1652	1656	0	1	1	2	2	2	3	3
·22	1660	1663	1667	1671	1675	1679	1683	1687	1690	1694	0	1	1	2	2	2	3	3
·23	1698	1702	1706	1710	1714	1718	1722	1726	1730	1734	0	1	1	2	2	2	3	3
·24	1738	1742	1746	1750	1754	1758	1762	1766	1770	1774	0	1	1	2	2	2	3	3
·25	1778	1782	1786	1791	1795	1799	1803	1807	1811	1816	0	1	1	2	2	2	3	3
·26	1820	1824	1828	1832	1837	1841	1845	1849	1854	1858	0	1	1	2	2	3	3	3
·27	1862	1866	1871	1875	1879	1884	1888	1892	1897	1901	0	1	1	2	2	3	3	3
·28	1905	1910	1914	1919	1923	1928	1932	1936	1941	1945	0	1	1	2	2	3	3	4
·29	1950	1954	1959	1963	1968	1972	1977	1982	1986	1991	0	1	1	2	2	3	3	4
·30	1995	2000	2004	2009	2014	2018	2023	2028	2032	2037	0	1	1	2	2	3	3	4
·31	2042	2046	2051	2056	2061	2065	2070	2075	2080	2084	0	1	1	2	2	3	3	4
·32	2089	2094	2099	2104	2109	2113	2118	2123	2128	2133	0	1	1	2	2	3	3	4
·33	2138	2143	2148	2153	2158	2163	2168	2173	2178	2183	0	1	1	2	2	3	3	4
·34	2188	2193	2198	2203	2208	2213	2218	2223	2228	2234	1	1	2	2	3	3	4	4
·35	2239	2244	2249	2254	2259	2265	2270	2275	2280	2286	1	1	2	2	3	3	4	4
·36	2291	2296	2301	2307	2312	2317	2323	2328	2333	2339	1	1	2	2	3	3	4	4
·37	2344	2350	2355	2360	2366	2371	2377	2382	2388	2393	1	1	2	2	3	3	4	4
·38	2399	2404	2410	2415	2421	2427	2432	2438	2443	2449	1	1	2	2	3	3	4	4
·39	2455	2460	2466	2472	2477	2483	2489	2495	2500	2506	1	1	2	2	3	3	4	5
·40	2512	2518	2523	2529	2535	2541	2547	2553	2559	2564	1	1	2	2	3	4	4	5
·41	2570	2576	2582	2588	2594	2600	2606	2612	2618	2624	1	1	2	2	3	4	4	5
·42	2630	2636	2642	2649	2655	2661	2667	2673	2679	2685	1	1	2	2	3	4	4	5
·43	2692	2698	2704	2710	2716	2723	2729	2735	2742	2748	1	1	2	3	3	4	4	5
·44	2754	2761	2767	2773	2780	2786	2793	2799	2805	2812	1	1	2	3	3	4	4	5
·45	2818	2825	2831	2838	2844	2851	2858	2864	2871	2877	1	1	2	3	3	4	5	5
·46	2884	2891	2897	2904	2911	2917	2924	2931	2938	2944	1	1	2	3	3	4	5	5
·47	2951	2958	2965	2972	2979	2985	2992	2999	3006	3013	1	1	2	3	3	4	5	5
·48	3020	3027	3034	3041	3048	3055	3062	3069	3076	3083	1	1	2	3	4	4	5	6
·49	3090	3097	3105	3112	3119	3126	3133	3141	3148	3155	1	1	2	3	4	4	5	6

	0	1	2	3	4	5	6	7	8	9	1	2	3	4	5	6	7	8	9
·50	3162	3170	3177	3184	3192	3199	3206	3214	3221	3228	1	1	2	3	4	4	5	6	7
·51	3236	3243	3251	3258	3266	3273	3281	3289	3296	3304	1	2	2	3	4	5	5	6	7
·52	3311	3319	3327	3334	3342	3350	3357	3365	3373	3381	1	2	2	3	4	5	5	6	7
·53	3388	3396	3404	3412	3420	3428	3436	3443	3451	3459	1	2	2	3	4	5	6	6	7
·54	3467	3475	3483	3491	3499	3508	3516	3524	3532	3540	1	2	2	3	4	5	6	6	7
·55	3548	3556	3565	3573	3581	3589	3597	3606	3614	3622	1	2	2	3	4	5	6	7	7
·56	3631	3639	3648	3656	3664	3673	3681	3690	3698	3707	1	2	3	3	4	5	6	7	8
·57	3715	3724	3733	3741	3750	3758	3767	3776	3784	3793	1	2	3	3	4	5	6	7	8
·58	3802	3811	3819	3828	3837	3846	3855	3864	3873	3882	1	2	3	4	4	5	6	7	8
·59	3890	3899	3908	3917	3926	3936	3945	3954	3963	3972	1	2	3	4	5	5	6	7	8
·60	3981	3990	3999	4009	4018	4027	4036	4046	4055	4064	1	2	3	4	5	6	6	7	8
·61	4074	4083	4093	4102	4111	4121	4130	4140	4150	4159	1	2	3	4	5	6	7	8	9
·62	4169	4178	4188	4198	4207	4217	4227	4236	4246	4256	1	2	3	4	5	6	7	8	9
·63	4266	4276	4285	4295	4305	4315	4325	4335	4345	4355	1	2	3	4	5	6	7	8	9
·64	4365	4375	4385	4395	4406	4416	4426	4436	4446	4457	1	2	3	4	5	6	7	8	9
·65	4467	4477	4487	4498	4508	4519	4529	4539	4550	4560	1	2	3	4	5	6	7	8	9
·66	4571	4581	4592	4603	4613	4624	4634	4645	4656	4667	1	2	3	4	5	6	7	9	10
·67	4677	4688	4699	4710	4721	4732	4742	4753	4764	4775	1	2	3	4	5	7	8	9	10
·68	4786	4797	4808	4819	4831	4842	4853	4864	4875	4887	1	2	3	4	6	7	8	9	10
·69	4898	4909	4920	4932	4943	4955	4966	4977	4989	5000	1	2	3	5	6	7	8	9	10
·70	5012	5023	5035	5047	5058	5070	5082	5093	5105	5117	1	2	4	5	6	7	8	9	11
·71	5129	5140	5152	5164	5176	5188	5200	5212	5224	5236	1	2	4	5	6	7	8	10	11
·72	5248	5260	5272	5284	5297	5309	5321	5333	5346	5358	1	2	4	5	6	7	9	10	11
·73	5370	5383	5395	5408	5420	5433	5445	5458	5470	5483	1	3	4	5	6	8	9	10	11
·74	5495	5508	5521	5534	5546	5559	5572	5585	5598	5610	1	3	4	5	6	8	9	10	12
·75	5623	5636	5649	5662	5675	5689	5702	5715	5728	5741	1	3	4	5	7	8	9	10	12
·76	5754	5768	5781	5794	5808	5821	5834	5848	5861	5875	1	3	4	5	7	8	9	11	12
·77	5888	5902	5916	5929	5943	5957	5970	5984	5998	6012	1	3	4	5	7	8	10	11	12
·78	6026	6039	6053	6067	6081	6095	6109	6124	6138	6152	1	3	4	6	7	8	10	11	13
·79	6166	6180	6194	6209	6223	6237	6252	6266	6281	6295	1	3	4	6	7	9	10	11	13
·80	6310	6324	6339	6353	6368	6383	6397	6412	6427	6442	1	3	4	6	7	9	10	12	13
·81	6457	6471	6486	6501	6516	6531	6546	6561	6577	6592	2	3	5	6	8	8	11	12	14
·82	6607	6622	6637	6653	6688	6683	6699	6714	6730	6745	2	3	5	6	8	9	11	12	14
·83	6761	6776	6792	6808	6823	6839	6855	6871	6887	6902	2	3	5	6	8	9	11	13	14
·84	6918	6934	6950	6966	6982	6998	7015	7031	7047	7063	2	3	5	6	8	10	11	13	15
·85	7079	7096	7112	7129	7145	7161	7178	7194	7211	7228	2	3	5	7	8	10	12	13	15
·86	7244	7261	7278	7295	7311	7328	7345	7362	7379	7396	2	3	5	7	8	10	12	13	15
·87	7413	7430	7447	7464	7482	7499	7516	7534	7551	7568	2	3	5	7	9	10	12	14	16
·88	7586	7603	7621	7638	7656	7674	7691	7709	7727	7745	2	4	5	7	9	11	12	14	16
·89	7762	7780	7798	7816	7834	7852	7870	7889	7907	7925	2	4	5	7	9	11	13	14	16
·90	7943	7962	7980	7998	8017	8035	8054	8072	8091	8110	2	4	6	7	9	11	13	15	17
·91	8128	8147	8166	8185	8204	8222	8241	8260	8279	8299	2	4	6	8	9	11	13	15	17
·92	8318	8337	8356	8375	8395	8414	8433	8453	8472	8492	2	4	6	8	10	12	14	15	17
·93	8511	8531	8551	8570	8590	8610	8630	8650	8670	8690	2	4	6	8	10	12	14	16	18
·94	8710	8730	8750	8770	8790	8810	8831	8851	8872	8892	2	4	6	8	10	12	14	16	18
·95	8913	8933	8954	8974	8995	9016	9036	9057	9078	9099	2	4	6	8	10	12	15	17	19
·96	9120	9141	9162	9183	9204	9226	9247	9268	9290	9311	2	4	6	8	11	13	15	17	19
·97	9333	9354	9376	9397	9419	9441	9462	9484	9506	9528	2	4	7	9	11	13	15	17	20
·98	9550	9572	9594	9616	9638	9661	9683	9705	9727	9750	2	4	7	9	11	13	16	18	20
·99	9772	9795	9817	9840	9863	9886	9908	9931	9954	9977	2	5	7	9	11	14	16	18	20

3.6 Further problems

Errors

1. The area of a triangle is given by $A = \frac{1}{2}bh$. If $b = 7.37$ cm and $h = 13.6$ cm, find the area of the triangle. [$50.12\ \text{cm}^2$]
2. Determine the velocity of a body v when $v = u + at$. The initial velocity $u = 14\ \text{m s}^{-1}$, the acceleration $a = 9.81\ \text{m s}^{-2}$ when the time t is 24 s. [$249\ \text{m s}^{-1}$]
3. The pressure p and volume V of a gas are constant, such that $pV = c$. Find c when $p = 105\,600$ Pa and $V = 0.43\ \text{m}^3$. [$45\,400\ \text{Pa m}^3$]

In Problems 4–10 state the type of error or errors which have been made if, indeed, there is an error present.

4. $73 \times 68.247 = 4\,982.03$
   [Accuracy error, add 'correct to 2 decimal places']
5. $18 \times 0.08 \times 6 = 86.4$ [Order of magnitude error]
6. $\dfrac{47.3}{9 \times 0.071} = 5\,301.2$ correct to 1 decimal place
   [Blunder]
7. $\dfrac{16.7 \times 0.02}{41} = 8.15 \times 10^{-3}$ correct to 3 significant figures [No error]
8. The force, $P = mf$, where m is the mass and a the acceleration. The value of P was found to be 1 775.61 N when m was 181 kg and f was $9.81\ \text{m s}^{-2}$.
   [Measured values, hence $P = 1\,776$ N]
9. $6.73 \div 0.006 = 4\,211.\dot{6}$ [Blunder]
10. $\dfrac{3.1 \times 0.008}{37.6 \times 0.347} = 9.100\,7 \times 10^{-4}$ [Blunder]

Squares, square roots and reciprocals

In problems 11–16, use four-figure tables to determine the values of the quantities shown.

11. (*a*) 3.18^2 (*b*) 5.476^2 (*c*) 8.888^2
    (*a*) [10.11] (*b*) [29.99] (*c*) [78.99]
12. (*a*) 839^2 (*b*) $0.029\,81^2$ (*c*) 67.42^2
    (*a*) [703 900] (*b*) [0.000 888 6] (*c*) [4 546]
13. (*a*) $\sqrt{1.072}$ (*b*) $\sqrt{7.779}$ (*c*) $\sqrt{10.72}$ (*d*) $\sqrt{77.79}$
    (*a*) [±1.035] (*b*) [±2.789] (*c*) [±3.274] (*d*) [±8.820]
14. (*a*) $\sqrt{7\,291}$ (*b*) $\sqrt{384\,800}$ (*c*) $\sqrt{0.171\,1}$ (*d*) $\sqrt{0.000\,271\,3}$
    (*a*) [±85.39] (*b*) [±620.3] (*c*) [±0.413 6] (*d*) [±0.016 47]
15. (*a*) $\dfrac{1}{5.281}$ (*b*) $\dfrac{1}{6.555}$ (*c*) $\dfrac{1}{9.132}$ (*d*) $\dfrac{1}{1.357}$
    (*a*) [0.189 4] (*b*) [0.152 6] (*c*) [0.109 5] (*d*) [0.735 3]
16. (*a*) $\dfrac{1}{927.4}$ (*b*) $\dfrac{1}{0.624}$ (*c*) $\dfrac{1}{1\,371}$ (*d*) $\dfrac{1}{0.000\,277}$
    (*a*) [0.001 079] (*b*) [1.603] (*c*) [0.000 729 9] (*d*) [3 610]

Logarithms

17. Find the value of:
(*a*) lg 3.824 (*b*) lg 271.8 (*c*) lg 38 140 (*d*) lg 63.8
(*a*) [0.582 6] (*b*) [2.434 3] (*c*) [4.581 4] (*d*) [1.804 8]

18. Find the value of:
(*a*) lg 9.214 (*b*) lg 0.627 (*c*) lg 0.000 23 (*d*) lg 0.009 27
(*a*) [0.964 5] (*b*) [$\bar{1}$.797 3] (*c*) [$\bar{4}$.361 7] (*d*) [$\bar{3}$.967 1]

19. Find the antilogarithms of:
(*a*) 0.147 1 (*b*) 0.462 9 (*c*) 0.923 7
(*a*) [1.403] (*b*) [2.903] (*c*) [8.389]

20. Find the antilogarithms of:
(*a*) 1.271 (*b*) 3.714 9 (*c*) 2.714
(*a*) [18.66] (*b*) [5 187] (*c*) [517.6]

21. Find the antilogarithms of:
(*a*) $\bar{1}$.472 8 (*b*) $\bar{3}$.814 (*c*) $\bar{2}$.9
(*a*) [0.297 0] (*b*) [0.006 516] (*c*) [0.079 43]

In problems 22–26 use logarithms to evaluate:

22. (*a*) 38.71 × 46.3 (*b*) 0.076 28 × 13.7 (*c*) 93.82 × 16.74 × 0.037 1
(*d*) 8.417 × 0.003 84 × 6 742
(*a*) [1 793] (*b*) [1.045] (*c*) [58.26] (*d*) [217.9]

23. (*a*) $\frac{28.41}{76.82}$ (*b*) $\frac{3.714}{0.039\ 7}$ (*c*) $\frac{743.8}{62.49}$ (*d*) $\frac{0.027\ 4}{837.2}$
(*a*) [0.369 8] (*b*) [93.56] (*c*) [11.91] (*d*) [3.273×10^{-5}]

24. (*a*) $\frac{1}{67.4}$ (*b*) $\frac{1}{7\ 628}$ (*c*) $\frac{1}{0.024\ 77}$ (*d*) $\frac{1}{7.241}$
(*a*) [0.014 84] (*b*) [1.311×10^{-4}] (*c*) [40.37] (*d*) [0.138 1]

25. (*a*) 23^2 (*b*) 67.4^3 (*c*) 824^2 (*d*) $0.024\ 7^4$ (*e*) $0.005\ 14^2$
(*a*) [529.0] (*b*) [3.063×10^5] (*c*) [6.789×10^5]
(*d*) [3.722×10^{-7}] (*e*) [2.642×10^{-5}]

26. (*a*) $\sqrt{7.284}$ (*b*) $\sqrt{23.84}$ (*c*) $\sqrt{4\ 977}$ (*d*) $\sqrt{0.624}$ (*e*) $\sqrt{0.039\ 28}$
(*f*) $\sqrt{0.002\ 91}$
(*a*) [±2.699] (*b*) [±4.883] (*c*) [±70.55] (*d*) [±0.790 0]
(*e*) [±0.198 2] (*f*) [±0.053 94]

Use logarithms in questions 27–53.

27. Find the voltage V when $V = IR$ (volts) given the current $I = 1.672$ amperes when the resistance $R = 27.48$ ohms. [45.95 V]

28. Find the volume V when $V = \frac{c}{p}$ (m^3) given the pressure $p = 6.384$ Pa when the constant $c = 453$. [70.96 m^3]

29. Find the distance s (m) when $s = \frac{1}{2}\,g\,t^2$ given the time $t = 0.047$ s when the acceleration $g = 9.81$ m s^{-2}. [0.010 84 m]

30. Find the distance s (m) when $s = \frac{1}{2}\,(u + v)t$ given the initial velocity

$u = 18.6$ m s^{-1} and the final velocity $v = 93.74$ m s^{-1} when the time $t = 0.098$ s. [5.505 m]

31. Find the energy E in joules when $E = \frac{1}{2} LI^2$ given the current $I = 3.841$ A when the inductance $L = 0.270$ H. [1.992 J]
32. Find the area A, m^2, of a triangle when $A = \frac{1}{2} bh$ given base length $b = 67.41$ m and height $h = 127.6$ m. [4 301 m^2]
33. Find the resistance R_t in ohms at temperature t° C when $R_t = R_o(1 + \alpha t)$ given the resistance at 0° C, $R_o = 240$ ohms and the temperature co-efficient of resistance $\alpha = 0.000\ 27/^\circ$C when $t = 87^\circ$C. [245.6]
34. The time in seconds, t, taken for sounds of average intensity to fall below limits of audibility in a room is given by $t = \dfrac{0.16\ V}{a}$ where V is the volume of the room in metre^3 and a, the number of absorption units, is given by
a = area of walls (m^2) × 0.02 + area of floor (in m^2) × 0.05 + area of ceiling (in m^2) × 0.02
Determine t for a room of length 32 m, width 12 m and height 6 m. [9.846 s]
35. The equilibrium constant, K, for the reaction $H_2 + CO_2 \rightleftharpoons H_2O + CO$ at 247° C is given by $K = \dfrac{(0.18)^2}{(1.82)^2}$. What is the value of K?
[0.009 78 or 9.78 × 10^{-3}]
36. If pH = lg $\dfrac{1}{C_{H^+}}$ where C_{H^+} is the concentration of hydrogen ions in g dm^{-3} determine pH when $C_{H^+} = 6.48 \times 10^{-4}$ g dm^{-3} [3.188]
37. Determine the safe load, p, which can be carried by a steel plate weakened by rivet holes, from the equation $p = f(b - nd)t$ given that $f = 150$ N mm^{-2}, $b = 112$ mm, $n = 4$, $d = 19$ mm, $t = 15$ mm
[81 000 N]
38. What is the volume of a cube of concrete of length 114.9 cm?
[1 517 000 cm^3 or 1.517 m^3]
39. The failing load of a material is given by
failing stress × area of section.
Determine the failing load of a concrete test cube of length 150 mm and failing stress of 0.027 8 kN mm^{-2}. [625.5 kN]
40. The pressure of a stack of 1 855 bricks on a ground area of 3.521 m by 1.249 m
$= \dfrac{1\ 855 \times 3.20 \times 9.81}{3.521 \times 1.249}$ N m^{-2}. Determine its value. [13 240 N m^{-2}]
41. The length, l, of the unit cell of titanium oxide is given by
$l^3 = \dfrac{63.90 \times 4.00}{5.536 \times 6.022 \times 10^{23}}$ cm^3. Determine l. [4.248 × 10^{-8} cm]
42. The density, d, of yellow lead oxide is given by
$d = \dfrac{223.2 \times 4.00}{6.022 \times 10^{23} \times 0.550 \times 0.472 \times 0.588 \times 10^{-21}}$ g cm^{-3}. Find d.
[9.713 g cm^{-3}]

43. The relative molecular mass, M, of urea is determined from an experiment which gives
$M = \frac{5.2 \times 100 \times 0.242}{25 \times 0.083}$. Find M. [60.65]
44. 74.60 g of potassium chloride react with 170.00 g of silver nitrate to form silver chloride. How much silver nitrate will react with 10.53 g of potassium chloride? [24.00 g]
45. Bronze contains 95% of copper, 4% of tin and 1% of zinc by mass. Brass contains 89% of copper, 10% of tin, and 1% of zinc by mass. In what proportions must bronze and brass be mixed to give an alloy containing 91% of copper? [1:2]
46. 18 560 bricks are required for a specific building job. If this includes 8% for cutting and wastage how many bricks are actually used for the job? [17 075]
47. Assuming a loss of bulk on mixing of 25%, how much sand and cement, mixed in a ratio of 3:1, will be required to give 147 m^3 of mortar? What will be the cost if sand costs £3.50 m^{-3} and cement costs £15.20 m^{-3}? [Sand = 137.8 m^3; cement = 45.94 m^3; £482, £698]
48. One tonne of gypsum plaster mixed with Type 1 sand in the ratio 2:3 will cover an area of plaster board of 300 m^2 to a thickness of 6 mm. How much plaster and sand will be required to cover 188 m^2? [Plaster = 0.626 7 tonne; sand = 0.940 0 tonne]
49. The ratio of the rise (vertical height) of a staircase to its going (horizontal distance) is 1 to 1.3. If the going of each step is 225 mm, what is the rise of a staircase of eight steps? [1.385 m]
50. It was found on analysis that a sample of water contained 47 parts per million by mass (ppm) of carbonates and bicarbonates, 8 ppm of mineral acids, 49 ppm of calcium and 29 ppm of magnesium. What is the percentage mass of these impurities in water expressed as a total percentage? [0.013 3%]
51. A brick has a mass of 3.200 kg when thoroughly wet and 2.800 kg when dry. What is the moisture content (percentage mass of the water to that of the dry brick) of the brick? [14.3%]
52. A Petrograd standard of timber has a volume of 4.67 m^3. How many standards are contained in a volume of 60.71 m^3? [13]
53. On strong heating 0.234 1 g of hydrated barium chloride loses 0.034 5 g of water of crystallisation. What is the percentage mass of water of crystallisation in the original salt? [14.74%]

In questions 54–60 use logarithms to evaluate the expressions shown.

54. $\frac{27.9 \times 83.4}{2.671}$ [871.2]
55. $\frac{63.81^2}{42.7^3}$ [0.052 30]
56. $\frac{8.191}{33.72 \times (0.714)^2}$ [0.476 5]
57. $\left(\frac{4.96 \times 18.31}{4.713}\right)^3$ [7 155]

58. $\dfrac{\sqrt{47.21} - \sqrt{0.071}}{15.6}$ taking the positive value of the square roots only.
[0.423 4]

59. $\dfrac{(3.817)^3}{(16.29)^2 - 14.3 \times 8.71}$ [0.394 9]

60. $\dfrac{0.671 \times 8.423^2}{\sqrt{17.84 + 43.57 \times 0.071\,8}}$ taking the positive value of the square root only. [6.475]

Calculator

Problems 11–16 and 22–60 can be repeated using a calculator.

Conversion tables and charts

61. Use Table 3.1 (page 73) to determine:
(*a*) the number of millimetres in 30 inches,
(*b*) a speed of 50 m.p.h. in km h^{-1},
(*c*) the number of miles in 177.1 km,

Table 3.9

HAVANT · Leigh Park · Eastern Road · SOUTHSEA (Clarence Pier)

Weekdays								
HAVANT, Market Parade	......	......	......	0718	0755	0825	0855	0925
Park Parade, Purbrook Way	0603	0633	0703	0733	0810	0840	0910	0940
Hulbert Road, Purbrook Way	0609	0639	0709	0739	0816	0846	0916	0946
Farlington, Rectory Avenue	0617	0647	0717	0747	0824	0854	0924	0954
Eastern Road, Airport Service Road	0624	0654	0724	0754	0831	0901	0931	1001
Stanley Avenue	0629	0659	0729	0759	0836	0906	0936	1006
Milton, St. Mary's Road	0636	0706	0736	0806	0843	0913	0943	1013
Station Street, Commercial Road	0646	0716	0746	0816	0853	0923	0953	1023
SOUTHSEA, Clarence Pier	......	......	......	0823	0900	0930	1000	1030
Guildhall	*	*	*	......	......	......	......	......
PORTSMOUTH, The Hard	0652	0722	0752	......	......	......	......	......

SOUTHSEA (Clarence Pier) · Eastern Road · Leigh Park · HAVANT

Weekdays								
PORTSMOUTH, The Hard	......	......	......	0705	0735	0805	......	......
Guildhall	......	......	......	......	......	......	......	......
SOUTHSEA, Clarence Pier	......	......	......	*	*	*	0833	0910
Station Street, Commercial Road	......	......	......	0711	0741	0811	0841	0918
Milton, St. Mary's Road	0551	0621	0651	0721	0751	0821	0851	0928
Stanley Avenue	0557	0627	0657	0727	0757	0827	0857	0934
Eastern Road, Airport Service Road	0602	0632	0702	0732	0802	0832	0902	0939
Farlington, Rectory Avenue	0609	0639	0709	0739	0809	0839	0909	0946
Hulbert Road, Purbrook Way	0617	0647	0717	0747	0817	0847	0917	0954
Park Parade, Stockheath Road	0623	0653	0723	0753	0823	0853	0923	1000
HAVANT, Market Parade	......	0708	0738	0808	0838	0908	0938	1015

(*d*) the number of kilograms in 44 lb,
(*e*) the number of litres in 20 gallons, and
(*f*) the number of gallons in 38 litres.

(*a*) [762 mm] (*b*) [80.5 km h^{-1}] (*c*) [110 miles]
(*d*) [20 lb] (*e*) [90.91 litres] (*f*) [8.36 gallons]

62. Use Table 3.2 (page 74) to determine:
(*a*) the pounds sterling 247.5 francs will buy,
(*b*) the number of Italian lira that can be exchanged for £15.50.
(*c*) the pounds sterling which can be bought for 512 Dm
(*d*) the number of Spanish pesetas which can be exchanged for £52.75, and
(*e*) the pounds sterling which can be exchanged for $209.

(*a*) [£22.50] (*b*) [35 650 lira] (*c*) [£107.56] (*d*) [10 022.5 pes.]
(*e*) [£110]

Service 24

0955	1025	1055		25	55		1825	1855	1925	2025	2125	2205
1010	1040	1110		40	10		1840	1910	1940	2040	2140	2220
1016	1046	1116	Then	46	16		1846	1916	1946	2046	2146	2226
1024	1054	1124	at	54	24		1854	1924	1954	2054	2154	2234
1031	1101	1131	these	01	31		1901	1931	2001	2101	2201	2241
1036	1106	1136	minutes	06	36	until	1906	1936	2006	2106	2206	2246
1043	1113	1143	past	13	43		1913	1943	2013	2113	2213	2253
1053	1123	1153	each	23	53		1923	1953	2023	2123	2223	2303
1100	1130	1200	hour	30	00		1930	2000	2030	2132†	....	....
....	....	....		....	....		....	....	....	2139	*	*
....	....	....		....	....		....	....	....	2145	2229‡	2309‡

Service 24

											●	
....		....	....		....	....	....	....	....	2200	2248	....
....		....	....		....	....	....	....	....	2206	....	....
0940	Then	10	40		1810	1840	1910	2010	2050	2215†	*	....
0948	at	18	48		1818	1848	1918	2018	2058	2223	2253	....
0958	these	28	58		1828	1858	1928	2028	2108	2233	2303	....
1004	minutes	34	06	until	1834	1904	1934	2034	2114	2239	2309	....
1009	past	39	09		1839	1909	1939	2039	2119	2244	....	....
1016	each	46	16		1846	1916	1946	2046	2126	2251	....	....
1024	hour	54	24		1854	1924	1954	2054	2134	2259	....	....
1030		00	30		1900	1930	2000	2100	2140	2305‡	....	....
1045		15	45		1915	....	2015	2115	2155	....	....	....

63. Three similar holidays are advertised. The prices are:
Italy 483 000 lira; Spain 40 850 pesetas; France 2 343 francs.
Use Table 3.2 (page 74) to determine the least expensive holiday.

[Italy £210, France £213, Spain £215]

64. Deduce the following information from the bus timetable shown in Table 3.9 (on pp. 92 and 93):

(*a*) At what time should a man catch a bus at Park Parade to enable him to be at Southsea by 8.45 a.m.?

(*b*) At what time should a man catch a bus at Clarence Pier to arrive at Park Parade by 4.15 p.m.?

(*c*) A lady leaves Havant at 8.55 a.m. and travels to Commercial Road. She needs 30 minutes for shopping before catching the next available bus back to Havant. What is the earliest time she can be back at Havant?

(*d*) A girl leaves Havant by the 6.55 p.m. bus and travels to Southsea. How long does the journey take? If she has to be back in Havant by 9.30 p.m., what time is the last bus that she should catch from Southsea? (*a*) [7.33 a.m.] (*b*) [3.10 p.m.] (*c*) [11.45 a.m.]
(*d*) [65 minutes; 8.10 p.m.]

65. A production schedule used in the manufacture of an engineering component is shown below. Stage 3 requires two men, the other stages requiring one man.

Stage	Start day	End day
1	1	3
2	3	8
3	2	9
4	5	8
5	5	11

Determine (*a*) the minimum time for the component to be completed,

(*b*) the number of man-days needed for the production of the component,

(*c*) the total labour cost if each man is paid £2.50 per hour and they each work a 7 hour day, and

(*d*) the percentage profit, based on the sale price, if materials and overheads cost £140 and the component is sold for £1 100.

(*a*) [11 days] (*b*) [36 man-days] (*c*) [£630] (*d*) [30%]

Section II

Algebra

Chapter 4

An introduction to algebra

4.1 Introduction

Algebra is that part of mathematics in which the relations and properties of numbers are investigated by means of general symbols. It is one of the most important concepts used in the understanding of mathematics and, in the form of general equations, is used in many areas of engineering and science.

Taking the simple case of a rectangle, whose area is found by multiplying the length by the breadth, this can be expressed algebraically as $A = l \times b$ where A represents the area, l the length and b the breadth. Hence $A = l \times b$ is a convenient general expression which is true for all positive values of l and b, when determining the area of a rectangle.

4.2 Basic notation

Problems involving numbers give a particular answer, for example $2 + 3 = 5$. This is called an **arithmetic equation.**

In algebra, results are generalised and letters are used which can stand for any number. The laws of indices introduced for numbers can be applied to show that

$$2^2 \times 2^3 = 2^{2+3}$$

Similarly

$$3^2 \times 3^3 = 3^{2+3}$$

and

$$10^2 \times 10^3 = 10^{2+3}$$

Since this result applies to all numbers, the general result is

$a^m \times a^n = a^{m+n}$

where a, m and n can each represent any number. This is an **algebraic equation.** A comparison between arithmetic and algebra is given in Table 4.1.

Table 4.1

Arithmetic	**Algebra**
Numbers 1, 2, 15, −6, . . . have fixed values	Letters a, b, p, $-y$, . . . can stand for any values
$(2 \times 3) + (4 \times 7)$ is an arithmetic expression	$2b + 4c$ is an algebraic expression. $2b$ means $2 \times b$, the $2b$ form being adopted since the letter x is often used in algebra
(2×3) is the first term of the expression and (4×7) the second term	$2b$ is the first term of the expression and $4c$ the second term

When a letter is used more than once in an algebraic expression, it must represent the same value throughout the expression.

4.3 Basic operations

In algebra addition (+), subtraction (−), multiplication (×) and division (÷) have the same meaning as in arithmetic.

In arithmetic $3 \times 5 = 5 + 5 + 5$
and $5^3 = 5 \times 5 \times 5$

Corresponding algebraic terms are

$3 \times a = a + a + a$

and

$a^3 = a \times a \times a$

The four basic operations in algebra are

Addition $2a + a = (a + a) + a = 3a$

Subtraction $2a - a = (a + a) - a = a$

Multiplication $2a \times a = 2 \times a \times a = 2a^2$

Division $2a \div a = \dfrac{2a}{a} = 2$ by cancellation

In arithmetic kilograms cannot be added to metres. In the same way, in algebra different letters represent different quantities and therefore cannot be added. If 'a' represents one number and 'b' another number, when $2a$ and b are added the result is written as

$2a + b$

When b is taken from $2a$, the result is written as

$2a - b$

Multiplying $2a$ by b is written as

$2a \times b = 2ab$,

and dividing $2a$ by b is written

$$2a \div b = \frac{2a}{b}$$

If b represents a number, then except when $b = 1$, b^2 will represent a different number. Hence terms containing b cannot be added or subtracted from those containing b^2. Thus the addition of b and b^2 is written as $b + b^2$. Subtraction of b from b^2 is written as $b^2 - b$. For multiplication and division, the terms can be simplified. Multiplication of b by b^2 becomes

$b \times b^2 = b \times b \times b = b^3$

Division of b by b^2 gives

$$b \div b^2 = \frac{b}{b^2} = \frac{b}{b \times b} = \frac{1}{b}$$

by cancellation.

Worked problems on basic operations

Problem 1. Find the value of $2ab - 3bc + abc$, when $a = 2$, $b = 5$, $c = 1$

Replacing a, b and c with numbers gives:

$$2 \times 2 \times 5 - 3 \times 5 \times 1 + 2 \times 5 \times 1 = 20 - 15 + 10$$
$$= \mathbf{15}$$

Problem 2. Find the value of $3a^2bc^3$ given $a = \frac{1}{2}$, $b = 2$ and $c = 1\frac{1}{2}$

Replacing a, b and c with numbers gives:

$$3 \times (\tfrac{1}{2})^2 \times 2 \times (\tfrac{3}{2})^3 = 3 \times \tfrac{1}{2} \times \tfrac{1}{2} \times 2 \times \tfrac{3}{2} \times \tfrac{3}{2} \times \tfrac{3}{2}$$
$$= \tfrac{81}{16}$$
$$= \mathbf{5\tfrac{1}{16}}$$

Problem 3. Find the sum of $3a$, $2a$, $-a$ and $-6a$

The sum of the positive terms is $3a + 2a = 5a$

The sum of the negative terms is $a + 6a = 7a$

Taking the sum of the negative terms from the sum of the positive terms gives

$5a - 7a$ or $-2a$. Thus
$3a + 2a + (-a) + (-6a) = -2a$.
The rule for this type of problem is to add all the positive terms, add all the negative terms and then to find the difference between these two sums.

Problem 4. Find the sum of $3a$, $2b$, c, $-2a$, $-3b$ and $4c$

Each letter must be dealt with individually.
For the a terms $+ 3a - 2a = a$
For the b terms $+ 2b - 3b = -b$
For the c terms $+ c + 4c = 5c$
Thus
$\mathbf{3a + 2b + c + (-2a) + (-3b) + 4c = a - b + 5c}$

The signs +, −, × and ÷ are called operators, meaning that they signify which operation has to be carried out. Adjacent operators in this type of problem follow the rule:
like signs give + overall (++ or −−)
unlike signs give − overall (+− or −+)
Thus the 4th term from problem 4, $+(-2a)$, becomes $-2a$ and the 5th term $+(-3b)$ becomes $-3b$. Therefore the problem can be written
$\mathbf{3a + 2b + c - 2a - 3b + 4c = a - b + 5c}$

Problem 5. Find the sum of $3a - b$, $2a + c$, $2b - 4d$ and $3b - a + 2d - 2c$

This problem can be laid out as shown, forming columns for the a's, b's, c's and d's. It should be remembered that $3a - b$ means $+ 3a - b$.

$+3a$	$-b$			
$+2a$		$+c$		
	$+2b$		$-4d$	
$-a$	$+3b$	$-2c$	$+2d$	
$\mathbf{4a}$	$\mathbf{+4b}$	$\mathbf{-c}$	$\mathbf{-2d}$	Adding

---

Problem 6. Subtract $2a + 4b - 3c$ from $a - 2b + 8c$

Check what is being subtracted from what, i.e.
From $a \quad -2b \quad +8c$
Take $2a \quad +4b \quad -3c$
Dealing with each term, the a terms become $a - 2a = -a$
the b terms give $-2b - (+4b) = -2b - 4b = -6b$
and the c terms give $8c - (-3c) = 8c + 3c = 11c$. Thus

a	$-2b$	$+8c$	
$2a$	$+4b$	$-3c$	
$-a$	$-6b$	$+11c$	Subtracting

An alternative approach is to think of all the signs of the 'take' expression as being changed and then to add the two expressions:

$$\begin{array}{lll} a & -2b & +8c \\ -2a & -4b & +3c \\ \hline -a & -6b & +11c \end{array} \quad \text{Adding}$$

Problem 7. Multiply $3ab$ by $2ab^2$

$3ab$ means $3 \times a \times b$
$2ab^2$ means $2 \times a \times b \times b$. Thus
$3ab \times 2ab^2 = 3 \times a \times b \times 2 \times a \times b \times b$
Grouping like terms
$3ab \times 2ab^2 = 3 \times 2 \times a \times a \times b \times b \times b$
$= 6 \times a^2 \times b^3$
$= 6a^2b^3$

Problem 8. Simplify $2pq \times 3p^2q^2 \times (-\frac{1}{2}pq^2)$

The product of two terms with like signs is positive.
The product of two terms with unlike signs is negative.
Thus $(+a) \times (-b) = -ab$ but $(+a) \times (+b) = ab$. Similarly $(-a) \times (-b) = ab$ and $(-a) \times (+b) = -ab$. Applying this general principle to the problem:
$2pq \times 3p^2q^2 = 6p^3q^3$ (like signs) and
$6p^3q^3 \times (-\frac{1}{2}pq^2) = -3p^4q^5$ (unlike signs). Hence
$2pq \times 3p^2q^2 \times (-\frac{1}{2}pq^2) = -3p^4q^5$

Problem 9. Multiply $3x + 4y - 5z$ by $x + y$

Each term in the first expression is multiplied by x, then each term in the first expression is multiplied by y and the two results are added. The usual layout is to keep similar terms in columns so that they are underneath one another as shown:

$$\begin{array}{llllll} & 3x & +4y & -5z & & \\ & x & +y & & & \\ \hline \text{Multiplying by } x \rightarrow & 3x^2 & +4xy & -5xz & & \\ \text{Multiplying by } y \rightarrow & & 3xy & & +4y^2 & -5yz \\ \hline \text{Adding} & 3x^2 & +7xy & -5xz & +4y^2 & -5yz \end{array}$$

Problem 10. Multiply $2p - 3q^2 + 4pq$ by $2p - 3q$

$$\begin{array}{lrrrrr}
 & 2p & -\ 3q^2 & +\ 4pq & & \\
 & 2p & -\ 3q & & & \\
\hline
\text{Multiplying by } 2p \rightarrow & 4p^2 & -\ 6pq^2 & +\ 8p^2q & & \\
\text{Multiplying by } -3q \rightarrow & & -\ 12pq^2 & & -\ 6pq & +\ 9q^3 \\
\hline
\text{Adding} & 4p^2 & -\ 18pq^2 & +\ 8p^2q & -\ 6pq & +\ 9q^3 \\
\hline
\end{array}$$

Problem 11. Simplify $2a \div 6ab$

This can be written as $\frac{2a}{6ab}$ and reduced by cancelling as in arithmetic. Thus

$$\frac{2a}{6ab} = \frac{{}^1\not{2} \times \not{a}^1}{{}_3\not{6} \times \not{a}_1 \times b} = \frac{1}{3b} \text{ by cancellation}$$

Problem 12. Simplify $\frac{3a^2bc^3}{9ab^2c^2}$

$$\frac{3a^2bc^3}{9ab^2c^2} = \frac{ac}{3b} \text{ by cancellation}$$

Problem 13. Express as single fractions: (a) $\frac{a}{c+d} + \frac{b}{c+d}$

(b) $\frac{x}{2x-1} - \frac{1-x}{-1+2x}$ (c) $\frac{4}{x+1} + \frac{x}{1+x} - 3.$

(a) $\frac{a}{c+d} + \frac{b}{c+d} = \frac{a+b}{c+d}$

(b) $\frac{x}{2x-1} - \frac{1-x}{-1+2x} = \frac{x}{2x-1} - \frac{1-x}{2x-1} = \frac{x-(1-x)}{(2x-1)} = \frac{2x-1}{2x-1} = \mathbf{1}$

(c) $\frac{4}{x+1} + \frac{x}{1+x} - 3 = \frac{4}{x+1} + \frac{x}{x+1} - \frac{3}{1} = \frac{4+x-3(x+1)}{x+1}$

$$= \frac{4+x-3x-3}{x+1} = \mathbf{\frac{1-2x}{x-1}}$$

Problem 14. Divide $x^3 + 3x^2y + 3xy^2 + y^3$ by $x + y$

In arithmetic, long division can be used to divide, say, 192 by 16.

$$
\begin{array}{r|l}
 & 12 \\ \hline
16 & 192 \\
 & \underline{16} \\
 & 32 \\
 & \underline{32} \\
 & \cdot\cdot
\end{array}
$$

Another way of doing this problem is

$$
\begin{array}{r|ccccc}
 & 6 & + & 5 & + & 0 \\ \hline
16 & 100 & + & 90 & + & 2 \\
 & \underline{96} & & \underline{80} & & \underline{0} \\
 & {}^{4}/_{16} & & {}^{10}/_{16} & & {}^{2}/_{16}
\end{array}
$$

That is $\frac{192}{16} = 6\frac{4}{16} + 5\frac{10}{16} + \frac{2}{16} = 12$

An approach something along the same lines is made in algebraic division.

$$
\begin{array}{r|l}
 & \overset{①}{x^2} + \overset{③}{2xy} + \overset{⑤}{y^2} \\ \hline
x + y & x^3 + 3x^2y + 3xy^2 + y^3 \\
 & \underline{x^3 + x^2y} \\
 & \qquad 2x^2y + 3xy^2 \\
 & \qquad \underline{2x^2y + 2xy^2} \\
 & \qquad\qquad xy^2 + y^3 \\
 & \qquad\qquad \underline{xy^2 + y^3} \\
 & \qquad\qquad \cdot \quad \cdot
\end{array}
$$

① x into x^3 goes x^2, put x^2 up

② $x^2(x + y)$ and subtracting

③ Bring down $3xy^2$. x into $2x^2y$ goes $2xy$, put $2xy$ up.

④ $2xy(x + y)$ and subtracting

⑤ Bring down y^3. x into xy^2 goes y^2, put y^2 up

⑥ $y^2(x + y)$ and subtracting

Thus $(x^3 + 3x^2y + 3xy^2 + y^3) \div (x + y) = x^2 + 2xy + y^2$

A check can be made on this answer by multiplying $x^2 + 2xy + y^2$ by $x + y$ and seeing if it is equal to $x^3 + 3x^2y + 3xy^2 + y^3$.

Problem 15. Simplify $\dfrac{a^3 - b^3}{a - b}$

$$
\begin{array}{lll}
 & a^2 + ab + b^2 & \\
 & a - b\,\overline{)\,a^3 + 0 + 0 - b^3} & a \text{ into } a^3 \text{ goes } a^2 \\
a^2(a-b) & \underline{a^3 - a^2b} & \\
\text{Subtract} & a^2b + 0 & a \text{ into } a^2b \text{ goes } ab \\
ab(a-b) & \underline{a^2b - ab^2} & \\
\text{Subtract} & ab^2 - b^3 & a \text{ into } ab^2 \text{ goes } b^2 \\
b^2(a-b) & \underline{ab^2 - b^3} & \\
\text{Subtract} & \underline{\cdot \quad \cdot} &
\end{array}
$$

Thus $\dfrac{a^3 - b^3}{a - b} = a^2 + ab + b^2$

The noughts shown in the $a^3 - b^3$ expression are not normally shown, but have been included to clarify the subtraction process and to keep similar terms in their respective columns.

Problem 16. Divide $6m^3 - 4m^2n + 2n^3$ by $2m - n$

$$
\begin{array}{l}
3m^2 - \tfrac{1}{2}mn - \tfrac{1}{4}n^2 \\
2m - n\,\overline{)\,6m^3 - 4m^2n \qquad\qquad\qquad + 2n^3} \\
\underline{6m^3 - 3m^2n} \\
 - m^2n \\
 \underline{- m^2n + \tfrac{1}{2}mn^2} \\
 - \tfrac{1}{2}mn^2 \\
 \underline{- \tfrac{1}{2}mn^2 + \tfrac{1}{4}n^3} \\
\phantom{2m-n\,)6m^3 - m^2n - \tfrac{1}{2}mn^2} - \tfrac{1}{4}n^3 + 2n^3
\end{array}
$$

Thus $\dfrac{6m^3 - 4m^2n + 2n^3}{2m - n} = \mathbf{3m^2 - \dfrac{mn}{2} - \dfrac{n^2}{4}}$ **and the remainder is** $\mathbf{(-\tfrac{1}{4}n^3 + 2n^3)}$ **or** $\mathbf{+\tfrac{7}{4}n^3}$.

Further problems on basic operations may be found in Section 4.7 (Problems 1–18), page 117.

4.4 Laws of indices

The laws of indices, introduced for arithmetic in Chapter 2, can now be generalised to apply to algebra. These rules are shown below in algebraic form, where a, m and n are any number.

$a^m = a \times a \times a \times \ldots$ to m terms
$a^n = a \times a \times a \times \ldots$ to n terms

Laws:

$$a^m a^n = a^{m+n} \qquad [I]$$

For example $a^3 a^5 = a^{3+5} = a^8$

$$\frac{a^m}{a^n} = a^{m-n} \qquad [II]$$

For example $\frac{a^6}{a^2} = a^{6-2} = a^4$

$$(a^m)^n = a^{mn} \qquad [III]$$

For example $(a^3)^5 = a^{3 \times 5} = a^{15}$

$$a^{\frac{m}{n}} = \sqrt[n]{a^m} \qquad [IV]$$

For example $a^{7/2} = \sqrt{a^7}$

$$a^{-n} = \frac{1}{a^n} \qquad [V]$$

For example $a^{-3} = \frac{1}{a^3}$

and expressing $a^4 \times a^{-7}$ with a positive index gives

$$a^{4-7} = a^{-3} \qquad \text{(by law [I])}$$
$$= \frac{1}{a^3} \qquad \text{(by law [V])}$$

$$a^0 = 1 \qquad [VI]$$

For example $\frac{a^2 a^3}{a^5} = \frac{a^{2+3}}{a^5}$ (by law [I])

$$= \frac{a^5}{a^5}$$
$$= a^{5-5} \qquad \text{(by law [II])}$$
$$= a^0$$
$$= 1 \qquad \text{(by law [VI])}$$

Worked problems on laws of indices

Problem 1. Simplify $a^2 \ b^3 \ c \times ab^2 \ c^3$ and evaluate when $a = 1$, $b = \frac{1}{2}$ and $c = 2$.

Grouping like terms this becomes $a^2 \times a \times b^3 \times b^2 \times c \times c^3$ and since $a = a^1$ and $c = c^1$, using law (1) this becomes

$$a^{(2+1)} \times b^{(3+2)} \times c^{(1+3)} = a^3 \times b^5 \times c^4$$
$$= a^3\, b^5\, c^4$$

When $a = 1$, $b = \frac{1}{2}$ and $c = 2$, $a^3\, b^5\, c^4 = (1)^3 (\frac{1}{2})^5 (2)^4$

$$= \frac{2^4}{2^5} = \frac{1}{2}$$

Problem 2. Simplify $p^{\frac{1}{2}}\, q^2\, r^{-3} \times p^{1/3}\, q^{\frac{1}{2}}\, r^{-\frac{1}{2}}$ and evaluate when $p = 64$, $q = \frac{1}{4}$ and $r = 1$.

This becomes by law (1), $p^{(1/2\,+\,1/3)}\, q^{(2\,+\,1/2)}\, r^{(-3-1/2)} = p^{5/6}\, q^{5/2}\, r^{-7/2}$

When $p = 64$, $q = 4$ and $r = 1$, $p^{5/6}\, q^{5/2}\, r^{-7/2} = (64)^{5/6} (\frac{1}{4})^{5/2} (1)^{-7/2}$

$$= \sqrt[6]{(64)^5}\, \sqrt{(\tfrac{1}{4})^5}\, (1) = (2)^5 (\tfrac{1}{2})^5 = \mathbf{1}.$$

Problem 3. Simplify $\dfrac{a^2\, b^3\, c^4}{ab^2\, c}$ and evaluate when $a = 4$, $b = 3$ and $c = \frac{1}{2}$.

Using the second law of indices

$$\frac{a^2}{a} = a^{2-1}, \frac{b^3}{b^2} = b^{3-2} \text{ and } \frac{c^4}{c} = c^{4-1}$$

Thus $\dfrac{a^2\, b^3\, c^4}{ab^2\, c} = a^{2-1}\, b^{3-2}\, c^{4-1} = \boldsymbol{abc^3}$

When $a = 4$, $b = 3$ and $c = \frac{1}{2}$, $ab\, c^3 = (4)\,(3)\,(\frac{1}{2})^3 = \dfrac{(4)\,(3)}{8} = \mathbf{1\frac{1}{2}}$.

Problem 4. Simplify $\dfrac{x^{\frac{1}{2}}\, y^2\, z^{1/3}}{x^{\frac{1}{4}}\, y^{\frac{1}{2}}\, z^{1/6}}$

Using law (II), this becomes

$x^{\frac{1}{2}-\frac{1}{4}}\, y^{2-\frac{1}{2}}\, z^{1/3\,-\,1/6}$ or $\boldsymbol{x^{\frac{1}{4}}\, y^{3/2}\, z^{1/6}}$

Problem 5. Simplify $\dfrac{x^2\, y^3 + x^{\frac{1}{2}}\, y}{x^{1/3} y^{1/3}}$

Algebraic expressions of the form $\frac{a+b}{c}$ can be split into $\frac{a}{c}+\frac{b}{c}$. Thus

$$\frac{x^2y^3 + x^{\frac{1}{2}}y}{x^{1/3}y^{1/3}} = \frac{x^2y^3}{x^{1/3}y^{1/3}} + \frac{x^{\frac{1}{2}}y}{x^{1/3}y^{1/3}}$$
$$= x^{2-1/3}y^{3-1/3} + x^{1/2-1/3}y^{1-1/3}$$
$$= x^{5/3}y^{8/3} + x^{1/6}y^{2/3}$$

Problem 6. Simplify $\frac{a^2b}{ab^2 - ab}$

The highest common factor of each of the three terms comprising the numerator and denominator is found. This is ab. Dividing each term by ab gives:

$$\frac{a^2b}{ab^2 - ab} = \frac{\frac{a^2b}{ab}}{\frac{ab^2}{ab} - \frac{ab}{ab}}$$
$$= \frac{a}{b-1}$$

Problem 7. Simplify $\frac{p^2q}{pq^2 - p^{\frac{1}{2}}q}$ and evaluate when $p = 9$, and $q = \frac{1}{2}$, taking positive values of square roots only.

A common factor approach must be made where addition or subtraction of terms occurs in the denominator. The highest common factor of the three terms is $p^{\frac{1}{2}}q$. Thus

$$\frac{p^2q}{pq^2 - p^{\frac{1}{2}}q} = \frac{\frac{p^2q}{p^{\frac{1}{2}}q}}{\frac{pq^2}{p^{\frac{1}{2}}q} - \frac{p^{\frac{1}{2}}q}{p^{\frac{1}{2}}q}} = \frac{p^{3/2}}{p^{\frac{1}{2}}q - 1}$$

When $p = 9$ and $q = \frac{1}{2}$, $\frac{p^{3/2}}{p^{\frac{1}{2}}q - 1} = \frac{(9)^{3/2}}{(9)^{\frac{1}{2}}(\frac{1}{2}) - 1} = \frac{\sqrt{(9)^3}}{[\sqrt{(9)}]\,(\frac{1}{2}) - 1}$

$$= \frac{3^3}{(3)\,(\frac{1}{2}) - 1} = \frac{27}{\frac{1}{2}} = 54$$

Problem 8. Simplify $(a^3)^{\frac{1}{2}}(b^2)^3$

Using law [III], the expression becomes $a^{3 \times \frac{1}{2}} b^{2 \times 3}$. Thus $(a^3)^{\frac{1}{2}}(b^2)^3 = a^{3/2} b^6$

Problem 9. Simplify $\dfrac{(m^2 n)^3}{(m^{\frac{1}{2}} n^{1/3})^4}$

The bracket indicates that each letter in the bracket must be raised to the power shown outside. Using law [III] the numerator becomes $m^{2 \times 3} n^{1 \times 3}$ or $m^6 n^3$ and the denominator becomes $m^{\frac{1}{2} \times 4} n^{\frac{1}{3} \times 4}$ or $m^2 n^{4/3}$. Hence

$$\frac{(m^2 n)^3}{(m^{\frac{1}{2}} n^{1/3})^4} = \frac{m^6 n^3}{m^2 n^{4/3}}$$

$$= m^{6-2} n^{3-4/3} \quad \text{by law [II]}$$

$$= m^4 n^{5/3}$$

Problem 10. Simplify $a^2 \sqrt{b} \sqrt{c^3} \times \sqrt{a} \sqrt[3]{b^2}\ c^4$ and evaluate when $a = \dfrac{1}{4}$, $b = 64$ and $c = 1$.

By law (IV) this can be written as

$$a^2\ b^{\frac{1}{2}}\ c^{3/2} \times a^{\frac{1}{2}}\ b^{2/3}\ c^4 = a^{5/2}\ b^{7/6}\ c^{11/2}$$

It is usual in mathematics to give the answer in the same form as the question.

Hence $a^{5/2}\ b^{7/6}\ c^{11/2} = \sqrt{a^5}\ \sqrt[6]{b^7}\ \sqrt{c^{11}}$

When $a = \dfrac{1}{4}$, $b = 64$ and $c = 1$, $\sqrt{a^5}\ \sqrt[6]{b^7}\ \sqrt{c^{11}} = \sqrt{(\frac{1}{4})^5}\ \sqrt[6]{(64)^7}\ \sqrt{(1)^{11}}$

$$= (\tfrac{1}{2})^5\ (2)^7\ (1) = \frac{2^7}{2^5} = 2^2 = 4$$

Problem 11. Simplify $(p^2 q)(p^{-3} q^{-2})$ expressing the answer with positive indices only.

Using law [I] $(p^2 q)(p^{-3} q^{-2}) = p^{-1} q^{-1}$

Applying law [V] $\quad p^{-1} q^{-1} = \dfrac{1}{pq}$

Problem 12. Simplify $\dfrac{p^2 q^4 r^{\frac{1}{2}}}{(p^{3/2} q^2 r^{3/2})^2}$

Using law [III], this becomes $\dfrac{p^2q^4r^{\frac{1}{2}}}{p^{3/2\times 2}q^{2\times 2}r^{3/2\times 2}} = \dfrac{p^2q^4r^{\frac{1}{2}}}{p^3q^4r^3}$

and by law [II] $\quad p^{2-3}q^{4-4}r^{\frac{1}{2}-3} = p^{-1}q^0r^{-5/2}$

By law [VI] $\quad q^0 = 1$. Hence

$$p^{-1}q^0r^{-5/2} = p^{-1}r^{-5/2} = \frac{1}{pr^{5/2}}$$

Problem 13. Simplify $\dfrac{(a^2b^{\frac{1}{2}})(\sqrt{a}\,\sqrt[3]{b^2})}{(a^5b^4)^{\frac{1}{2}}}$

By laws [III] and [IV] this expression becomes $\dfrac{(a^2b^{\frac{1}{2}})(a^{\frac{1}{2}}b^{2/3})}{a^{5/2}b^{4/2}}$

$= a^{2+\frac{1}{2}-5/2}b^{\frac{1}{2}+2/3-2}$ (laws [I] and [II])

$= a^0b^{-5/6}$

$= \dfrac{1}{b^{5/6}}$ (laws [V] and [VI])

Further problems on the laws of indices may be found in Section 4.7 (Problems 17–32), page 117.

4.5 Brackets and factorisation

The use of brackets

The addition of the sum of b and c to a may be written as

$a + (b + c)$

The subtraction of the sum of b and c from a is written as

$a - (b + c)$

The multiplication of twice the sum of a and b by the difference between c and d is written as

$2(a + b)(c - d)$

The division of the square of the sum of a and b by twice the sum of c and d is written as

$\dfrac{(a+b)^2}{2(c+d)}$

Removing brackets

An addition sign in front of a bracket does not change the signs within the bracket when they are removed, thus

$(a + b) + (c + d) = a + b + c + d$

and

$(a + b) + (c - d) = a + b + c - d$

A subtraction sign in front of a bracket results in all terms within the bracket being multiplied by -1, thus having their signs changed when the brackets are removed. Thus

$(a + b) - (c + d) = a + b - c - d$

and

$(a + b) - (c - d) = a + b - c + d$

A term in front of or behind a bracket indicates that each term within the bracket is multiplied by that term. Hence

$$a(b + c) = ab + ac$$

and

$d(a + b) - 2(c - d) = ad + bd - 2c + 2d$

and

$$(3a + b)d = 3ad + bd$$

One bracketed expression immediately followed by another bracketed expression indicates that the contents of one bracket should be multiplied by the contents of the other bracket when they are being removed. Thus, using algebraic multiplication techniques,

$(a + b)(c + d) = ac + ad + bc + bd$

A bracketed expression raised to some index indicates that the contents of the bracket must be multiplied by itself the number of times indicated by the power, thus

$$\begin{aligned}(p + q)^2 = (p + q)(p + q) &= p^2 + pq + pq + q^2 \\ &= p^2 + 2pq + q^2\end{aligned}$$

using algebraic multiplication techniques. When an expression contains more than one type of bracket, it is usual to remove the innermost bracket first and work outwards clearing one bracket at a time. For an expression containing two brackets

$$\begin{aligned}6\left\{3y - 4(2z + 4)\right\} &= 6\left\{3y - 8z - 16\right\} \\ &= 18y - 48z - 96\end{aligned}$$

and for an expression containing three brackets:

$$\begin{aligned}4\left[2 + a\left\{c - 3(a + 2)\right\}\right] &= 4\left[2 + a\left\{c - 3a - 6\right\}\right] \\ &= 4\left[2 + ac - 3a^2 - 6a\right] \\ &= 8 + 4ac - 12a^2 - 24a\end{aligned}$$

Factorisation

When two or more terms in an algebraic expression contain a common factor, then this factor can be shown outside of a bracket. Thus

$$4ax + 2ay = 2a \cdot 2x + 2a \cdot y$$
$$= 2a(2x + y)$$

(The product point $\cdot$ is used here instead of $\times$ to indicate multiplication.) This process is called **factorisation**, $2a$ and $(2x + y)$ being the factors of the expression $4ax + 2ay$. In the expression

$$ax + 3a + 2bx + 6b$$

a is common to the first two terms and $2b$ to the last two terms. Thus the expression can be written as

$$a(x + 3) + 2b(x + 3)$$

Also $(x + 3)$ is a common factor of both the first and second terms and so

$$a(x + 3) + 2b(x + 3) = (x + 3)(a + 2b)$$

Then $(x + 3)$ and $(a + 2b)$ are the factors of the expression

$$ax + 3a + 2bx + 6b$$

Worked problems on brackets and factorisation

Problem 1. Remove the brackets and simplify the expression
$(2a + b) + 3(b + c) - 2(c - d)$

Both b and c in the second bracket have to be multiplied by 3 and c and $-d$ in the third bracket by -2 when the brackets are removed. Thus
$(2a + b) + 3(b + c) - 2(c - d) = 2a + b + 3b + 3c - 2c + 2d$
and collecting together the like terms gives $2a + 4b + c + 2d$, that is
$(2a + b) + 3(b + c) - 2(c - d) = \mathbf{2a + 4b + c + 2d}$

Problem 2. Simplify $x^2 - (3x - xy) - x(4y + x)$

When the brackets are removed, both $3x$ and $-xy$ in the first bracket must be multiplied by -1 and both $4y$ and x in the second bracket by $-x$. Thus
$x^2 - (3x - xy) - x(4y + x) = x^2 - 3x + xy - 4xy - x^2$
Grouping like terms gives $(x^2 - x^2) - 3x + (xy - 4xy)$
or
$-3x - 3xy$
Since $-3x$ is a common factor, this can be factorised as $-3x(1 + y)$. Thus
$x^2 - (3x - xy) - x(4y + x) = \mathbf{-3x(1 + y)}$
When forming a bracket in this way, a check should be made by multiplying out to see that the signs have been correctly determined.

Problem 3. Remove the brackets from the expression $(2p - q)(p + q^2)$

Each term in the second bracket has to be multiplied by each term in the first bracket. Thus

$(2p - q)(p + q^2) = 2p(p + q^2) - q(p + q^2)$
$= 2p^2 + 2pq^2 - pq - q^3$

Alternatively, algebraic multiplication can be used.

Problem 4. Remove the brackets from the expression
$-3\left[a^2 - 2(b + c) + b^2\right]$

Removing the inner bracket by multiplying b and c by -2 gives
$-3\left[a^2 - 2b - 2c + b^2\right]$
Removing the outer bracket by multiplying all terms within this bracket by -3 gives
$-3a^2 + 6b + 6c - 3b^2$

Then

$-3\left[a^2 - 2(b + c) + b^2\right] = \mathbf{-3a^2 + 6b + 6c - 3b^2}$

Problem 5. Remove the brackets and simplify the expression
$4a - \left[2\left\{(2a - 4b) - 2(a + 5b)\right\} + a\right]$

Starting by removing the innermost brackets gives
$4a - \left[2\left\{2a - 4b - 2a - 10b\right\} + a\right]$
The $\{\ \}$ brackets are now the inner ones and are removed next, giving

$4a - [4a - 8b - 4a - 20b + a]$
Multiplying all terms in the square brackets by -1 gives
$4a - 4a + 8b + 4a + 20b - a$
Grouping like terms together results in
$4a - \left[2\left\{(2a - 4b) - 2(a + 5b)\right\} + a\right]$ **being equal to** $\mathbf{3a + 28b}$

Problem 6. Simplify $m(3m - 2n) - 2m(3m + 4n)$

Multiplying each term in the first bracket by m and each term in the second bracket by $-2m$ gives
$3m^2 - 2mn - 6m^2 - 8mn = -3m^2 - 10mn$
Since $-m$ is a common factor of both terms, the expression can be written as $-m(3m + 10n)$, thus
$m(3m - 2n) - 2m(3m + 4n) = \mathbf{-m(3m + 10n)}$

Problem 7. Factorise (*a*) $pq - 2pr$ (*b*) $3a^2 + 15ab^2$
(*c*) $5x^2y - 10xy^2 + 25xy$

For each part of this problem, the highest common factor of the terms will become one of the factors. Thus

(*a*) $pq - 2pr = \mathbf{p(q - 2r)}$

(*b*) $3a^2 + 15ab^2 = \mathbf{3a(a + 5b^2)}$

(*c*) $5x^2y - 10xy^2 + 25xy = \mathbf{5xy(x - 2y + 5)}$

Problem 8. Factorise $ax + bx - by - ay$

The first two terms have a common factor of x and the last two terms a common factor of y. Thus

$ax + bx - by - ay = x(a + b) + y(-a - b)$
$= x(a + b) - y(a + b)$

since the second bracket also has a common factor of -1. Both terms now have a common factor of $(a + b)$, hence

$ax + bx - by - ay = \mathbf{(a + b)(x - y)}$

Problem 9. Factorise $3ax - 2ay + 3bx - 2by$

Selecting a as a common factor of the first two terms and b as a common factor of the last two terms gives

$a(3x - 2y) + b(3x - 2y)$

Since $(3x - 2y)$ is a common factor of both these terms, the expression becomes $(3x - 2y)(a + b)$. Thus

$3ax - 2ay + 3bx - 2by = \mathbf{(3x - 2y)(a + b)}$

Alternatively, selecting $3x$ as the common factor of the first and third terms and $-2y$ as the common factor of the second and last terms gives

$3x(a + b) - 2y(a + b) = (3x - 2y)(a + b)$

the same result as for selecting a and b as the factors.

Problem 10. Factorise $x^3 - 2x^2 - x + 2$

Selecting x^2 as a common factor of the first two terms gives

$x^3 - 2x^2 - x + 2 = x^2(x - 2) - x + 2$

and by making -1 a factor of the third and fourth terms the expression becomes $x^2(x - 2) - 1(x - 2)$

which is usually written $x^2(x - 2) - (x - 2)$

But $(x - 2)$ is now a common factor, hence

$x^3 - 2x^2 - x + 2 = \mathbf{(x^2 - 1)(x - 2)}$

Further problems on brackets and factorisation may be found in Section 4.7 (Problems 33–45), page 118.

4.6 Fundamental laws and precedence

The basic laws introduced in arithmetic are generalised in algebra. These are given below, where a, b and c represent any three numbers.

$a + (b + c) = (a + b) + c$ [I]

For example

$$2m + (3m + 4m) = 2m + 7m = 9m$$
and $$(2m + 3m) + 4m = 5m + 4m = 9m$$
thus $$2m + (3m + 4m) = (2m + 3m) + 4m = 9m$$

$a(bc) = (ab)c$ [II]

For example

$$2m(3m \cdot 4m) = 2m(12m^2) = 24m^3$$
and $$(2m \cdot 3m)4m = (6m^2)4m = 24m^3$$
thus $$2m(3m \cdot 4m) = (2m \cdot 3m)4m = 24m^3$$

Laws (I) and (II) are called associative laws.

$(a + b) = (b + a)$ [III]

For example

$$(2m + 3m) = 5m$$
and $$(3m + 2m) = 5m$$
thus $$(2m + 3m) = (3m + 2m) = 5m$$

$ab = ba$ [IV]

For example

$$2m \cdot 3m = 6m^2$$
and $$3m \cdot 2m = 6m^2$$
thus $$2m \cdot 3m = 3m \cdot 2m = 6m^2$$

Laws (III) and (IV) are called commutative laws.

$a(b + c) = ab + ac$ [V]

For example

$$2m(3m + 4m) = 2m(7m) = 14m^2$$
and $$2m \cdot 3m + 2m \cdot 4m = 6m^2 + 8m^2 = 14m^2$$
thus $$2m(3m + 4m) = 2m \cdot 3m + 2m \cdot 4m = 14m^2$$

Law (V) is called the distributive law.

The rules of precedence which apply to arithmetic governing the order in which operations are performed apply equally to algebraic expressions. The order is **B**rackets, **O**f, **D**ivision, **M**ultiplication, **A**ddition and **S**ubtraction, that is, governed by **BODMAS.** (As with arithmetic, the word 'of' appearing in an expression should be changed to a multiplication sign after brackets have been removed.)

For example, to simplify the expression

$(5a + 3a) - 4a \times 6a \div 7a$

the bracketed terms must first be evaluated, then the division performed, followed by the multiplication and finally the subtraction. Thus

$$
\begin{aligned}
(5a + 3a) - 4a \times 6a \div 7a &= 8a - 4a \times 6a \div 7a && \text{(B)} \\
&= 8a - 4a \times \frac{6a}{7a} && \text{(D)} \\
&= 8a - 4a \times \frac{6}{7} \\
&= 8a - \frac{24a}{7} && \text{(M)} \\
&= \frac{56a - 24a}{7} = \frac{32a}{7} && \text{(S)}
\end{aligned}
$$

Worked problems on fundamental laws and precedence

Problem 1. Simplify $3y + 7y \times 2y - y$

The order of precedence indicates that the multiplication must be done before addition and subtraction, hence

$$
\begin{aligned}
3y + 7y \times 2y - y &= 3y + 14y^2 - y \\
&= 14y^2 + 2y \\
&= 2y(7y + 1)
\end{aligned}
$$

Problem 2. Simplify $(3y + 7y) \times 2y - y$

The order is brackets, multiplication and subtraction, hence

$$
\begin{aligned}
(3y + 7y)2y - y &= 10y \cdot 2y - y \\
&= 20y^2 - y \\
&= y(20y - 1)
\end{aligned}
$$

Problem 3. Simplify $3y + 7y \times (2y - y)$

The order is brackets, multiplication and addition, thus

$$
\begin{aligned}
3y + 7y(2y - y) &= 3y + 7y \times y \\
&= 3y + 7y^2 \\
&= y(3 + 7y)
\end{aligned}
$$

Problem 4. Simplify $6k \div 3k + 4k - 2k$

The order is division, addition and subtraction, thus

$$6k \div 3k + 4k - 2k = \frac{6k}{3k} + 4k - 2k$$

$= 2 + 4k - 2k$
$= 2 + 2k = \mathbf{2(1 + k)}$

Problem 5. Simplify $6k \div (3k + 4k) - 2k$

The order is brackets, division and subtraction, hence

$$6k \div (3k + 4k) - 2k = 6k \div 7k - 2k$$
$$= \frac{6}{7} - 2k$$

Problem 6. Simplify $6k \div (3k + 4k - 2k)$

The order is brackets and then division, thus

$$6k \div (3k + 4k - 2k) = 6k \div 5k$$
$$= \frac{6}{5}$$

Problem 7. Simplify $4m + 2m \times 3m + m \div 5m - 6m$

Applying BODMAS gives the order of precedence as division, multiplication, addition and finally subtraction. Thus

$$4m + 2m \times 3m + m \div 5m - 6m = 4m + 2m \times 3m + \frac{m}{5m} - 6m$$
$$= 4m + 6m^2 + \frac{1}{5} - 6m$$
$$= \mathbf{6m^2 - 2m + \frac{1}{5}}$$

Problem 8. Simplify $(4m + 2m)3m + m \div 5m - 6m$

The order is brackets, division, multiplication, addition and subtraction, thus

$$(4m + 2m)3m + m \div 5m - 6m = 6m \times 3m + m \div 5m - 6m$$
$$= 6m \times 3m + \frac{m}{5m} - 6m$$
$$= 18m^2 + \frac{1}{5} - 6m$$
$$= \mathbf{6m(3m - 1) + \frac{1}{5}}$$

Problem 9. Simplify $4m + 2m \times 3m + m \div (5m - 6m)$

The order is brackets, division, multiplication and addition. Thus

$$4m + 2m \times 3m + m \div (5m - 6m) = 4m + 2m \times 3m + m \div -m$$

$$= 4m + 2m \times 3m + \frac{m}{-m}$$

Now $\frac{m}{-m} = \frac{1}{-1}$ and multiplying numerator and denominator by -1 gives $\frac{1 \times -1}{-1 \times -1} = -1$, hence

$$4m + 2m \times 3m + \frac{m}{-m} = 4m + 6m^2 - 1$$

$$= \mathbf{2m(3m + 2) - 1}$$

Problem 10. Simplify $(4m + 2m)(3m + m) \div (5m - 6m)$

The order is brackets, division and multiplication, thus

$$(4m + 2m)(3m + m) \div (5m - 6m) = 6m \times 4m \div -m$$

$$= 6m \times \frac{4m}{-m}$$

$$= \frac{24m^2}{-m}$$

$$= \frac{24m}{-1} \times \frac{-1}{-1} = \mathbf{-24m}$$

Problem 11. Simplify $(2y - 5) \div 4y + 6 \times 7 - 3y$

The bracket around the $2y - 5$ shows that both $2y$ and -5 have to be divided by $4y$, and to remove the bracket the expression is written in fraction form. Thus

$$(2y - 5) \div 4y + 6 \times 7 - 3y = \frac{2y - 5}{4y} + 6 \times 7 - 3y$$

$$= \frac{2y - 5}{4y} + 42 - 3y$$

$$= \frac{2y - 5 + 168y - 12y^2}{4y}$$

$$= \mathbf{\frac{170y - 5 - 12y^2}{4y}}$$

Problem 12. Simplify $\frac{1}{4}$ of $3y + 8y(5y - 2y)$

Applying BODMAS, the expression becomes $\frac{1}{4}$ of $3y + 8y(3y)$
and changing 'of' to ×, this becomes $\frac{1}{4} \times 3y + 8y \times 3y$
or $\frac{3}{4}y + 24y^2$ that is $\mathbf{y(\frac{3}{4} + 24y)}$

Further problems on the order of precedence may be found in the following Section (4.7) (Problems 46–57), page 119.

4.7 Further problems

Basic operations

1. Find the sum of $-4c$, $3c$, $-12c$ and $6.5c$ [$-6.5c$]
2. Simplify $\frac{3}{2}p + 8p - \frac{3}{5}p - \frac{p}{3}$ [$8\frac{17}{30}p$]
3. Find the sum of $4p$, $7q$, $-16p$, $3r$, $\frac{1}{2}p$ and $-3q$ [$-11\frac{1}{2}p + 4q + 3r$]
4. Add together $3x + 4y + z$, $-2x + y - 5z$ and $5x - 3y - 10z$ [$6x + 2y - 14z$]
5. From $3x - 2y + z$ subtract $2x - 3y - 4z$ [$x + y + 5z$]
6. Subtract $\frac{1}{2}p + \frac{1}{3}q - \frac{1}{4}r$ from $\frac{1}{3}p - \frac{1}{4}q - \frac{1}{5}r$ [$-\frac{1}{6}p - \frac{7}{12}q + \frac{1}{20}r$]
7. From $0.3a - 0.4b + 6z$ subtract $0.5a + 0.6b - 0.8z$ [$-0.2a - b + 6.8z$]
8. Multiply $4xyz$ by $3x^2y^3z$ [$12x^3y^4z^2$]
9. Simplify $3a^2b \times 4ac^2 \times -2abc$ [$-24a^4b^2c^3$]
10. Multiply $3p - 6pq + 2q^2$ by $p - q$ [$3p^2 - 6p^2q + 8pq^2 - 3pq - 2q^3$]
11. Multiply $3x - 4y$ by $x^2 + 3xy + 2y^3$ [$3x^3 + 5x^2y + 6xy^3 - 12xy^2 - 8y^4$]
12. Simplify $\frac{abc}{3a^2b^3c^4}$ $\left[\frac{1}{3ab^2c^3}\right]$
13. Simplify $\frac{3xy^2z^3}{24xy^3z}$ $\left[\frac{z^2}{8y}\right]$
14. Divide $a^2 - 2ab + b^2$ by $a - b$ [$a - b$]
15. Express $\frac{3a}{a-2} - \frac{a}{a-2}$ as a single fraction. $\left[\frac{2a}{a-2}\right]$
16. Express $\frac{2x}{1+x} + \frac{5}{x+1} - 3$ as a single fraction. $\left[\frac{2-x}{1+x}\right]$
17. Divide $a^3 + b^3$ by $a + b$ [$a^2 - ab + b^2$]
18. Divide $x^6 - y^6$ by $x^2 - xy + y^2$ [$x^4 + x^3y - xy^3 - y^4$]

Laws of indices

19. Simplify $p^2 q^3 r \times pq^2 r^5$ and evaluate when $p = 3, q = 2$ and $r = \frac{1}{2}$. [$p^3 q^5 r^6$; $13\frac{1}{2}$]
20. Simplify $\frac{x^{\frac{1}{2}} y^2 z^{3/2}}{y^{3/2} z}$ and evaluate when $x = 2, y = 2$ and $z = 9$. [$x^{\frac{1}{2}} y^{\frac{1}{2}} z^{\frac{1}{2}}$; ± 6]
21. Simplify $a^{\frac{1}{4}} b^2 c^{-2} \times b^{-3/2} c^3$ and evaluate when $a = 16$, $b = \frac{1}{9}$ and $c = 6$. [$a^{\frac{1}{4}} b^{\frac{1}{2}} c$; ± 4]

In problems 22 to 32, simplify the expressions given.

22. $\dfrac{p^{1/5}q^{1/4}r^{1/3}}{p^{1/6}q^{-\frac{1}{2}}r^{-2}}$ $\quad [p^{1/30}q^{3/4}r^{7/3}]$

23. $\dfrac{k^2l + k^{\frac{1}{2}}l^{3/2}}{k^3l^4}$ $\quad \left[\dfrac{1}{kl^3} + \dfrac{1}{(kl)^{5/2}}\right]$

24. $\dfrac{x^3y^{\frac{1}{2}}}{xy - x^2y^3}$ $\quad \left[\dfrac{x^2}{y^{\frac{1}{2}} - xy^{5/2}}\right]$

25. $\dfrac{(abc)^3}{(a^2b^{\frac{1}{2}}c^3)^{\frac{1}{2}}}$ $\quad [a^2b^{11/4}c^{3/2}]$

26. $(a^2)^3(b^{\frac{1}{2}})^{1/3}$ $\quad [a^6b^{1/6}]$

27. $x^2(\sqrt{y})\sqrt[3]{z^2} \times (\sqrt[3]{x})y^2z^3$ $\quad [x^{7/3}y^{5/2}z^{11/3}]$

28. $a^2b^3c \times a^{-3}b^{-5}c^2$, expressing the answer with positive indices only $\left[\dfrac{c^3}{ab^2}\right]$

29. $\dfrac{p^{1/3}q^2r^{1/5}}{(pq^{\frac{1}{2}}r^3)^{1/3}}$ $\quad [q^{11/6}r^{-4/5}]$

30. $\dfrac{k^3(\sqrt{l})\,m^{3/2} \times l\sqrt[3]{m^2}}{k^{1/3}\sqrt{l^3}}$ $\quad [k^{8/3}m^{13/6}]$

31. $(x^{-2/3}y^{\frac{1}{2}}z^3)^{\frac{1}{2}} \div (x^{2/3}y^2z^{1/3})^{1/3}$ expressing the answer with positive indices only $\quad \left[\dfrac{z^{25/18}}{x^{5/9}y^{5/12}}\right]$

32. $\dfrac{a^2b^{\frac{1}{2}}c^{-1/4} \times (ab)^{1/4}}{\sqrt[3]{a^2}\sqrt{b^3}c^{\frac{1}{2}}}$ $\quad [a^{19/12}b^{-3/4}c^{-3/4}]$

Brackets and factorisation

In problems 33 to 40, remove the brackets and simplify where possible.

33. (*a*) $(p + 2q) + (3p - q)$ (*b*) $(a - 6b) + (3b - 2c)$
(*c*) $(3x + 4y) - (x - 2y)$
(*a*) $[4p + q]$ (*b*) $[a - 3b - 2c]$ (*c*) $[2x + 6y]$

34. (*a*) $(a + b - c) + (2a - 4b - 3c)$ (*b*) $(3p - 4q + r) - (2p + 3q - 6r)$
(*c*) $(4x - z) - (y + 2x - 4z)$
(*a*) $[3a - 3b - 4c]$ (*b*) $[p - 7q + 7r]$ (*c*) $[2x - y + 3z]$

35. (*a*) $3(p - q) - 2(q - p)$ (*b*) $6(a - 4b) - 4(2a + b)$
(*c*) $3(p + 2q - r) - 4(r - 2q + p)$
(*a*) $[5p - 5q]$ (*b*) $[-2a - 28b]$ (*c*) $[-p + 14q - 7r]$

36. (*a*) $(k + 2l)(k + l)$ (*b*) $(p + q)(2p - 3q)$ (*c*) $2(x - 3y)(x - y + 2z)$
(*a*) $[k^2 + 3kl + 2l^2]$ (*b*) $[2p^2 - pq - 3q^2]$
(*c*) $[2x^2 - 8xy + 4xz + 6y^2 - 12yz]$

37. (*a*) $(a + b)^2$ (*b*) $(a - 2b)^2$ (*c*) $3(2a - b)^2$ (*d*) $(a - b)^3$
(*a*) $[a^2 + 2ab + b^2]$ (*b*) $[a^2 - 4ab + 4b^2]$ (*c*) $[12a^2 - 12ab + 3b^2]$
(*d*) $[a^3 - 3a^2b + 3ab^2 - b^3]$

38. (*a*) $k^2(l + m) - l(m^2 - k)$ (*b*) $3p(q + r) + 4r(p - q)$
(*c*) $2a(p + q) - 3p(2a - 2q)$
(*a*) $[k^2l + k^2m - lm^2 + lk]$ (*b*) $[3pq + 7pr - 4qr]$
(*c*) $[-4ap + 2aq + 6pq]$

39. (*a*) $3a + \{b - (2a + b)\}$ (*b*) $4p - \{2q + 3(p - q) + 6p\}$
(*c*) $3x + 4\{2x - (4x + 3)\}$
(*a*) $[a]$ (*b*) $[-5p + q]$ (*c*) $[-5x - 12]$

40. (*a*) $3 - 4\{(a + b) - (a - b)^2\}$ (*b*) $-4\{(a - 2b + c) - (3c - a)\}$
(*c*) $\{(2p + 4q) - (4q - r)\} - 3pq + 2qr$
(*a*) $[4a^2 + 4b^2 - 8ab - 4a - 4b + 3]$ (*b*) $[-8a + 8b + 8c]$
(*c*) $[2p + r - 3pq + 2qr]$

For problems 41 to 45, factorise

41. (*a*) $ab + ac$ (*b*) $p^2 + pq$ (*c*) $6x - 18$
(*a*) $[a(b + c)]$ (*b*) $[p(p + q)]$ (*c*) $[6(x - 3)]$

42. (*a*) $2m^2 + 10mn$ (*b*) $15ab^2 - 30a^2b$ (*c*) $21x^2y^2 - 28xy$
(*a*) $[2m(m + 5n)]$ (*b*) $[15ab(b - 2a)]$ (*c*) $[7xy(3xy - 4)]$

43. (*a*) $16a^3b^3 - 48a^2b^2$ (*b*) $ab^2 + bc - b^3$ (*c*) $3p^2q - 6pq^2 + 15pq$
(*a*) $[16a^2b^2(ab - 3)]$ (*b*) $[b\{c + b(a - b)\}]$ (*c*) $[3pq(p - 2q + 5)]$

44. (*a*) $ax + bx + a + b$ (*b*) $ay + by + az + bz$ (*c*) $ep - fp + eq - fq$
(*a*) $[(a + b)(x + 1)]$ (*b*) $[a + b)(y + z)]$ (*c*) $[(e - f)(p + q)]$

45. (*a*) $ax - ay - by + bx$ (*b*) $ep + fp - fq - eq$ (*c*) $ak + al - bk - bl$
(*a*) $[(a + b)(x - y)]$ (*b*) $[(e + f)(p - q)]$ (*c*) $[(a - b)(k + l)]$

Order of precedence

In problems 46 to 57, simplify

46. $3a \div 4a + 7a$ $[7a + \frac{3}{4}]$

47. $3a \div (4a + 7a)$ $[\frac{3}{11}]$

48. $4m - 3m \times 5m + 2m$ $[3m(2 - 5m)]$

49. $(4m - 3m)5m + 2m$ $[m(5m + 2)]$

50. $4m - 3m(5m + 2m)$ $[m(4 - 21m)]$

51. $3x + 4 \div 7x + 2 \times 5 - 2x$ $\left[x + \dfrac{4}{7x} + 10\right]$

52. $(3x + 4) \div 7x + 2 \times 5 - 2x$ $\left[\dfrac{3x + 4}{7x} + 10 - 2x\right]$

53. $3x + 4 \div 7x + 2(5 - 2x)$ $\left[\dfrac{4}{7x} - x + 10\right]$

54. $2 \div x + 3 \div (x + 1)$ $\left[\dfrac{5x + 2}{x(x + 1)}\right]$

55. $a^2 - 3ab \times 2a \div 6b + ab$ $[ab]$

56. $(x + 2)(x - 3) \div (2x - 6)$ $[\frac{1}{2}(x + 2)]$

57. $6k \div (4kl + 8l \times 2k) - 2lk \div l - k$ $\left[3\left(\dfrac{1}{10l} - k\right)\right]$

Chapter 5

Simple equations

5.1 Solution of simple equations

An **equation** is simply a statement that two quantities are equal. For example,

1 m = 100 cm
0°C = 273.15 K

Often, however, an equation contains an unknown quantity which is represented by a symbol, say x. Simple linear equations are those in which an unknown quantity is raised only to the power 1, i.e. $x^1 = x$ (an equation of the **first degree** is another term used to describe such equations). Terms such as x^2, x^3 etc. do not appear in simple equations. (The word 'simple' in this context does not necessarily mean 'easy' – it is merely the term given to an equation with an unknown of power unity.)

To **'solve an equation'** means 'to find the value of the unknown'. When the unknown is found, its value is said **'to satisfy'** the given equation, or, it is the **'solution of the equation'**. With simple linear equations there is only one value of the unknown which satisfies the equation. (It will be seen later that simultaneous equations and quadratic equations can have more than one value for the unknown.) For example, if $x + 2 = 5$, x represents the unknown value. There is only one value of x which is correct for this equation, i.e. x must be equal to 3. Thus, $x = 3$ **is the 'solution of the equation'**. Another way of expressing this fact is to say that $x = 3$ **satisfies the equation** $x + 2 = 5$. There is no other possible value of x which satisfies the given equation.

An **identity** is a relationship which is true for all values of the unknown, whereas an equation is only true for particular values of the unknown. For example, $2x - 5 = 1$ is an equation since it is only true when $x = 3$, whereas

$2x \equiv 6x - 4x$ is an identity since it is true for **all** values of x. (Note that "$\equiv$" means "is identical to").

What is the difference between an algebraic expression and an equation? The answer to this question is shown by the following:

$\frac{3x - 6}{5x - 10}$ is an algebraic expression, but $\frac{3x - 6}{5x - 10} = 3$ is an equation,

i.e. **it contains an 'equals' sign.**

The most important thing when dealing with simple equations is to remember that the equality of a given equation must be maintained. For example, if the algebraic expression to the left of the 'equals' sign is multiplied by 2, then the algebraic expression to the right of the 'equals' sign must also be multiplied by 2. Similarly if 3 is subtracted from the left-hand side (L.H.S.) of the 'equals' sign then 3 must also be subtracted from the right-hand side (R.H.S.). **The equality of the equation must be maintained.**

The solution of simple linear equations has many practical applications and some of these are shown later, but firstly, to ensure a firm understanding of the various types of simple linear equations, the student should work through the following graded worked examples in order.

The general objective with each of the following problems is to get the unknown, x, by itself on the left-hand side of the equation.

Worked problems on simple equations

Problem 1. Solve the equation $3x = 15$

Dividing each side of the equation by 3 gives

$$\frac{3x}{3} = \frac{15}{3}$$

(Note, the same operation has been applied to both L.H.S. and R.H.S. of the equation so the equality has been maintained.)

Therefore $x = 5$

Thus $x = 5$ is the solution to the equation.

The equation should always be checked. This is accomplished by substituting the solution into the **original equation.**

Check:

Since L.H.S. $= 3x$

L.H.S. $= (3)(5) = 15$

R.H.S. $= 15$

Hence the solution $x = 5$ is correct. Solutions to simple equations can always be checked in this manner.

Problem 2. Solve the equation $\frac{2x}{3} = 4$

The L.H.S. is a fraction and this can be removed by multiplying both sides of the equation by 3.

Hence $3\left(\frac{2x}{3}\right) = 3(4)$

The 3 in the numerator and the denominator of the L.H.S. will 'cancel'.

Therefore $2x = 12$

Dividing both sides of the equation by 2 gives $x = 6$

Check:

L.H.S. $= \frac{2}{3}(6) = 4$

R.H.S. $= 4$

Hence the solution $x = 6$ is correct.

Problem 3. Solve $x - 4 = 9$

Adding 4 to both sides of the equation gives

$x - 4 + 4 = 9 + 4$

Hence $x = 9 + 4$

Therefore $x = 13$

Check:

L.H.S. $= 13 - 4 = 9$

R.H.S. $= 9$

Hence the solution $x = 13$ is correct.

It is often easier to look upon the above procedure as follows:

To solve $x - 4 = 9$

The '-4' on the L.H.S. of the equation may be moved to the R.H.S. of the equation **but the sign must be changed.** That is, when the '-4' on the L.H.S. is moved across the 'equals' sign to the R.H.S. it becomes a '$+4$'.

Therefore $x - 4 = 9$

becomes $x \quad = 9 + 4$

i.e. $\quad x \quad = 13$

Problem 4. Solve $x + 2 = 7$

As in Problem 3, the equivalent term, '$+2$', on the L.H.S. may be moved across the 'equals' sign to the R.H.S., but the positive sign must be changed to a negative sign.

Hence $\quad x + 2 = 7$

becomes $x \quad = 7 - 2$

i.e. $\quad x \quad = 5$

This, of course, should be checked.

Sometimes terms containing the unknown quantity appear on both sides of an equation.

Problem 5. Solve $7x + 1 = 2x + 6$

In such cases as this, the terms containing x are grouped on one side of the equation and the remaining terms grouped on the other side of the equation. When this is done, changing from one side of an equation to the other must be accompanied by a change of sign.

Hence $7x + 1 = 2x + 6$

Therefore $7x - 2x = 6 - 1$

i.e. $5x = 5$

Therefore $\frac{5x}{5} = \frac{5}{5}$

or $x = 1$

Check:

L.H.S. $= 7(1) + 1 = 8$

R.H.S. $= 2(1) + 6 = 8$

Hence the solution $x = 1$ is correct.

Problem 6. Solve $3 - 4x = 2x - 15$

In order to keep the term in x positive the unknowns can be moved to the R.H.S.

Hence $3 + 15 = 2x + 4x$

or $18 = 6x$

Therefore $\frac{18}{6} = \frac{6x}{6}$

i.e $x = 3$

Check:

L.H.S. $= 3 - 4(3) = -9$

R.H.S. $= 2(3) - 15 = -9$

Hence the solution $x = 3$ is correct.

If the unknown quantities had been grouped on the L.H.S. instead of the R.H.S. the resulting equation would have been solved as follows:

$3 - 4x = 2x - 15$

Therefore $-4x - 2x = -15 - 3$

i.e. $-6x = -18$

Therefore $\frac{-6x}{-6} = \frac{-18}{-6}$

$x = 3$

Obviously the same result has been achieved. However, it is suggested that if at all possible, students should try to work with positive values of the unknown, until adept at this kind of work.

Further problems similar to 1 to 6 on simple equations may be found in Section 5.3 (Problems 1–21), page 134.

Problem 7. Solve $4(x-4)=2$

This equation contains a bracket and before anything else this bracket must be removed. Hence

$$4x-16=2$$

Therefore
$$4x=2+16$$
$$4x=18$$
$$\frac{4x}{4}=\frac{18}{4}$$
$$x=4.5$$

Check:
L.H.S. $=4(4.5-4)=4(0.5)=2$
R.H.S. $=2$
Hence the solution $x=4.5$ is correct.

Problem 8. Solve $3(2x-6)+2(x+3)=4(x-5)$

Hence
$$6x-18+2x+6=4x-20$$
Therefore
$$6x+2x-4x=-20+18-6$$
$$4x=-8$$
$$\frac{4x}{4}=\frac{-8}{4}$$
$$x=-2$$

Check:
L.H.S. $=3(-4-6)+2(-2+3)$
$=3(-10)+2(1)=-30+2=-28$
R.H.S. $=4(-2-5)=4(-7)=-28$
Hence the solution $x=-2$ is correct.

Further problems similar to 7 and 8, on simple equations involving brackets may be found in Section 5.3 (Problems 22–35), page 134.

Worked problems on simple equations involving fractions

Problem 9. Solve $\frac{4}{x}=\frac{2}{7}$

This is achieved by multiplying each term of the equation by the lowest common multiple (L.C.M.) of the denominators. If

$$\frac{4}{x}=\frac{2}{7}$$

the L.C.M. of the denominators, i.e. the lowest algebraic expression that both x and 7 will divide into, is $7x$.

Hence $7x\left[\frac{4}{x}\right] = 7x\left[\frac{2}{7}\right]$

In the L.H.S. the x can be 'cancelled'. In the R.H.S. the 7 can be 'cancelled'.
Hence $7(4) = x(2)$
i.e. $28 = 2x$
or $x = \mathbf{14}$

Check:

L.H.S. $= \frac{4}{14} = \frac{2}{7}$

R.H.S. $= \frac{2}{7}$

Hence the solution $x = 14$ is correct.

Problem 10. Solve $\frac{2x}{5} + \frac{4}{7} = 3 - \frac{2x}{3}$

The L.C.M. of the denominators is 105. Thus

$105\left(\frac{2x}{5}\right) + 105\left(\frac{4}{7}\right) = 105\,(3) - 105\left(\frac{2x}{3}\right)$

Terms may then be 'cancelled' where appropriate. Thus

$$21(2x) + 15(4) = 105(3) - 35(2x)$$

Therefore
$$42x + 60 = 315 - 70x$$
$$42x + 70x = 315 - 60$$
$$112x = 255$$
$$x = \mathbf{\frac{255}{112} = 2\frac{31}{112}}$$

Obviously it is more arithmetically difficult to check such an answer as this, but it should still be done.

Check:

L.H.S. $= \frac{2}{5}\left[\frac{255}{112}\right] + \frac{4}{7}$

$= \frac{51}{56} + \frac{4}{7} = \frac{51 + 32}{56} = \frac{83}{56}$

R.H.S. $= 3 - \frac{2}{3}\left[\frac{255}{112}\right]$

$= 3 - \frac{85}{56} = \frac{168 - 85}{56} = \frac{83}{56}$

Hence the solution $x = 2\frac{31}{112}$ is correct.

Problem 11. Solve $\frac{4}{x-2} = \frac{5}{3x+4}$

The L.C.M. of the denominators is $(x-2)(3x+4)$. Thus

$$(x-2)(3x+4)\left[\frac{4}{x-2}\right] = (x-2)(3x+4)\left[\frac{5}{3x+4}\right]$$

The next step is to cancel where appropriate. Hence

$$\cancel{(x-2)}(3x+4)\left[\frac{4}{\cancel{x-2}}\right] = (x-2)\cancel{(3x+4)}\left[\frac{5}{\cancel{3x+4}}\right]$$

Therefore

$$\begin{aligned}(3x+4)4 &= (x-2)5\\ 12x+16 &= 5x-10\\ 12x-5x &= -10-16\\ 7x &= -26\\ x &= \frac{-26}{7}\\ x &= -3\tfrac{5}{7}\end{aligned}$$

Check:

$$\text{L.H.S.} = \frac{4}{-3\frac{5}{7}-2} = \frac{4}{-5\frac{5}{7}} = \frac{4}{-\frac{40}{7}} = -\frac{4(7)}{40}$$

$$= -\frac{28}{40} = -\frac{7}{10}$$

$$\text{R.H.S.} = \frac{5}{3\left[\frac{-26}{7}\right]+4} = \frac{5}{\frac{-78}{7}+4} = \frac{5}{\frac{-78+28}{7}}$$

$$= \frac{5}{\frac{-50}{7}} = \frac{-5(7)}{50} = \frac{-35}{50} = \frac{-7}{10}$$

Hence the solution $x = -3\frac{5}{7}$ is correct.

Further problems similar to 9 to 11, on simple equation involving fractions may be found in Section 5.3 (Problems 36–70), page 135.

Worked problems involving squares and square roots

Problem 12. Solve $\sqrt{x} = 5$

Whenever square root signs are involved with the unknown, both sides of the equation must be squared.

Hence $(\sqrt{x})^2 = (5)^2$

Therefore $x = 25$

Check:
L.H.S. $= \sqrt{25} = 5$
R.H.S. $= 5$
Hence the solution $x = 25$ is correct.

Problem 13. Solve $2\sqrt{x} = 6$

It is often easier to arrange the equation so that the term involving the square root of the unknown is on its own.

Hence $\dfrac{2\sqrt{x}}{2} = \dfrac{6}{2}$

Therefore $\sqrt{x} = 3$

Squaring both sides gives $x = 9$

Check:
L.H.S. $= 2\sqrt{9} = 2(3) = 6$
R.H.S. $= 6$
Hence the solution $x = 9$ is correct.

Problem 14. Solve $\dfrac{\sqrt{x}+1}{\sqrt{x}} = 4$

To remove the fraction, each term must be multiplied by $\sqrt{x}$.

Hence $\sqrt{x}\left[\dfrac{\sqrt{x}+1}{\sqrt{x}}\right] = \sqrt{x}(4)$

Therefore
$$\sqrt{x} + 1 = 4\sqrt{x}$$
$$1 = 4\sqrt{x} - \sqrt{x}$$
$$1 = 3\sqrt{x}$$
$$\frac{1}{3} = \frac{3\sqrt{x}}{3}$$
$$\frac{1}{3} = \sqrt{x}$$

Squaring both sides gives $\left[\dfrac{1}{3}\right]^2 = x$

Therefore $x = \dfrac{1}{9}$

Check:

$$\text{L.H.S.} = \frac{\sqrt{\left[\frac{1}{9}\right]} + 1}{\sqrt{\left[\frac{1}{9}\right]}} = \frac{\frac{1}{3} + 1}{\frac{1}{3}} = \frac{\frac{4}{3}}{\frac{1}{3}} = 4$$

R.H.S. $= 4$

Hence the solution $x = \dfrac{1}{9}$ is correct.

The following two examples involves a power of the unknown and hence they are not strictly simple equations. However, they are included for completeness.

Problem 15. Solve $x^2 = 16$

Whenever a square of the unknown is involved, the square root of both sides is taken

Hence $\quad \sqrt{x^2} = \sqrt{16}$

Therefore $\quad x = 4$

However, $x = -4$ is also a solution of the equation because $(-4) \times (-4) = +16$. Therefore it should be noted that whenever the square root of a number is found there are always **two** answers, one positive and the other negative.

The solution of this problem is written as $x = \pm 4$

Problem 16. Solve $\dfrac{2}{9} = \dfrac{16}{3x^2}$

The L.C.M. of the denominator is $9x^2$.

Hence $\quad 9x^2\left[\dfrac{2}{9}\right] = 9x^2\left[\dfrac{16}{3x^2}\right]$

Therefore $\quad 2x^2 = 48$

$$x^2 = 24$$

$$x = \sqrt{24}$$

$$\mathbf{x = \pm 4.899}$$

Check:

L.H.S. $= \dfrac{2}{9}$

R.H.S. $= \dfrac{16}{3(24)} = \dfrac{2}{3(3)} = \dfrac{2}{9}$

Hence the solution $x = \pm 4.889$ is correct.

Further problems similar to 12 to 16, on simple equations involving squares and square roots, may be found in Section 5.3 (Problems 71–81), page 136.

Summary of important points in the solution of simple equations

1. The equality of an equation must be maintained at every step.
2. Fractions are normally undesirable and are therefore removed. This is accomplished by multiplying each term by the lowest common multiple of the denominators.
3. Brackets are normally undesirable and are therefore removed at the earliest opportunity by simply 'multiplying out'.
4. The unknown terms must be grouped on one side of the equation and all other terms on the other side.
5. When the square root of the unknown is involved it is best to isolate that term on one side of the equation, and then square both sides.
6. When the square of the unknown is involved it is best to isolate that term on one side of the equation, and then take the square root of both sides.

When taking square roots there are always two answers, one positive and the other negative.

7. All solutions must be checked.
8. Do not try to take more than one step at a time.

5.2 Practical problems

There are many different practical applications where the solution of simple linear equations are needed. Various worked problems are shown below.

From now on it will be assumed that the student will check all solutions by substitution into the original equation.

Worked problems on practical applications of simple equations

---

Problem 1. A copper wire of length, l = 2 km has a resistance of 4 Ω and a resistivity, ρ, of 17.2×10^{-6} Ω mm. Find the cross-sectional area, a, of the wire given that

$$R = \frac{\rho l}{a}$$

$$\text{Therefore } 4 = \frac{(17.2 \times 10^{-6})(2\,000 \times 10^{3})}{a}$$

$$= \frac{34.4}{a}$$

$$4a = 34.4$$

$$a = \frac{34.4}{4} = 8.6 \text{ mm}^2$$

Hence the cross-sectional area of the wire, a, is **8.6 mm²**.

Problem 2. The temperature coefficient of resistance, α, may be calculated from the formula $R_t = R_o\,(1 + \alpha t)$. Find α, given $R_t = 0.726$, $R_o = 0.6$, and $t = 50$

As $R_t = R_o\,(1 + \alpha t)$ then

$$0.726 = 0.6\,(1 + \alpha\,(50))$$

$$0.726 = 0.6 + \alpha(50)\,(0.6)$$

$$0.726 - 0.6 = 30\alpha$$

$$0.126 = 30\alpha$$

$$\alpha = \frac{0.126}{30}$$

$$\alpha = \mathbf{0.004\,2}$$

Problem 3. When three resistors in an electrical circuit are connected in parallel the total resistance, R_T, is found from the formula

$$\frac{1}{R_T} = \frac{1}{R_1} + \frac{1}{R_2} + \frac{1}{R_3}$$

Find the total resistance when resistors of 4 Ω, 6 Ω and 12 Ω are connected in parallel.

Therefore $\frac{1}{R_T} = \frac{1}{4} + \frac{1}{6} + \frac{1}{12}$

The L.C.M. of the denominators is 12 R_T.

$$\text{Therefore } 12\,R_T\left[\frac{1}{R_T}\right] = 12\,R_T\left[\frac{1}{4}\right] + 12\,R_T\left[\frac{1}{6}\right] + 12\,R_T\left[\frac{1}{12}\right]$$

$$12 = 3\,R_T + 2\,R_T + R_T$$

$$12 = 6\,R_T$$

$$\frac{12}{6} = \frac{6\,R_T}{6}$$

$$R_T = 2\ \Omega$$

Alternatively, the fractions on the R.H.S. could be added first, as shown below:

$$\frac{1}{R_T} = \frac{1}{4} + \frac{1}{6} + \frac{1}{12} = \frac{3+2+1}{12}$$

$$\frac{1}{R_T} = \frac{6}{12} = \frac{1}{2}$$

Taking the reciprocal of both sides gives $R_T = 2\ \Omega$

Problem 4. A formula used to calculate power, P, in a d.c. electrical circuit is $P = \frac{V^2}{R}$ where V is the supply voltage and R is the resistance of the circuit. If the resistance is 6.25 Ω and the power measured is 100 watts find the supply voltage.

$$P = \frac{V^2}{R}$$

$$\text{Therefore} \quad 100 = \frac{V^2}{6.25}$$

$$6.25(100) = 6.25\left[\frac{V^2}{6.25}\right]$$

$$625 = V^2$$

$$V = \sqrt{625}$$

$$V = \pm 25 \text{ volts}$$

Problem 5. A rectangular box with square ends has its length 10 cm greater than its breadth and the total length of its edges is 1.52 m. Find the width of the box.

Let x cm = breadth = width of the box. The box is shown in Fig. 5.1.

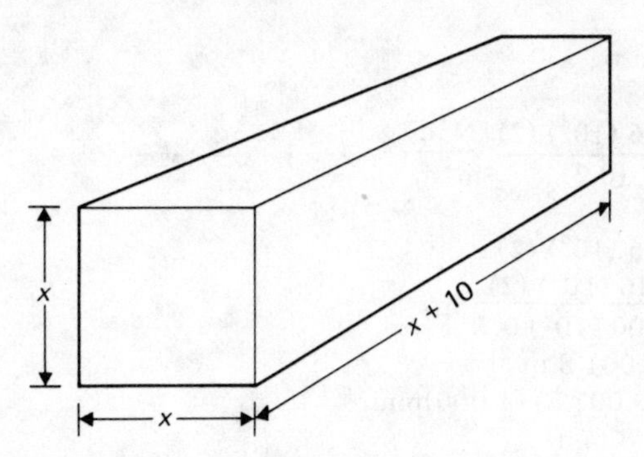

Fig. 5.1

Then the length of the box is $(x + 10)$cm and the lengths of the edges is $2(4x) + 4(x + 10)$ cm.

Therefore $152 \text{ cm} = (8x + 4x + 40)\text{cm}$

$$152 - 40 = 12x$$
$$112 = 12x$$
$$x = 9\tfrac{1}{3} \text{ cm}$$

Hence the width of the box = **$9\frac{1}{3}$ cm**

Problem 6. A number of people agree to pay equally for the use of a hire-tool, and each pays £1.50. If there had been two more people in the party each would have paid £1.00. Find how many people there were in the party and the cost of the hire-tool.

Let the number of people in the party be x. Therefore the cost of the hire-tool for x people at £1.50 each is £$1.5x$.

Two more people in the party would make $(x + 2)$ people and it would now cost £1.00 each. So the cost of hire-tool for $(x + 2)$ people at £1.00 each is £1 $(x + 2)$.

Therefore $1.5x = x + 2$

$$1.5x - x = 2$$
$$0.5x = 2$$
$$x = \frac{2}{0.5} = 4$$

Hence there are **4 people in the party.**

Cost of hire-tool = **£$1.5x$ = £1.5(4) = £6**

Problem 7. The extension, x m, of a mild steel tie bar of length l m and cross-sectional area A m^2 when carrying a load F N is given by the Modulus of Elasticity, $E = \dfrac{Fl}{Ax}$

If: $E = 200 \times 10^9 \text{ N m}^{-2}$, $F = 36 \times 10^6$ N, $A = 0.2 \text{ m}^2$, and $l = 2$ m, find the extension of the tie bar in mm.

$$E = \frac{Fl}{Ax}$$

Therefore $200\,(10^9)\text{ N m}^{-2} = \dfrac{36\,(10^6)\,(2)}{(0.2)\,x}\ \dfrac{\text{N m}}{\text{m}^2\,\text{m}}$

(Note: the unit x is in metres.)

Therefore $200\,(10^9)\,(0.2)\ x = 36\,(10^6)\,(2)$

$$x = \frac{36\,(10^6)\,(2)}{200\,(10^9)\,0.2}\ \text{m}$$

$$x = 0.001\,8\text{ m}$$

$$x = (0.001\,8)\,(1\,000)\text{mm}$$

$$x = 1.8\text{ mm}$$

Hence the extension of the tie bar is **1.8 mm**

Problem 8. A formula for distance, s metres, travelled in time t seconds is given by $s = ut + \frac{1}{2}at^2$ where u is the initial velocity in m s^{-1} and a is the acceleration in m s^{-2}. Find the acceleration of a body if it travels 150 m in 5 s with an initial velocity of 5 m s^{-1}.

$$s = ut + \tfrac{1}{2}at^2$$

$s = 150$, $u = 5$, and $t = 5$

Therefore $\quad 150 = (5)\,(5) + \frac{1}{2}a(5)^2$

$$150 = 25 + \frac{25}{2}a$$

$$150 - 25 = \frac{25}{2}a$$

$$125 = \frac{25}{2}a$$

$$125(2) = 25a$$

$$250 = 25a$$

$$\frac{250}{25} = a$$

$$a = 10$$

Hence the acceleration of the body is **10** m s^{-1}

Problem 9. A formula relating initial and final states of pressures, p, volumes, V, and absolute temperatures, T, of an ideal gas, is

$$\frac{p_1V_1}{T_1} = \frac{p_2V_2}{T_2}$$

If $p_1 = 2\,063 \times 10^3$, $p_2 = 101.3 \times 10^3$, $T_1 = 323$, $T_2 = 273$ and $V_1 = 0.5$, find V_2.

Therefore $\dfrac{2\,063\,(10^3)\,0.5}{323} = \dfrac{101.3\,(10^3)\,V_2}{273}$

$$(273)\,(323)\left[\frac{2\,063\,(10^3)\,0.5}{323}\right] = (273)\,(323)\left[\frac{101.3 \times 10^3}{273}\right] V_2$$

Cancelling gives: $273\,(2\,063)\,10^3\,(0.5) = 323\,(101.3)\,10^3\,V_2$

Dividing both sides by $(323)\,(101.3)\,10^3$ gives

$$\frac{(273)\,(2\,063)\,10^3\,(0.5)}{(323)\,(101.3)\,(10^3)} = \frac{(323)\,(101.3)\,(10^3)\,V_2}{(323)\,(101.3)\,10^3}$$

$$V_2 = \frac{273\,(2\,063)\,10^3\,0.5}{323\,(101.3)\,10^3}$$

Hence $V_2 =$ **8.6**

Problem 10. A welder is paid £1.50 per hour for a basic 40 hour week, but overtime is paid for at one-and-a-half times this rate. Find how many hours he has to work in a week to earn £80.25.

Basic rate per hour = £1.50
Rate for overtime per hour = 1.5(£1.50)
= £2.25

Let the number of overtime hours worked = x

Therefore $40\,(1.5) + x(2.25) = 80.25$

$$60 + 2.25x = 80.25$$
$$2.25x = 80.25 - 60$$
$$2.25x = 20.25$$
$$x = \frac{20.25}{2.25}$$
$$x = 9$$

Thus 9 hours overtime would have to be worked to earn £80.25 per week. Hence total number of hours worked is **49**.

Problem 11. When two resistors R_1 and R_2 are connected in parallel the voltage across R_2 may be found from

$$V_2 = \left[\frac{R_1 R_2}{R_1 + R_2}\right] I$$

where I is the total circuit current. Find R_2 given $I = 14\,A$, $R_1 = 5\,\Omega$ and $V_2 = 20$ V.

$$V_2 = \left[\frac{R_1 R_2}{R_1 + R_2}\right] I$$

Therefore $$20 = \left[\frac{5\,R_2}{5 + R_2}\right] 14$$

$$20(5 + R_2) = (5\,R_2)\,14$$
$$100 + 20\,R_2 = 70\,R_2$$
$$100 = 70\,R_2 - 20\,R_2$$
$$100 = 50\,R_2$$
$$R_2 = \frac{100}{50} = 2\,\Omega$$

Hence $R_2 = 2\,\Omega$

Further problems similar to 1 to 11 on the practical type of simple equations may be found in the following Section (5.3) (Problems 82–110), page 137.

5.3 Further problems

Simple equations

Solve the following equations:

1. $a + 3 = 5$ [2]
2. $x - 2 = 9$ [11]
3. $3s = 12$ [4]
4. $7 - q = 4$ [3]
5. $2t + 2 = 1$ [$-\frac{1}{2}$]
6. $\frac{r}{3} = 5$ [15]
7. $3x + 4 = 8$ [$\frac{4}{3}$]
8. $7p + 12 = 5$ [−1]
9. $3b - 4b + 2b = 7$ [7]
10. $14 - 3c = 8$ [2]
11. $3u - 2 = 7u - 6$ [1]
12. $6m + 11 = 25 - m$ [2]
13. $3p - 18 = 8p + 22$ [−8]
14. $0.4a = 1.6$ [4]
15. $2.6x - 0.4 = 0.9x + 1.3$ [1]
16. $2d - 1 - 3d = 4d - 21$ [4]
17. $2a - 4 = 3a - 2 - 5a$ [$\frac{1}{2}$]
18. $3x + 1 - 4x = 0$ [1]
19. $0 = 2 + 4b$ [$-\frac{1}{2}$]
20. $2r + 5 - 3r - 2 = 0$ [3]
21. $18x - 5 + x = 9x + 2 - 7$ [0]

Equations involving brackets

22. $3(x - 1) = 6$ [3]
23. $4(x + 2) = 14$ [$\frac{3}{2}$]
24. $3(e - 1) - 4(2e + 3) = 15$ [−6]
25. $5(g + 2) = 29 + 3(g - 5)$ [2]

26. $5(b-3)=2b$ [5]
27. $4(3-2x)-5=-3(x-2)$ [$\frac{1}{5}$]
28. $3(2y-1)-2=17$ [$3\frac{2}{3}$]
29. $4(3x-4)-(x+1)=10$ [$2\frac{5}{11}$]
30. $0=2(3a-5)-5$ [$2\frac{1}{2}$]
31. $0=2(b-2)-3(3b+1)$ [-1]
32. $5(2x+3)=6(x+2)-2(x-4)$ [$\frac{5}{6}$]
33. $20+8(3c-2)=(c-1)+3(2c-4)$ [-1]
34. $16-(p+2)=3(p-7)+10$ [$6\frac{1}{4}$]
35. $8+5(d-1)-6(d-3)=3(4-d)$ [$-4\frac{1}{2}$]

Equations involving fractions

36. $\frac{1}{3}a+2=5$ [9]
37. $\frac{3}{4}x+2=\frac{1}{3}x+\frac{1}{2}$ [$-3\frac{3}{5}$]
38. $\frac{1}{3}(2a-1)+4=\frac{1}{2}$ [$-4\frac{3}{4}$]
39. $3-\frac{1}{3}(2y+1)=\frac{3}{4}$ [$2\frac{7}{8}$]
40. $\frac{1}{2}(2f-3)+\frac{1}{3}(f-4)=0$ [$2\frac{1}{8}$]
41. $0=\frac{2}{3}(2m-1)-\frac{3}{4}(m+5)$ [$7\frac{4}{7}$]
42. $\frac{1}{3}(3a-2)-\frac{1}{4}(5a+1)+\frac{1}{5}(2a-7)=0$ [$15\frac{4}{9}$]
43. $-\frac{1}{9}(3x-4)+\frac{1}{5}(4x-1)=\frac{1}{15}-\frac{1}{3}(x+5)$ [$-2\frac{11}{36}$]
44. $\dfrac{t}{3}-\dfrac{t}{5}=2$ [15]
45. $\dfrac{V}{5}+\dfrac{V}{3}=\dfrac{47}{30}-\dfrac{V}{4}$ [2]
46. $\dfrac{m}{3}+3-\dfrac{m}{6}=1-\dfrac{m}{3}$ [-4]
47. $\dfrac{5}{x}=10$ [$\frac{1}{2}$]
48. $2t+\frac{1}{2}=\frac{4}{5}t-\frac{3}{10}$ [$-\frac{2}{3}$]
49. $\dfrac{3}{w}=\dfrac{9}{5}$ [$1\frac{2}{3}$]
50. $\dfrac{5x}{9}-\dfrac{2x}{5}=\dfrac{1}{15}$ [$\frac{3}{7}$]
51. $\dfrac{1}{3a}+\dfrac{1}{4a}=\dfrac{3}{20}$ [$3\frac{8}{9}$]
52. $\dfrac{c+3}{4}=2+\dfrac{c-3}{5}$ [13]
53. $\dfrac{3}{2}+\dfrac{3y}{20}=\dfrac{-(y-6)}{12}+\dfrac{2y}{15}$ [-10]
54. $\dfrac{5-d}{4}=\dfrac{d}{5}+\dfrac{7}{20}$ [2]
55. $4-a=\dfrac{2a+6}{5}$ [2]
56. $\dfrac{f-2}{f-3}=4$ [$3\frac{1}{3}$]

57. $\dfrac{3}{y-2} = \dfrac{4}{y+4}$ [20]

58. $\dfrac{2}{a-5} = \dfrac{3}{a-1}$ [13]

59. $\dfrac{4}{3b+2} = \dfrac{7}{5(b-3)}$ [−74]

60. $\dfrac{3}{4} - \dfrac{1}{2a-1} = \frac{1}{2}$ [$2\frac{1}{2}$]

61. $\dfrac{1}{3x-2} + \dfrac{1}{5x+3} = 0$ [$-\frac{1}{8}$]

62. $\dfrac{2b}{2b-3} = \dfrac{3b-1}{3b+2}$ [$\frac{1}{5}$]

63. $\dfrac{1}{3(y-2)} - \dfrac{2}{5(y+3)} = 0$ [27]

64. $\dfrac{3x+23}{3x+12} = \dfrac{4}{3}$ [7]

65. $(c-3)(2c+6) = 2c(c-18)$ [$\frac{1}{2}$]

66. $\dfrac{1-a}{a+1} = 9$ [$-\frac{4}{5}$]

67. $\dfrac{x-5}{x-2} = \frac{3}{4}$ [14]

68. $\dfrac{a}{4} - \dfrac{a+3}{2} = \dfrac{a+6}{5}$ [−6]

69. $\dfrac{x}{6} - \dfrac{3(x+4)}{2} = 2(x-3)$ [0]

70. $\dfrac{2f+1}{3} - 2 = \dfrac{2f-3}{4} - \dfrac{1-f}{5}$ [$-21\frac{1}{2}$]

Equations involving squares and square roots

71. $4\sqrt{a} = 12$ [9]

72. $2\sqrt{x} = \frac{5}{3}$ [$\frac{25}{36}$]

73. $6 = \sqrt{\left(\dfrac{x}{2}\right)} - 1$ [98]

74. $4 = \sqrt{\dfrac{3}{x}} + 2$ [$\frac{3}{4}$]

75. $\dfrac{2\sqrt{b}}{1-\sqrt{b}} = 4$ [$\frac{4}{9}$]

76. $8 = 4\sqrt{\left(\dfrac{a}{2} - 1\right)}$ [10]

77. $24 = \dfrac{6}{a^2}$ [$\pm\frac{1}{2}$]

78. $\dfrac{4}{3} = \sqrt{\left[\dfrac{x+2}{x-2}\right]}$ [$7\frac{1}{7}$]

79. $16 = \dfrac{e^2}{9}$ [±12]

80. $\frac{8}{x} = \frac{2x}{25}$ [±10]

81. $\frac{9}{2} = 4 + \frac{2}{x^2}$ [±2]

Practical problems

82. A formula used for calculating resistance of an electrical cable is $R = \frac{\rho l}{a}$. Given $R = 0.112$, $l = 100$, $a = 25$, find the value of ρ. [0.028]
83. Ohms Law states that $V = IR$. Find R when $I = 0.5$ and $V = 200$. [400]
84. If force, F N, mass, m kg and acceleration, a m s^{-2} are linked by the formula $F = m\, a$ find the acceleration when a force of 2 kN is applied to a mass of 1 000 kg. [2 m s^{-2}]
85. Kinetic energy $= \frac{1}{2} mv^2$ where m is the mass in kg and v is the velocity in m s^{-1}. Find the mass of a body projected vertically upwards with velocity 20 m s^{-1} if the kinetic energy is 1 000 joules. [5 kg]
86. Two variables x and y are related by $y = ax + \frac{b}{x}$, where a and b are constants. If $y = 13$, $x = 3$ and $a = 3\frac{1}{3}$ find b. [9]
87. If $PV = mRT$ find R given $P = 100$, $T = 298$, $V = 2.00$ and $m = 0.322$ [2.084]
88. When 3 resistors, R_1, R_2 and R_3 in an electrical circuit are connected in parallel, the total resistance, R_T, is found from the formula

$$\frac{1}{R_T} = \frac{1}{R_1} + \frac{1}{R_2} + \frac{1}{R_3}$$

(*a*) Find the total resistance when $R_1 = 5$, $R_2 = 10$, and $R_3 = 30$ [3]

(*b*) Find R_2 given $R_T = 1$, $R_1 = 2$ and $R_3 = 6$ [3]

89. If $\frac{1}{R_T} = \frac{1}{R_1} + \frac{1}{R_2} + \frac{1}{R_3} + \frac{1}{R_4}$ find R_4 given $R_1 = 5$, $R_2 = 4.5$, $R_3 = 3.7$ and $R_T = 1.12$ [5]
90. Given that $\frac{p}{p+1} = \frac{p+1}{p+5}$ find the numerical value of $\frac{2p+1}{2p+6}$ [$\frac{1}{4}$]
91. Four sparking plugs and a fan belt cost £2.90. If the fan belt costs 15p more than a sparking plug find the cost of each component. [55p, 70p]
92. If $R = \frac{R_1 R_2}{R_1 + R_2}$ find R_2 given that $R_1 = 3$ and $R = 2$ [6]
93. Ohms Law can be represented by $I = \frac{V}{R}$ where I is the current in amperes, V is the voltage in volts and R is the resistance in ohms.

(*a*) A 25 ohm resistor is connected across a voltage of 12.5 V. Find the current flowing. [0.5 A]

(*b*) A current of 5 amperes is flowing in a circuit of 50 ohms. What is the voltage? [250 V]

(*c*) An electric soldering iron takes a current of 0.25 A, at 240 V. Find the resistance of the element. [960 ohms]

94. If $R_2 = R_1(1 + \alpha t)$ find α given $R_1 = 3.5$, $R_2 = 3.65$ and $t = 42.8$ [0.001]

95. If $\dfrac{P_1 V_1}{T_1} = \dfrac{P_2 V_2}{T_2}$ find T_2 given that $P_1 = 2\,000 \times 10^3$, $V_1 = 5$, $T_1 = 300$, $P_2 = 10^6$, and $V_2 = 15$ [450]

96. If $s = ut + \frac{1}{2} ft^2$ find u given that $s = 500$, $t = 4$ and $f = -15$ [155]

97. A rectangle has a length of 15 cm and a width w cm. When its width is reduced by 2 cm its area becomes 45 cm^2. Find the original width and area of the rectangle. [5 cm, 75 cm^2]

98. If $v^2 = u^2 + 2\,as$ find u given $v = 25$, $a = -50$ and $s = 2.75$ [30]

99. A rectangular box with square ends has its length 3 m greater than its breadth and the total length of its edges is 21 m. Find the width of the box. [$\frac{3}{4}$ m]

100. If $I = \dfrac{V - E}{R + r}$ find r given $I = 0.5$, $R = 3$, $V = 10$ and $E = 8$ [1]

101. Two joiners and four mates earn £296 between them for fitting a kitchen. If a joiner earns £19 more than a mate find their wages for the job. [£62, £43]

102. If $t = 2\pi\sqrt{\dfrac{w}{Sg}}$ find the value of S given $t = 0.3196$, $w = 4$, and $g = 32.2$ [48]

103. If the area of a circle is given by $A = \pi r^2$ find r when the area is 15.0 cm^2. [2.185]

104. A rectangular metal plate is 20 cm long. A strip 4 cm wide is cut from one end and a second strip of 2 cm is cut from the other. The remainder weighs 168 g. Find the width of the plate if 1 cm^2 of it weighs 1.5 g. [8 cm]

105. Evaluate R if $\dfrac{1}{R-1} = \dfrac{3}{1-t} - \dfrac{4}{1+t}$ and $t = 0.6$ [1.2]

106. The relation between the temperature on a Fahrenheit thermometer and that on a Celcius thermometer is expressed by the formula $F = \frac{9}{5} C + 32$. Express 27.5°C in Fahrenheit degrees. [81.5]

107. The stress f in the material of a thick cylinder is given by the formula $\dfrac{D}{d} = \sqrt{\left(\dfrac{f+p}{f-p}\right)}$ Calculate the stress given $D = 19.5$, $d = 9.75$ and $p = 1\,500$ [2 500]

108. The perimeter of a rectangle is 14 cm. The length of a diagonal is given by $\sqrt{(w^2 + L^2)}$ where w and L are the width and length of the rectangle respectively. If the length is 1 cm greater than the width find the dimensions of the rectangle and the length of the diagonal. [3 cm by 4 cm; 5 cm]

109. Sorel cement contains 36.5% magnesium chloride, 19.5% magnesium oxide and 44% water by mass. How much of the magnesium compounds is contained in a cement containing 48 g water?
[magnesium compounds = 61.1 g]

110. An alloy contains 70% by weight of copper, the remainder being zinc. How much copper must be mixed with 40 kg of this alloy to give an alloy containing 80% copper? [20 kg]

Chapter 6

Simultaneous equations

6.1 Methods of solving simultaneous equations

It often happens that the solution of a problem involves the use of more than one unknown.

Consider the equation $2x + 3y = 11$. There are two unknowns involved, namely x and y.

If $2x + 3y = 11$
then $2x = 11 - 3y$
Therefore $x = \dfrac{11 - 3y}{2}$

This does not give the actual numerical value of x, but a result which involves the other unknown y.

If a value of y is known then a corresponding value of x can be found.

For example, if $y = 1$, $x = \dfrac{11 - 3(1)}{2}$

$$x = \frac{11 - 3}{2} = \frac{8}{2} = 4$$

Similarly, if $y = 2$, $x = \dfrac{11 - 3(2)}{2}$

$$= \frac{11 - 6}{2} = \frac{5}{2} = 2\tfrac{1}{2}$$

and if $y = 3$, $x = \dfrac{11 - 3(3)}{2}$

$$= \frac{11 - 9}{2} = \frac{2}{2} = 1$$

This process can be continued indefinitely, so that for every value of x there is a corresponding value of y, and there is an endless number of possibilities. If we need, for some reason, only **one** pair of values for x and y then some further information must be available in order that the correct pair can be found.

In many engineering/science applications there exists two or more unknown quantities, each of which have only one value: the value of one unknown depends on the value of the other unknown. In such cases it is found that one equation is insufficient to determine the value of each of the unknowns. If, as in the example $2x + 3y = 11$, there are two quantities x and y having one value each, then another equation is needed in order to find this unique pair of values.

Similarly, if a problem contains three quantities, each of which has only one value, then three equations are needed in order to evaluate the three quantities. If as stated

$2x + 3y = 11$ [equation (1)]

and **simultaneously**, the same quantities x and y also satisfy the equation:

$4x + 2y = 10$, [equation (2)]

then there is only one value of x and one value of y which satisfies both equations. Such equations are called **simultaneous equations,** the word 'simultaneous' meaning 'existing or taking place at the same time'.

In this book we will deal only with simultaneous linear equations in two unknowns.

There are two methods of solving such equations.

Method 1: By substitution

Let $2x + 3y = 11$ (1)
and $4x + 2y = 10$ (2)

Then from equation (1), $x = \dfrac{11 - 3y}{2}$

Let this expression for x be substituted into equation (2). Thus

$$4\left[\frac{11 - 3y}{2}\right] + 2y = 10$$

This is now a simple equation in y and may be solved as shown in Chapter 5. Multiplying both sides of the equation by 2 gives

$4(11 - 3y) + 4y = 20$

Removing brackets gives

$44 - 12y + 4y = 20$

Rearranging gives

$$44 - 20 = 12y - 4y$$
$$24 = 8y$$
Hence $y = 3$

This value of y may be substituted into either equation (1) or equation (2). (The result should be the same in both cases.)

Substituting in equation (1) gives

$$2x + 3(3) = 11$$

Therefore $2x + 9 = 11$
$$2x = 11 - 9$$
$$2x = 2$$
Hence $x = 1$

A check may be obtained by substituting in equation (2) for x and y. Thus

$4(1) + 2(3) = 10$
$4 + 6 = 10$
L.H.S. = R.H.S.

Therefore **the solution of the simultaneous equations $2x + 3y = 11$ and $4x + 2y = 10$ is $x = 1$ and $y = 3$.** These are the only pair of values that satisfies both equations.

Method 2: By elimination

Again

let $2x + 3y = 11$ (1)
and $4x + 2y = 10$ (2)

If equation (1) is multiplied throughout by 2 the resulting equation will be

$4x + 6y = 22$ (3)

The reason that equation (1) is multiplied by 2 is that the coefficient of x (i.e. the number multiplying x) in equation (2) and equation (3) is now the same. Sometimes it is necessary to multiply both equations by constants chosen so that the coefficients of x or y in each equation become the same.

Equation (2) can now be subtracted from equation (3). Thus

$4x + 6y = 22$ (3)
$4x + 2y = 10$ (2)

Subtracting $0 + 4y = 12$

Hence $4y = 12$
Therefore $y = 3$

This value of y may now be substituted in equation (1) or equation (2) exactly as in method 1 to find the value of x.

It will be found from experience that in many cases method 1, that of substitution, is unnecessarily cumbersome, so that method 2, the elimination procedure will be employed.

6.2 Procedure for solving any simultaneous linear equation in two unknowns by the method of elimination

The procedure will be outlined using the following problem.

Problem. Solve the simultaneous equations $3x + 4y = 5$ and $12 + 2x - 5y = 0$

1. The two equations should be written down with similar unknown terms under each other. It is usual to put the two unknown terms on the left-hand side of the equation with the constant term on the right-hand side.
Thus $3x + 4y = 5$
and $2x - 5y = -12$

2. The two equations are numbered. This merely makes later reference to particular equations easier. Thus

$$3x + 4y = 5 \qquad (1)$$
$$2x - 5y = -12 \qquad (2)$$

3. The equations must be multiplied in such a way that the coefficient of one unknown becomes the same in both equations, i.e. either the coefficients of x, or the coefficients of y, have to be made equal. The choice is arbitrary: it is made with convenience and ease as the prime considerations.

If, in this particular problem, the x coefficients are to be made equal, then equation (1) would have to be multiplied by 2 and equation (2) would have to be multiplied by 3. Thus, multiplying equation (1) by 2 gives

$$6x + 8y = 10$$

and multiplying equation (2) by 3 gives:

$$6x - 15y = -36$$

4. Any newly-formed equations should be numbered. Thus

$$6x + 8y = 10 \qquad (3)$$
$$6x - 15y = -36 \qquad (4)$$

5. By adding or subtracting newly-formed equations one unknown may be entirely eliminated, leaving a simple equation containing the other unknown.

If the sign of the common coefficients is the **same** in both equations (i.e. both positive or both negative) then one equation is **subtracted** from the other.

If the sign of the common coefficients is **different** in each equation (i.e. one coefficient is positive and the other negative) then the two equations are **added.**

In equations (3) and (4) the coefficients and sign of the x term are the same in both, i.e. +6. Thus equation (4) is subtracted from equation (3)

$$6x + 8y = 10 \qquad (3)$$
$$6x - 15y = -36 \qquad (4)$$
Subtracting $$0 + 23y = 46$$

It is very important to note that

$8y$ take away $-15y$ is the same as

$8y - -15y$, which is the same as

$8y + 15y = +23y$.

Similarly 10 take away -36 is the same as

$10 - -36$, which is the same as

$10 + 36 = 46$.

If mistakes are going to be made in solving simultaneous equations it is usually at this stage, where the rules of arithmetic must be strictly adhered to.

6. The new equation formed by the above subtraction contains one unknown only, i.e. y. It is therefore necessary to solve the simple equation

$23y = 46$

Dividing both sides of the equation by 23 gives

$y = 2$

7. The solution $y = 2$ is substituted into either of the **original** equations, i.e. equation (1) or equation (2), which is then solved for the other unknown, x. Once again the choice is arbitrary.

Substituting $y = 2$ in equation (1) gives

$$3x + 4(2) = 5$$

Therefore
$$\begin{aligned} 3x + 8 &= 5 \\ 3x &= 5 - 8 \\ 3x &= -3 \\ x &= -1 \end{aligned}$$

8. It is essential to check that the pair of values found (i.e. $x = -1$ and $y = 2$) satisfy **both** of the **original** equations.

It is important to note that stages 7 and 8, the substitution and checking stages, must be done in the original and not the derived equations.

From above, $x = -1$ and $y = 2$ satisfies equation (1). Checking in equation (2)

$$\begin{aligned} \text{L.H.S.} = 2x - 5y &= 2(-1) - 5(2) \\ &= -2 - 10 \\ &= -12 \\ \text{R.H.S.} &= -12 \end{aligned}$$

Thus the solution of the simultaneous equations $3x + 4y = 5$ and $12 + 2x - 5y = 0$ is $x = -1$ and $y = 2$. These are the **only** pair of values that satisfies **both** equations.

In Chapter 8 it will be shown that equations such as the two used in the above problem are straight lines when plotted graphically. The above solution of the two equations represents the point where the two lines intersect, and two straight lines can intersect at only one point.

Problem 1. Solve $8x - 3y = 39$ (1)
$7x + 5y = -4$ (2)

When equation (1) is multiplied by 5 and equation (2) is multiplied by 3 the coefficient of y in each equation is numerically the same, i.e. 15. However, in this case, they are of opposite sign. Thus, equation (1) multiplied by 5 gives
$40x - 15y = 195$ (3)
and equation (2) multiplied by 3 gives
$21x + 15y = -12$ (4)
To eliminate the unknown y, equation (3) is added to equation (4). Thus
$61x = 183$
Dividing both sides of the equation by 61 gives $\boldsymbol{x = 3}$
Substituting in equation (1) gives
$8(3) - 3y = 39$
Therefore $24 - 3y = 39$
$- 3y = 39 - 24$
$- 3y = 15$
Dividing both sides by -3 gives
$$y = \frac{15}{-3}$$
$$\boldsymbol{y = -5}$$
Checking in equation (2)
L.H.S. $= 7(3) + 5(-5)$
$= 21 - 25$
$= -4 =$ R.H.S.
Thus the solution of the given equations is $\boldsymbol{x = 3}$ **and** $\boldsymbol{y = -5}$

Problem 2. Solve $4x - 18 = 3y$
$1 + x + 2y = 0$

Rearranging gives
$4x - 3y = 18$ (1)
$x + 2y = -1$ (2)
Multiplying equation (2) by 4 gives
$4x + 8y = -4$ (3)
Subtracting equation (3) from equation (1) gives
$-11y = 22$
$$\text{Therefore } y = \frac{22}{-11}$$
$$\boldsymbol{y = -2}$$
Substituting in equation (1) gives
$4x - 3(-2) = 18$

Therefore $4x + 6 = 18$

$$4x = 18 - 6$$
$$4x = 12$$
$$x = 3$$

Checking in equation (2)

L.H.S. $= (3) + 2(-2)$

$= 3 - 4$

$= -1 =$ R.H.S.

Thus the solution is $\boldsymbol{x = 3, y = -2}$.

Further problems on simple simultaneous equations shown in worked Problems 1 and 2 may be found in Section 6.4 (Problems 1–15), page 156.

Worked problem on simultaneous equations involving fractions

Problem 3. Solve $\frac{a}{6} - \frac{b}{8} = 0$ (1)

$$\frac{a}{3} + \frac{b}{4} = 6 \qquad (2)$$

It is usually easier to rid the equations of fractions before the procedure for solving simultaneous equations is attempted. This is achieved by multiplying each equation by the lowest common multiple of the denominators. Multiplying equation (1) by 24 gives

$$24\left[\frac{a}{6}\right] - 24\left[\frac{b}{8}\right] = 24(0)$$

Therefore $4a - 3b = 0$ (3)

Multiplying equation (2) by 12 gives

$$12\left[\frac{a}{3}\right] + 12\left[\frac{b}{4}\right] = 12(6)$$

Therefore $4a + 3b = 72$ (4)

The usual procedure for solution is now adopted using equations (3) and (4).

Therefore $4a - 3b = 0$ (3)

$$4a + 3b = 72 \qquad (4)$$

By coincidence the coefficients of 'a' are the same in both equations. Also the coefficients of 'b' are the same although the sign is different. Adding equations (3) and (4) gives

$8a = 72$

$a = 9$

Substituting in equation (3) gives

$$4(9) - 3b = 0$$

Therefore $36 - 3b = 0$

$$36 = 3b$$
$$b = 12$$

Checking in equation (4) gives

L.H.S. $= 4(9) + 3(12)$
$= 36 + 36$
$= 72 =$ R.H.S.

Whenever fractions are involved in simultaneous equations a final check must be made in the original equations. Thus, in equation (1)

L.H.S. $= \frac{9}{6} - \frac{12}{8}$
$= 1\frac{1}{2} - 1\frac{1}{2}$
$= 0 =$ R.H.S.

And in equation (2)

L.H.S. $= \frac{9}{3} + \frac{12}{4}$
$= 3 + 3$
$= 6 =$ R.H.S.

Thus the solution is $\boldsymbol{a = 9, b = 12}$.

Further problems on simultaneous equations involving fractions as shown in worked Problem 3 may be found in Section 6.4 (Problems 16–23), page 156.

Worked problems on simultaneous equations involving decimal fractions

---

Problem 4. Solve

$1.2x - 1.8y = -21$ (1)
$2.5x + 0.6y = 65$ (2)

Many practical examples of simultaneous equations will contain decimal fractions. It is generally easier to rid the equations of their decimal fractions and in Problem 4 this is achieved by multiplying throughout by 10.

Thus, multiplying equation (1) by 10 gives

$12x - 18y = -210$ (3)

and multiplying equation (2) by 10 gives

$25x + 6y = 650$ (4)

Equations (3) and (4) do not contain decimal fractions, and are thus solved as follows:

Multiplying equation (4) by 3 gives

$75x + 18y = 1\,950$ (5)

Adding equations (3) and (5) gives

$87x = 1\,740$

Therefore $x = \frac{1\,740}{87}$

$\boldsymbol{x = 20}$

Substituting in equation (1)

$1.2(20) - 1.8y = -21$

Therefore $24 - 1.8y = -21$

$$24 + 21 = 1.8y$$
$$45 = 1.8y$$
$$y = \frac{45}{1.8}$$
$$\mathbf{y = 25}$$

The solution $x = 20$ and $y = 25$ must now be substituted into equation (2), to serve as a check.

Check:

In equation (2), L.H.S. $= 2.5x + 0.6y$

$$= 2.5(20) + 0.6(25)$$
$$= 50 + 15$$
$$= 65 = \text{R.H.S.}$$

Thus the solution is $\mathbf{x = 20}$ **and** $\mathbf{y = 25}$.

The following worked problem is included because it shows clearly that the use of a calculator is often invaluable. The procedure does not differ in any way to that previously adopted. Numbers can become large and cumbersome in practical applications and this is unavoidable.

Problem 5. Solve $0.460b + 0.150d = 1.990$ (1)

$$1.21b - 0.536d = 4.304 \quad (2)$$

If equation (1) is multiplied by 100 and equation (2) by 1 000 the decimal fractions will be changed into integers. However, if a calculator is used, this changing of decimal fractions to integers is an unnecessary step.

If equation (1) is multiplied by 0.536 and equation (2) by 0.15 the coefficients of d will be the same, albeit of opposite sign. Thus

$$(0.536)(0.460)b + (0.536)(0.150)d = (0.536)(1.990) \quad (3)$$
$$(0.150)(1.21)b - (0.150)(0.536)d = (0.150)(4.304) \quad (4)$$

Therefore $0.246\ 56b + 0.0804d = 1.066\ 64$ (3)

$$0.181\ 50b - 0.0804d = 0.645\ 60 \quad (4)$$

Adding equations (3) and (4) gives

$$0.428\ 06b = 1.712\ 24$$

Therefore $b = \dfrac{1.712\ 24}{0.428\ 06}$

$$b = 4$$

Substituting in equation (1) gives

$$(0.46)4 + 0.15d = 1.99$$

Therefore $1.84 + 0.15d = 1.99$

$$0.15d = 1.99 - 1.84$$
$$0.15d = 0.15$$
$$d = 1$$

Checking in equation (2)

L.H.S. $= (1.21)(4) - (0.536)(1)$

$= 4.840 - 0.536$

$= 4.304 =$ R.H.S.

Thus the solution is $\boldsymbol{b = 4}$ **and** $\boldsymbol{d = 1}$.

Further problems on simultaneous equations involving decimal fractions as shown in worked Problems 4 and 5 may be found in Section 6.4 (Problems 24–32), page 157.

Worked problem on simultaneous equations involving reciprocals

---

Problem 6. Solve $\frac{3}{a} + \frac{6}{b} = 4$ (1)

$\frac{9}{a} - \frac{12}{b} = -3$ (2)

When the reciprocal of the unknowns are included throughout, the following substitution should be made before continuing with the adopted procedure. Let

$\frac{1}{a} = x$ and $\frac{1}{b} = y$

Equations (1) and (2) then become

$3x + 6y = 4$ (3)

$9x - 12y = -3$ (4)

Multiplying equation (3) by 3 gives

$9x + 18y = 12$ (5)

Subtracting equation (4) from equation (5) gives

$30y = 15$

Therefore $y = \frac{15}{30}$

$\boldsymbol{y = \frac{1}{2}}$

Substituting in equation (3) gives

$3x + 6(\frac{1}{2}) = 4$

Therefore $3x + 3 = 4$

$3x = 4 - 3$

$3x = 1$

$\boldsymbol{x = \frac{1}{3}}$

But $x = \frac{1}{a}$ or $a = \frac{1}{x}$

Therefore $a = \frac{1}{\frac{1}{3}}$

$\boldsymbol{a = 3}$

Also $y = \frac{1}{b}$ or $b = \frac{1}{y}$

Therefore $b = \frac{1}{\frac{1}{2}}$

$$b = 2$$

A check must be made in the original two equations. Substituting $a = 3$ and $b = 2$ in equation (1) gives

L.H.S. $= \frac{3}{3} + \frac{6}{2}$

$= 1 + 3$

$= 4 =$ R.H.S.

Substituting in equation (2) gives

L.H.S. $= \frac{9}{3} - \frac{12}{2}$

$= 3 - 6$

$= -3 =$ R.H.S.

Thus the solution is **$a = 3$ and $b = 2$.**

Further problems on simultaneous equations involving reciprocals as shown in worked Problem 6 may be found in Section 6.4 (Problems 33–39), page 157.

Worked problem on simultaneous equations involving more difficult fractions

---

Problem 7. Solve $\frac{f-1}{3} + \frac{g+2}{2} = 3$ (1)

$\frac{1-f}{6} + \frac{4-g}{2} = \frac{1}{2}$ (2)

Before equations (1) and (2) can be simultaneously solved the fractions need to be removed. Thus, multiplying equation (1) by 6 gives

$$6\left[\frac{f-1}{3}\right] + 6\left[\frac{g+2}{2}\right] = 6(3)$$

Therefore $2(f-1) + 3(g+2) = 18$

$2f - 2 + 3g + 6 = 18$

$2f + 3g = 18 + 2 - 6$

$2f + 3g = 14$ (3)

Multiplying equation (2) by 6 gives

$$6\left[\frac{1-f}{6}\right] + 6\left[\frac{4-g}{2}\right] = 6(\tfrac{1}{2})$$

Therefore $(1-f) + 3(4-g) = 3$

$1 - f + 12 - 3g = 3$

$-f - 3g = 3 - 1 - 12$

$-f - 3g = -10$

This equation can be multiplied throughout by -1 to give:

$$f + 3g = 10 \qquad (4)$$

Thus the initial problem containing fractions becomes simply the solution of the simultaneous equations

$$2f + 3g = 14 \qquad (3)$$

$$f + 3g = 10 \qquad (4)$$

Subtracting equation (4) from equation (3) gives

$f = 4$

Substituting in equation (3) gives

$$(2)(4) + 3g = 14$$

Therefore $8 + 3g = 14$

$$3g = 14 - 8$$

$$3g = 6$$

$$g = 2$$

Checking in equation (4)

L.H.S. $= 4 + 3(2)$

$= 4 + 6$

$= 10 =$ R.H.S.

For a complete check, the solutions $f = 4$, $g = 2$ should be substituted into equations (1) and (2). Thus, substituting in equation (1)

L.H.S. $= \dfrac{4-1}{3} + \dfrac{2+2}{2}$

$= \dfrac{3}{3} + \dfrac{4}{2}$

$= 1 + 2$

$= 3 =$ R.H.S.

Substituting in equation (2)

L.H.S. $= \dfrac{1-4}{6} + \dfrac{4-2}{2}$

$= \dfrac{-3}{6} + \dfrac{2}{2}$

$= -\frac{1}{2} + 1$

$= \frac{1}{2} =$ R.H.S.

Thus, the solution is $f = 4$ and $g = 2$

Further problems on simultaneous equations involving more difficult fractions as shown in worked Problem 7 may be found in Section 6.4 (Problems 40–45), page 157.

6.3 Simultaneous equations involved in practical problems

In many practical problems it is necessary to firstly form two equations from given data, and then to solve using the procedure of Section 6.1.

Worked problems on solving practical problems

The following worked problems are typical of some of the practical problems that may be encountered by scientists, technicians and engineers.

Problem 1. In a system of pulleys, the effort P required to raise a load W is given by $P = aW + b$, where a and b are constants. If $W = 50$ when $P = 10$ and $W = 80$ when $P = 13$ find the values of a and b.

Substituting $W = 50$ and $P = 10$ in the given equation gives

$$10 = 50a + b \qquad (1)$$

Substituting $W = 80$ and $P = 13$ in the given equation gives

$$13 = 80a + b \qquad (2)$$

Thus two equations have been produced from the given data. Subtracting equation (1) from equation (2) gives

$$3 = 30a$$

Therefore $a = \dfrac{3}{30}$

$$\mathbf{a = \frac{1}{10}}$$

Substituting in equation (1) gives

$$10 = 50\left[\frac{1}{10}\right] + b$$

Therefore $10 = 5 + b$

$$10 - 5 = b$$

$$\mathbf{b = 5}$$

Checking in equation (2)

$$\text{R.H.S.} = 80\left[\frac{1}{10}\right] + 5$$

$$= 8 + 5$$

$$= 13 = \text{L.H.S.}$$

Thus the solution is $\mathbf{a = \frac{1}{10}}$ **and** $\mathbf{b = 5}$.

Problem 2. By using Kirchhoffs laws in a certain electrical network the currents I_1 and I_2 (in amperes) are connected by the equations

$$0.2I_1 + 1.6I_2 = 8.6 \qquad (1)$$

$$1.8I_1 - 2.2I_2 = -5.6 \qquad (2)$$

Find the currents I_1 and I_2.

Multiplying both equations by 10 changes the decimal fractions into integers. Hence,

$$2I_1 + 16I_2 = 86 \qquad (3)$$

$$18I_1 - 22I_2 = -56 \qquad (4)$$

Multiplying equation (3) by 9 gives

$$18I_1 + 144I_2 = 774 \qquad (5)$$

Subtracting equation (4) from equation (5) gives

$$166I_2 = 830$$

Therefore $I_2 = \frac{830}{166}$

$$I_2 = 5$$

Substituting in equation (3) gives

$$2I_1 + 16(5) = 86$$

Therefore $2I_1 + 80 = 86$

$$2I_1 = 86 - 80$$

$$2I_1 = 6$$

$$I_1 = \frac{6}{2}$$

$$I_1 = 3$$

Checking in equation (1)

L.H.S. $= (0.2)(3) + (1.6)(5)$

$= 0.6 + 8.0$

$= 8.6 =$ R.H.S.

Checking in equation (2)

L.H.S. $= (1.8)(3) - (2.2)(5)$

$= 5.4 - 11$

$= -5.6 =$ R.H.S.

Thus the solution is $I_1 = 3$A and $I_2 = 5$A.

Problem 3. The distance s m of a body travelling with constant acceleration a m s^{-2} in a straight line is given by $s = ut + \frac{1}{2}at^2$ where u m s^{-1} is the initial velocity and t s the time. Calculate the initial velocity and the acceleration given that $s = 36$ m at $t = 2$ s and $s = 124$ m when $t = 4$ s.

Substituting $s = 36$ and $t = 2$ in the given equation gives

$$36 = 2u + \tfrac{1}{2}a(2)^2$$

Therefore $36 = 2u + 2a$ (1)

Substituting $s = 124$ and $t = 4$ in the given equation gives

$$124 = 4u + \tfrac{1}{2}a(4)^2$$

Therefore $124 = 4u + 8a$ (2)

Multiplying equation (1) by 2 gives

$$72 = 4u + 4a \quad (3)$$

Subtracting equation (3) from equation (2) gives

$$52 = 4a$$

Therefore $a = \frac{52}{4}$

$$a = 13$$

Substituting in equation (1) gives

$$36 = 2u + 2(13)$$

Therefore $36 = 2u + 26$

$$36 - 26 = 2u$$
$$10 = 2u$$
$$u = \frac{10}{2}$$
$$u = 5$$

Checking in equation (2)

R.H.S. $= 4(5) + 8(13)$

$= 20 + 104$

$= 124 =$ L.H.S.

Thus the solution is that the initial velocity is **5 m s^{-1}** and the acceleration is **13 m s^{-2}**.

Problem 4. An engineer and five technicians together earn £3 552 per month, whilst three engineers and nine technicians together earn £7 200 per month. Find how much an engineer and how much a technician earn per month.

Let E represent the salary of an engineer and T that of a technician.

Thus $E + 5T = 3\,552$ (1)

and $3E + 9T = 7\,200$ (2)

Multiplying equation (1) by 3 gives

$3E + 15T = 10\,656$ (3)

Subtracting equation (2) from equation (3) gives

$$6T = 3\,456$$

Therefore $T = \frac{3\,456}{6}$

$$T = 576$$

Substituting in equation (1) gives

$E + 5\,(576) = 3\,552$

Therefore $E + 2\,880 = 3\,552$

$E = 3\,552 - 2\,880 = 672$

Checking in equation (2):

L.H.S. $= 3(672) + 9(576)$

$= 2\,016 + 5\,184$

$= 7\,200 =$ R.H.S.

Thus the solution is that an **engineer earns £672 per month** and a **technician £576 per month.**

Problem 5. The resistance R ohms of a length of wire at t°C is given by the formula

$R = R_0(1 + \alpha t)$
where R_0 is the resistance at 0°C. Find α and R_0 if R = 30 ohms at 60°C and R = 32 ohms at 100°C.

Substituting R = 30 and t = 60 in the equation gives

$$30 = R_0(1 + 60\alpha) \quad (1)$$

Substituting R = 32 and t = 100 in the equation gives

$$32 = R_0(1 + 100\alpha) \quad (2)$$

If equation (1) is divided by equation (2) the resulting equation will involve only one unknown, α. Thus

$$\frac{30}{32} = \frac{R_0(1 + 60\alpha)}{R_0(1 + 100\alpha)}$$

Therefore $\dfrac{30}{32} = \dfrac{(1 + 60\alpha)}{(1 + 100\alpha)}$

The L.C.M. of the denominators is $32(1 + 100\alpha)$. Therefore, multiplying both sides of the equation by $32(1 + 100\alpha)$ gives

$$32(1 + 100\alpha)\frac{30}{32} = 32(1 + 100\alpha)\frac{(1 + 60\alpha)}{(1 + 100\alpha)}$$

Therefore $30(1 + 100\alpha) = 32(1 + 60\alpha)$

$$30 + 3\,000\alpha = 32 + 1\,920\alpha$$

$$3\,000\alpha - 1\,920\alpha = 32 - 30$$

$$1\,080\alpha = 2$$

$$\alpha = \frac{2}{1\,080}$$

$$\boldsymbol{\alpha = \frac{1}{540}} \text{ or } \mathbf{0.001\,85}$$

Substituting in equation (1) gives

$$30 = R_0\left(1 + \frac{1}{540}(60)\right)$$

Therefore $30 = R_0\left[1 + \dfrac{1}{9}\right]$

$$30 = \frac{10}{9}R_0$$

$$R_0 = \frac{9}{10}(30)$$

$$\boldsymbol{R_0 = 27}$$

Checking in equation (2)

$$\text{R.H.S.} = 27\left[1 + \frac{100}{540}\right]$$

$$= 27\left[1 + \frac{10}{54}\right]$$

$$= 27\left[\frac{64}{54}\right]$$

$$= 32 = \text{L.H.S.}$$

Thus the solution is $\boldsymbol{\alpha}$ **= 0.001 85** and $\boldsymbol{R_0}$ **= 27 ohms.**

 Further problems on simultaneous equations obtained from practical data as shown in worked Problems 1 to 5 may be found in the following Section (6.4) (Problems 46–60), page 158.

6.4 Further problems

Solve the following simultaneous equations for the unknowns and verify the results.

Simple simultaneous equations

1. $x + y = 5$
$x - y = 2$
$[x = 3\frac{1}{2}, y = 1\frac{1}{2}]$

2. $a - b = 8$
$a + b = 12$
$[a = 10, b = 2]$

3. $2s + 3t = 5$
$s + t = 2$
$[s = 1, t = 1]$

4. $d + e = 3$
$3d + 2e = 7$
$[d = 1, e = 2]$

5. $3g - 2h = 7$
$g + 2h = 5$
$[g = 3, h = 1]$

6. $x + 3y = 11$
$x + 2y = 8$
$[x = 2, y = 3]$

7. $4x - 3y = 18$
$x + 2y = -1$
$[x = 3, y = -2]$

8. $3m - 2n = -4.5$
$4m + 3n = 2.5$
$[m = -\frac{1}{2}, n = 1\frac{1}{2}]$

9. $7a - 4b = 37$
$6a + 3b = 51$
$[a = 7, b = 3]$

10. $3x = 2y$
$4x + y = -11$
$[x = -2, y = -3]$

11. $4c = 2 - 5d$
$3d + c + 3 = 0$
$[c = 3, d = -2]$

12. $4x - 3y = 3$
$3x + 5y = 111$
$[x = 12, y = 15]$

13. $3a + 4b - 5 = 0$
$12 = 5b - 2a$
$[a = -1, b = 2]$

14. $6m - 19 = 3n$
$13 = 5m + 6n$
$[m = 3, n = -\frac{1}{3}]$

15. $4a - 6b + 2.5 = 0$
$7a - 5b + 0.25 = 0$
$[a = \frac{1}{2}, b = \frac{3}{4}]$

Simultaneous equations involving fractions

16. $\frac{a}{3} + \frac{b}{4} = 6$
$\frac{a}{6} - \frac{b}{8} = 0$
$[a = 9, b = 12]$

17. $s + t = 17$
$\frac{s}{5} - \frac{t}{7} = 1$
$[s = 10, t = 7]$

18. $\frac{a}{2} - 11 = -2b$
$\frac{3}{5}b = 9 - 3a$
$[a = 2, b = 5]$

19. $\frac{3}{2}p - 2q = \frac{1}{2}$
$p + \frac{3}{2}q = 6$
$[p = 3, q = 2]$

20. $\frac{2}{5}x - \frac{y}{4} = \frac{17}{20}$
$\frac{x}{4} + \frac{2}{7}y = \frac{13}{7}$
$[x = 4, y = 3]$

21. $\frac{c}{5} + \frac{d}{3} = \frac{43}{30}$
$\frac{c}{9} - \frac{d}{6} = -\frac{1}{12}$
$[c = 3, d = 2\frac{1}{2}]$

22. $y - \frac{x}{12} = 1$
$x + \frac{y}{4} = \frac{25}{2}$
$[x = 12, y = 2]$

23. $\frac{a}{8} + \frac{5}{2} = b$
$13 - \frac{b}{3} - 3a = 0$
$[a = 4, b = 3]$

Simultaneous equations involving decimal fractions

24. $1.2a - 1.8b = -21$
$2.5a + 0.6b = 65$
$[a = 20, b = 25]$

25. $2.5x + 0.45 - 3y = 0$
$1.6x + 0.8y - 0.8 = 0$
$[x = 0.30, y = 0.40]$

26. $1.2p + q = 1.8$
$p - 1.2q = 3.94$
$[p = 2.50, q = -1.20]$

27. $0.5b - 1.2c = -13$
$0.8b + 0.3c = 12.5$
$[b = 10, c = 15]$

28. $6.9 + 1.7y = 0.9x$
$4.3 + 1.3x + 2.3y = 0$
$[x = 2.0, y = -3.0]$

29. $1.4x - 3.2y = 7.06$
$-2.1x + 6.7y = -12.87$
$[x = 2.30, y = -1.20]$

30. $-0.5f + 0.4g = 0.7$
$1.2f - 0.3g = 3.6$
$[f = 5.0, g = 8.0]$

31. $1.25d - 0.75e = 1$
$0.25d + 1.25e = 17$
$[d = 8.0, e = 12.0]$

32. $0.1p + 0.2q = -0.2$
$1.5p - 0.4q = 10.6$
$[p = 6.0, q = -4.0]$

Simultaneous equations involving reciprocals

33. $\frac{3}{x} - \frac{2}{y} = 0$
$\frac{1}{x} + \frac{4}{y} = 14$
$[x = \frac{1}{2}, y = \frac{1}{3}]$

34. $\frac{2}{a} - \frac{3}{b} = -8$
$\frac{4}{a} + \frac{5}{b} = 50$
$[a = \frac{1}{5}, b = \frac{1}{6}]$

35. $\frac{3}{r} - \frac{4}{s} = \frac{5}{6}$
$\frac{1}{r} + \frac{3}{s} = 1$
$[r = 2, s = 6]$

36. $\frac{1}{2a} + \frac{3}{5b} = 7$
$\frac{4}{a} + \frac{1}{2b} = 13$
$[a = \frac{1}{2}, b = \frac{1}{10}]$

37. $\frac{3}{h} - \frac{2}{k} = \frac{1}{2}$
$\frac{5}{h} + \frac{3}{k} = \frac{29}{12}$
$[h = 3, k = 4]$

38. $\frac{10}{a} - \frac{4}{b} = 3$
$\frac{6}{a} + \frac{8}{b} = 7$
$[a = 2, b = 2]$

39. $\frac{5}{m} + \frac{4}{n} = 1.3$
$\frac{3}{m} - \frac{7}{n} = -1.1$
$[m = 10, n = 5]$

Simultaneous equations involving more difficult fractions

40. $\frac{a-1}{3} + \frac{b+2}{2} = 3$
$\frac{1-a}{6} + \frac{4-b}{2} = \frac{1}{2}$
$[a = 4, b = 2]$

41. $\frac{f-2}{3} + \frac{k-1}{4} = \frac{13}{12}$
$\frac{2-f}{2} + \frac{3+k}{3} = \frac{11}{6}$
$[f = 3, k = 4]$

42. $\frac{4}{a-b} = \frac{16}{21}$
$\frac{3}{a+b} = \frac{4}{9}$
$[a = 6, b = \frac{3}{4}]$

43. $\frac{2x+1}{5} - \frac{1-4y}{2} = \frac{5}{2}$
$\frac{1-3x}{7} + \frac{2y-3}{5} + \frac{32}{35} = 0$
$[x = 2, y = 1]$

44. If $5p - \frac{2}{q} = 11$ and $p + \frac{3}{q} = 9$ find the value of $\frac{pq+1}{q}$ [5]

45. When the numerator of a certain fraction is increased by 3 and the denominator by 1, its value changes to $1\frac{1}{4}$; but when the numerator is increased by 4 and the denominator decreased by $1\frac{1}{2}$, it is equal to 4. Find the original fraction [$\frac{2}{3}$]

46. In an engineering process two variables x and y are related by $y = ax + \frac{b}{x}$ where a and b are constants. Find a and b if $y = 13$ when $x = 3$ and $y = 12$ when $x = 2$ [$a = 3$, $b = 12$]
47. When an effort E newtons is applied to a machine it is found that the resistance R newtons can be overcome and that E and R are connected by the formula: $E = a + bR$. An effort of 3.5 N overcomes a resistance of 5 N while an effort of 5.3 N overcomes a resistance of 8 N. Find a and b and the effort required to overcome a resistance of 10 N.
[$a = 0.50$, $b = 0.60$, $E = 6.5$ N].
48. When Kirchhoffs Laws are applied to a two mesh circuit the following equations are produced
$3.0 = 0.10I_1 + (I_1 - I_2)$
$-2.0 = 0.05I_2 - (I_1 - I_2)$
Find I_1 and I_2 [$I_1 = 7.42$ A, $I_2 = 5.16$ A]
49. 500 Cotswold bricks and 80 breeze blocks cost £84.50. Another order of 220 bricks and 50 breeze blocks cost £43.10. Find the cost of one breeze block and the cost, per hundred, of bricks. [40p, £10–50]
50. The law connecting friction F and load L for an experiment to find the friction force between two surfaces is of the type $F = aL + b$ where a and b are constants. When $F = 6.0$ N, $L = 7.5$ N and when $F = 2.7$ N, $L = 2.0$ N. Find the values of a and b and the value of F when $L = 18$ N.
[$a = 0.60$, $b = 1.5$, $F = 12.3$ N]
51. The distance s metres of a vehicle travelling with constant acceleration a m s^{-2} in a straight line is given by: $s = ut + \frac{1}{2}at^2$ where u is the initial velocity and t the time. Calculate the initial velocity and the acceleration given that $s = 51.75$ m when $t = 3$ s and $s = 118.75$ m when $t = 5$ s.
[$u = 7.50$ m s^{-1}, $a = 6.50$ m s^{-2}]
52. The weekly wage bill for three instructors and thirty apprentices in a mechanical training workshop is £1 761. In an electrical training centre the wage bill is £1 080 for two instructors and eighteen apprentices. If the five instructors each receive the same pay, and the forty-eight apprentices each receive the same pay, calculate the weekly wage of an instructor and the weekly wage of an apprentice [Instructor £117, apprentice £47]
53. $y = mx + c$ is the equation of a straight line of slope m and intercept on the y-axis of c. If a line passes through the point where $x = 3$ and $y = 2\frac{1}{2}$, and also through the point where $x = 5\frac{1}{2}$, and $y = 3\frac{3}{4}$ find the slope and intercept of the straight line [$m = \frac{1}{2}$, $c = 1$]
54. The resistance R ohms of a length of wire at t°C is given by the formula $R = R_0(1 + \alpha t)$ where R_0 ohms is the resistance at 0°C and α is a constant.
(*a*) If $R = 12$ ohms at 20°C, find R at 60°C. Assume $\alpha = 0.004\,2$. Find also the resistance at 0°C.
(*b*) If $R = 25$ ohms at 50°C and $R = 26$ ohms at 90°C find α and R_0.
(*a*) [$R_{60} = 13.86$ ohms, $R_0 = 11.07$ ohms] (*b*) [$R_0 = 23.75$ ohms, $\alpha = \frac{1}{950}$ or 0.001 05]

55. Applying Kirchhoffs Laws to a two-mesh network gives the following equations:

$14 = 0.2I_1 + 2I_1 + 8(I_1 - I_2)$

$0 = -8(I_1 - I_2) + 2I_2 + 10I_2$

Find the values of I_1 and I_2 [$I_1 = 2$, $I_2 = 0.8$]

56. A garage is supplied with two orders for tyres. If 2 radial tyres and 3 crossply tyres cost £107.50, and 5 radials and 7 crossplies cost £259, find the cost of each type of tyre. [Radial £24.50, crossply £19.50]

57. Equations connecting parallel resistances in an electrical circuit are

$$\frac{4}{R_1} + \frac{6}{R_2} + \frac{9}{R_3} = 6$$

$$\frac{15}{R_1} + \frac{11}{R_2} + \frac{2}{R_3} = 8\frac{1}{12}$$

If $R_2 = R_3$, find the values of R_1, R_2 and R_3 [$R_1 = 4$, $R_2 = R_3 = 3$]

58. The compound $C_y H_z$ reacts with oxygen in proportions obeyed by the equations

$44y - 9z = 60$

$32y + 8z = 160$

Determine y and z and hence the formula of $C_y H_z$ [$C_3 H_8$]

59. The molar heat capacity of a solid compound is given by the equation $c = a + bT$ where c is measured in JK^{-1} $mole^{-1}$ and T in K. When c is 41, T is 100 K and when c is 108.5, T is 350 K. Find a and b [$a = 14$, $b = 0.27$]

60. x g of cartridge brass (70% copper, 30% zinc, by mass) and y g of naval brass (62% copper, 37% zinc, 1% tin, by mass) when fused together give a new alloy containing 342.6 g of copper and 165.6 g of zinc. Find x and y. [$x = 330$, $y = 180$]

Chapter 7

Evaluation and transposition of formulae

7.1 Evaluation of formulae

The statement $I = \frac{V}{R}$ is said to be a **formula** for I in terms of V and R.

I, V and R are called **symbols.**

The single term on the left-hand side is called the **subject of the formula.**

Similarly, $A = 2\pi rh + 2\pi r^2$ is a formula for A in terms of r and h, and A is the subject of the formula.

There are two methods widely used to evaluate formulae:

1. Use of tables (for example, logarithms, squares, roots, reciprocals, and so on)
2. Calculators.

Worked problems on the evaluation of formulae

In most practical branches of science and engineering the evaluation of formulae is essential. Below are some worked problems typical of the types of formulae met in day to day calculations.

The actual method of calculation is not shown, but it is useful to be able to evaluate formulae by both of the available methods.

The use of brackets in place of multiplication signs is employed throughout.

Problem 1. The surface area, A, of a hollow cone is given by the formula $A = \pi rl$. Find the surface area in square centimetres, taking π as 3.14, when

$r = 4.0$ cm and $l = 9.0$ cm.

Substituting the values for symbols in the formula $A = \pi rl$ gives:

$A = (3.14)(4.0)\ (9.0)\ \text{cm}^2$

$= 113\ \text{cm}^2$

Hence the surface area, $A =$ **113 cm²**.

Problem 2. In an electrical circuit the voltage V volts is given by $V = IR$. Find the voltage correct to 4 significant figures when $I = 7.240$ amperes and $R = 12.57$ ohms.

$V = IR$

Therefore $V = (7.240)(12.57)$ volts

$= 91.01$ volts

Hence the voltage, $V =$ **91.01 volts.**

Problem 3. A formula used for calculating velocity v m s^{-1} is given by $v = u + at$. If $u = 12.47$ m s^{-1}, $a = 5.46$ m s^{-2} and $t = 4.92$ s find v correct to 2 decimal places.

$v = u + at$

Therefore $v = 12.47 + (5.46)(4.92)$

$= 12.47 + 26.86$

$= 39.33$ m s^{-1}

Hence the velocity, $v =$ **39.33 m s^{-1}**.

Problem 4. The area A m^2 of a circle is given by $A = \pi r^2$. Find the area correct to 2 decimal places given $\pi = 3.142$ and $r = 4.156$ m.

$A = \pi r^2$

Therefore $A = (3.142)(4.156)^2\ \text{m}^2$

$= (3.142)(17.27)\ \text{m}^2$

$= 54.27\ \text{m}^2$

Hence the area, $A =$ **54.27 m²**.

Problem 5. Power P watts in an electrical circuit may be expressed by the formula $P = \dfrac{V^2}{R}$. Evaluate the power correct to 2 decimal places, given that $V = 24.62$ volts and $R = 45.21$ ohms.

$P = \dfrac{V^2}{R}$

Therefore $P = \dfrac{(24.62)^2}{45.21}$ watts

$= \dfrac{606.14}{45.21}$ watts $= 13.41$ watts

Hence the power, $P =$ **13.41 watts.**

Problem 6. The volume V cm^3 of a right circular cone is given by the formula $V = \frac{1}{3}\pi r^2 h$. Given $\pi = 3.142$, $r = 5.637$ cm and $h = 16.41$ cm find the volume in standard form correct to 3 significant figures.

$$V = \frac{1}{3}\pi r^2 h$$

$$\text{Therefore } V = \frac{1}{3}(3.142)(5.637)^2\ (16.41)\ \text{cm}^3$$

$$= \frac{1}{3}(3.142)(31.78)(16.41)\ \text{cm}^3$$

$$= \frac{1\,638}{3}\ \text{cm}^3$$

$$= 546\ \text{cm}^3$$

Hence the volume V = **5.46×10^2 cm^3.**

Problem 7. If force F newtons is given by

$$F = \frac{G\, m_1 m_2}{d^2}$$

where m_1 and m_2 are masses, d their distance apart and G a constant, find the force given $G = 6.67 \times 10^{-11}$ Nm^{2}kg^{-2}, $m_1 = 8.43$ kg, $m_2 = 17.2$ kg and $d = 24.2$ m. Express the answer in standard form to 3 significant figures.

$$F = \frac{G\, m_1 m_2}{d^2}$$

$$= \frac{(6.67)(10^{-11})\ \text{Nm}^2\text{kg}^{-2}(8.43)\ \text{kg}(17.2)\ \text{kg}}{(24.2)^2\ \text{m}^2}$$

$$= \frac{(6.67)(8.43)(17.2)(10^{-11})\ \text{N}}{585.6}$$

$$= \frac{(967.1)(10^{-11})}{585.6}\ \text{N}$$

$$= 1.651 \times 10^{-11}\ \text{N}$$

$$= 1.65 \times 10^{-11}\ \text{N to 3 significant figures}$$

Hence the force, F = **1.65×10^{-11} N.**

Problem 8. The time t seconds of swing of a simple pendulum is given by the formula

$$t = 2\pi \sqrt{\left[\frac{l}{g}\right]}$$

Find the time, correct to 3 decimal places, given $\pi = 3.142$, $l = 13.0$ m and $g = 9.81$ m s^{-2}.

$$t = 2\pi\sqrt{\left[\frac{l}{g}\right]}$$

$$t = (2)(3.142)\sqrt{\left[\frac{13.0 \text{ m}}{9.81 \text{ ms}^{-2}}\right]}$$

$= (2)(3.142)\sqrt{(1.325 \text{ s}^2)}$
$= (2)(3.142)(1.151) \text{ s}$
$= 7.234 \text{ s}$

Hence the time t = **7.234 s.**

Problem 9. A formula for resistance variation with temperature is $R = R_0(1 + \alpha t)$ Given R_0 = 15.42 ohms, α = 0.002 70 and t = 78.4°C evaluate R ohms, correct to 2 decimal places.

$R = R_0(1 + \alpha t)$
$= 15.42\,[1 + (0.002\,7)(78.4)]$ ohms
$= 15.42\,[1 + 0.211\,7]$ ohms
$= 15.42\,[1.211\,7]$ ohms

Hence R = **18.68 ohms.**

Further problems on evaluation of formulae may be found in Section 7.3 (Problems 1–23), page 167.

7.2 Transposition of formulae

Formulae play a very important part in mathematics, science and engineering, as by them it is possible to give a concise, accurate and generalised statement of laws.

From Section 7.1 it can be seen that some formulae contain several symbols. Usually, one of the symbols is isolated on one side of the equation and is called the subject of the formula. Sometimes a symbol other than the subject is required to be calculated. In such circumstances it is usually easiest to rearrange the formula to make a new subject before numbers are substituted for symbols. This rearranging process is called **transposing the formula** or simply **transposition.** Transposition should be treated carefully because an error can result in a statement that is not true, with possibly serious effects in certain branches of industry.

Basically the rules used for transposition of formulae are the same as the rules used for the solution of simple equations (see Chapter 5).

Worked problems on transposition of formulae

In the following graded worked problems four types of transposition have been categorised (Case A to Case D); however, it will become evident that many practical examples will require the application of several of these categories in order to transpose a formula.

The objective, in transposition, is to obtain the required new subject on its own on the left-hand side (L.H.S.) of the equation. Therefore, as a first step in any transposition, the equation, if necessary, should be changed around so that the side containing the required new subject is on the left.

Problem 1. Transpose $a = b + c + d$ to make c the subject.

The first step is to change the equation around so that the new subject, i.e. c, is on the L.H.S. Thus
$b + c + d = a$
Subtracting $(b + d)$ from both sides of the equation gives:

$$(b + c + d) - (b + d) = a - (b + d)$$

Therefore $b + c + d - b - d \quad = a - b - d$

$$c = a - b - d.$$

It will be clear from a knowledge of the solution of simple equations that what has actually happened in the above example is that $+ b$ has been moved from the L.H.S. to the R.H.S., with the necessary change of sign. Similarly $+ d$ has been changed from the L.H.S. to the R.H.S. with a change of sign.

Problem 2. If $x + y = a - b + c$ express b as the subject.

$$a - b + c = x + y$$

Therefore $-b \quad = x + y - a - c$

Obtaining a negative subject instead of the required positive subject is not an uncommon occurrence. In such circumstances both sides of the equation should be multiplied by -1.
Hence $(-1)(-b) = (-1)(x + y - a - c)$
Therefore $\quad +b = -x - y + a + c$
The result of multiplying each side of the equation by -1 has simply changed all of the signs in the equation, i.e. the negative signs have become positive and vice versa.
Therefore $b = a + c - x - y$

When a number of symbols having different signs are involved in a formula it is usual to express the positive symbols before the negative ones. That is, rather than $b = -x - y + a + c$, it is more conventional (although no more correct) to express the formula as $b = a + c - x - y$.

Case B. Formulae involving products

Problem 3. Transpose $V = IR$ to make I the subject.

$IR = V$

On the L.H.S. R is multiplying the required new subject, I. The opposite operation to multiplication is division. Hence, if both sides of the equation are divided by R, then

$$\frac{IR}{R} = \frac{V}{R}$$

Therefore $I = \dfrac{V}{R}$

Problem 4. If a body falls freely through a height h the velocity is given by the formula $V^2 = 2gh$. Express this formula with h as the subject.

$2gh = V^2$

Dividing both sides of the equation by $2g$ gives

$$\frac{2gh}{2g} = \frac{V^2}{2g}$$

Therefore $h = \dfrac{V^2}{2g}$

Case C. Formulae containing fractions

Problem 5. If $K = \dfrac{M}{V}$, rearrange the equation to make M the subject.

$\dfrac{M}{V} = K$

On the L.H.S. the prospective new subject, M, is divided by V. The opposite operation to division is multiplication. Hence, if both sides of the equation are multiplied by V then

$$V\left[\frac{M}{V}\right] = V(K)$$

Therefore $M = VK$

Note that, as in the solution of simple equations, to rid an equation of a fraction, both sides of the equation needs to be multiplied by the lowest common multiple of the denominators, in the above case, V.

Problem 6. If $a = \dfrac{F}{m}$ rearrange the equation to make m the subject.

$\dfrac{F}{m} = a$

Multiplying both sides by m gives

$$m\left[\frac{F}{m}\right] = m(a)$$

Therefore $F = ma$

The required new subject is now on the R.H.S. Thus

$ma = F$

Dividing both sides by a gives

$$\frac{ma}{a} = \frac{F}{a}$$

Therefore $m = \dfrac{F}{a}$

Problem 7. Transpose $R = \frac{\rho l}{a}$ to make (i) a and (ii) ρ the subject.

(i) $\frac{\rho l}{a} = R$

Multiplying both sides by a gives

$$a\left[\frac{\rho l}{a}\right] = a(R)$$

Therefore $\rho l = aR$

$$aR = \rho l$$

Dividing both sides by R gives

$$\frac{Ra}{R} = \frac{\rho l}{R}$$

Therefore $\boldsymbol{a = \frac{\rho l}{R}}$

(ii) $\frac{\rho l}{a} = R$

Multiplying both sides by a gives

$$a\left[\frac{\rho l}{a}\right] = a(R)$$

Therefore $\rho l = aR$

Dividing both sides by l gives

$$\frac{\rho l}{l} = \frac{aR}{l}$$

Therefore $\boldsymbol{\rho = \frac{aR}{l}}$

Problem 8. If $v = u + \frac{ft}{m}$ transpose the equation to make t the subject.

$$u + \frac{ft}{m} = v$$

From Case A: $\frac{ft}{m} = v - u$

From Case C: $m\left[\frac{ft}{m}\right] = m(v - u)$

Therefore $ft = m(v - u)$

From Case B: $\frac{ft}{f} = \frac{m(v - u)}{f}$

Therefore $\boldsymbol{t = \frac{m}{f}(v - u)}$

Case D. Formulae containing the required new subject in a bracket

---

Problem 9. Transpose the equation $s = \frac{1}{2}(u + v)t$ to make v the subject.

$\frac{1}{2}(u + v)t = s$

From Case C: $2(\frac{1}{2})(u + v)t = 2s$

Therefore $(u + v)t = 2s$

From Case B: $\frac{(u + v)t}{t} = \frac{2s}{t}$

Therefore $u + v = \frac{2s}{t}$

From Case A: $v = \frac{2s}{t} - u.$

Note that the R.H.S. $\frac{2s}{t} - u = \frac{2s - ut}{t}$

Therefore $v = \frac{2s - ut}{t}$

Both of the above answers for v are correct, the latter being the neater version.

Problem 10. Transpose the equation $L_2 = L_1(1 + \alpha\theta)$ to make α the subject.

$L_1(1 + \alpha\theta) = L_2$

Whenever the new subject is contained within a bracket it is advisable to isolate the bracket on the L.H.S. Thus, dividing both sides of the equation by L_1 gives

$$\frac{L_1(1 + \alpha\theta)}{L_1} = \frac{L_2}{L_1}$$

Therefore $1 + \alpha\theta = \frac{L_2}{L_1}$

$$\alpha\theta = \frac{L_2}{L_1} - 1$$

$$\alpha\theta = \frac{L_2 - L_1}{L_1}$$

Dividing both sides by θ gives

$$\frac{\alpha\theta}{\theta} = \frac{L_2 - L_1}{L_1\theta}$$

Therefore $\boldsymbol{\alpha = \frac{L_2 - L_1}{L_1\theta}}$

Further problems on transposition of formulae may be found in the following Section (7.3) (Problems 24–59), page 169.

7.3 Further problems

Evaluation of formulae

1. The area A of a rectangle is given by the formula $A = lb$. Evaluate the area when l is 14.21 cm and b is 7.46 cm [$A = 106.0\ \text{cm}^2$]

2. The circumference C of a circle is given by the formula $C = 2\pi r$. Find the circumference given $\pi = 3.14$ and $r = 6.20$ cm [$C = 38.94$ cm]
3. The area A of a triangle may be evaluated by using the formula $A = \frac{1}{2}bh$. Find the area when $b = 7.0$ mm and $h = 12.5$ mm [$A = 43.75\ \text{mm}^2$]
4. A formula used when dealing with gas laws is $R = \frac{pV}{T}$. Evaluate R when p is $2\,000\ \frac{\text{kN}}{\text{m}^2}$, V is 5 m^3 and T is 200 K $\left[\frac{50\ \text{kN m}}{\text{K}} = \frac{50\ \text{kJ}}{\text{K}}\right]$
5. In the following formula evaluate X when $p = 751$, $a = 5$ and $t = 21$: $X = p\left[1 + \frac{at}{100}\right]$ [1 540]
6. The voltage drop, V, in an electrical circuit is given by the formula $V = E - IR$. Evaluate the voltage drop when $E = 4.81$ volts, $I = 0.680$ amperes and $R = 5.37$ ohms [$V = 1.158$ volts]
7. $a = \frac{uv}{u + v}$. Evaluate a when $u = 10.93$ and $v = 7.380$ [$a = 4.405$]
8. When a number of cells are connected together the current I amperes of the combination is given by the formula $I = \frac{nE}{R + nr}$. Evaluate the current when $n = 40$, $E = 2.3$ volts, $R = 2.9$ ohms and $r = 0.60$ ohms [$I = 3.42$ amperes]
9. The power P watts of an electrical circuit is given by the formula $P = \frac{V^2}{R}$. Find the power when V is 14.8 volts and R is 19.5 ohms [$P = 11.23$ watts]
10. If $W = 8.20$, $v = 10.0$ and $g = 9.81$ evaluate B given that $B = \frac{Wv^2}{2g}$ [$B = 41.79$]
11. If $F = \frac{1}{2}m(v^2 - u^2)$ find F when $m = 17.0$, $v = 14.8$ and $u = 9.24$ [$F = 1\,136$]
12. The time t seconds of oscillation for a simple pendulum is given by $t = 2\pi\sqrt{\frac{l}{g}}$. Evaluate the time of oscillation when π is 3.142, $l = 61.42$ m and $g = 9.81\ \frac{\text{m}}{\text{s}^2}$ [$t = 15.72$ s]

Evaluate the subject of each of the following formulae shown in Problems 13–23.

13. $S = 2\pi r^2 + 2\pi rh$ when $\pi = 3.14$, $r = 2.42$ and $h = 3.47$ [89.51]
14. $E = \frac{1}{2}LI^2$ when $L = 3.55$ and $I = 0.490$ [0.426]
15. $A = \frac{V}{100}\left[Q - \frac{mV^2}{g}\right]$ when $Q = 50.28$, $m = 17$, $V = 5.0$ and $g = 9.81$ [0.348]
16. $I = \frac{V}{\sqrt{(R^2 + X^2)}}$ when $V = 240$, $R = 10.0$ and $X = 15.0$ [13.31]
17. $a = \frac{T(1 - S)}{S(T + 1)}$ when $T = 2.31$ and $S = 2.62$ [$-0.431\,5$]

18. $X = T\left(1 + \dfrac{b}{10}\right)^n$ when $T = b = n = \dfrac{3}{4}$ [0.791 8]

19. $A = \sqrt{s(s-a)(s-b)(s-c)}$ where $s = \dfrac{a+b+c}{2}$ and $a = 2.0$, $b = 3.0$ and $c = 4.0$ [2.91]

20. $y = \sqrt{\left(\dfrac{ak}{p} - \dfrac{b}{q}\right)}$ when $a = 0.72$, $k = 0.620$, $p = 0.003\ 00$, $b = 41.36$ and $q = 0.480$ [7.914]

21. $Z = \sqrt{\left[R^2 + \left(wL - \dfrac{1}{wc}\right)^2\right]}$ when $R = 24$, $L = 0.30$, $w = 352$ and $c = 6.5 \times 10^{-5}$ [66.4]

22. $S = ut + \frac{1}{2}at^2$ when $u = 10.1$, $t = 5.20$ and $a = -3.80$ [1.144]

23. $a = \dfrac{T}{\sqrt{(ku^2 - lv^2)}}$ when $T = 17.42$, $k = 0.073\ 0$, $u = 5.46$, $l = 0.089\ 0$ and $v = 3.21$ [$\pm$15.52]

Transposition of formulae

Make the symbol indicated in round brackets the subject of each of the formulae shown in Problems 24–54 and express each in its simplest form.

24. $k + l = m - n + p$ (m)
$[m = k + l + n - p]$

25. $a + 2b = c$ (b)
$\left[b = \dfrac{c-a}{2}\right]$

26. $2abc = d$ (c)
$\left[c = \dfrac{d}{2ab}\right]$

27. $c = 2\pi r$ (r)
$\left[r = \dfrac{c}{2\pi}\right]$

28. $5 - 2bc = p + q$ (c)
$\left[c = \dfrac{5-p-q}{2b}\right]$

29. $y = mx + c$ (m)
$\left[m = \dfrac{y-c}{x}\right]$

30. $pV = c$ (p)
$\left[p = \dfrac{c}{V}\right]$

31. $A = \pi rl$ (l)
$\left[l = \dfrac{A}{\pi r}\right]$

32. $I = PRT$ (R)
$\left[R = \dfrac{I}{PT}\right]$

33. $I = \dfrac{V}{R}$ (R)
$\left[R = \dfrac{V}{I}\right]$

34. $P = \dfrac{RT}{V}$ (V)
$\left[V = \dfrac{RT}{P}\right]$

35. $S = \dfrac{a}{1-r}$ (r)
$\left[r = \dfrac{S-a}{S}\right]$

36. $V = \dfrac{\pi}{12}d^2h$ (h)
$\left[h = \dfrac{12V}{\pi d^2}\right]$

37. $a = \dfrac{wL^2}{4d}$ (d)
$\left[d = \dfrac{wL^2}{4a}\right]$

38. $F = \dfrac{9}{5}C + 32$ (C)
$\left[C = \dfrac{5}{9}(F - 32)\right]$

39. $T = \dfrac{\lambda(x - C)}{C}$ (x)

$\left[x = \dfrac{C(T + \lambda)}{\lambda}\right]$

40. $Y = \dfrac{Fl}{Ax}$ (l)

$\left[l = \dfrac{Axy}{F}\right]$

41. $\dfrac{a + b}{b} = \dfrac{3p}{q}$ (q)

$\left[q = \dfrac{3bp}{a + b}\right]$

42. $A = 2\pi r(r + h)$ (h)

$\left[h = \dfrac{A}{2\pi r} - r\right]$

43. $A = B - 5.6C + 2$ (C)

$\left[C = \dfrac{2 + B - A}{5.6}\right]$

44. $W = aq(x - t)$ (x)

$\left[x = t + \dfrac{W}{aq}\right]$

45. $P = \dfrac{N - a}{4b}$ (N)

$[N = 4bP + a]$

46. $B = \dfrac{8(D - d)}{L}$ (d)

$\left[d = D - \dfrac{LB}{8}\right]$

47. $I = \dfrac{bd^3}{12}$ (b)

$\left[b = \dfrac{12I}{d^3}\right]$

48. $y = \dfrac{Ml^2}{8EI}$ (I)

$\left[I = \dfrac{Ml^2}{8Ey}\right]$

49. $R_2 = R_1(1 + \alpha t)$ (t)

$\left[t = \dfrac{R_2 - R_1}{R_1\alpha}\right]$

50. $\dfrac{1}{R} = \dfrac{1}{R_1} + \dfrac{1}{R_2}$ (R_1)

$\left[R_1 = \dfrac{RR_2}{R_2 - R}\right]$

51. $I = \dfrac{E - e}{R + r}$ (r)

$\left[r = \dfrac{E - e - IR}{I}\right]$

52. $y^2 = 4a\left(x + \dfrac{c^2}{4a}\right)$ (x)

$\left[x = \dfrac{y^2 - c^2}{4a}\right]$

53. $S = \dfrac{n}{2}\left[2a + (n - 1)d\right]$ (d)

$\left[d = \dfrac{2(S - an)}{n(n - 1)}\right]$

54. $A = b\left(\dfrac{1}{1 - d} - r\right)$ (d)

$\left[d = \dfrac{A + b(r - 1)}{A + br}\right]$

55. If P is the safe load which may be carried by a steel plate weakened by rivet holes then $P = f(b - nd)\,t$. Make f, the safe working stress in the steel and then n, the number of rivet holes, the subject of the formula.

$\left[f = \dfrac{P}{(b - nd)t};\ n = \dfrac{fbt - p}{fdt}\right]$

56. The modulus of elasticity of a structural material (E) is given by the formula $E = \dfrac{Wl}{Ax}$. Make x the subject of the formula $\left[x = \dfrac{Wl}{EA}\right]$

57. Van der Waals equation for the pressure of a real gas (p) is $\left(p + \dfrac{a}{v^2}\right)\left(V - b\right) = RT$. Make p the subject of the formula

$$\left[p = \frac{RT}{V - b} - \frac{a}{v^2} \right]$$

58. The observed growth yield, Y, of an organism is given by $\frac{1}{Y} = \frac{m}{\mu} + \frac{1}{Y_G}$.

Make Y_G, the true growth yield, the subject of the equation

$$\left[Y_G = \frac{\mu Y}{\mu - mY} \right]$$

59. The viscosity coefficient of a liquid (η) is given by the equation $\eta = \frac{\pi P r^4 t}{8vl}$.

Make v the subject of the formula $\left[v = \frac{\pi P r^4 t}{8\eta l} \right]$

Chapter 8

Direct and inverse proportionality and straight line graphs

8.1 Direct and inverse proportionality

An expression such as $y = 2x$ contains two variables. For every value of x there is a corresponding value of y. The variable x is called the **independent variable** and y is called the **dependent variable.** The formula used for the circumference c of a circle is $c = 2\pi r$, where r is the radius of the circle. Changes in the radius cause a corresponding change in the circumference, thus r is the independent variable and c is the dependent variable.

Direct proportion

When an increase or decrease in an independent variable leads to an increase or decrease of the same proportion in the dependent variable then this is termed **direct proportion.** If $y = 2x$, then y is directly proportional to x, which may be written as $y \propto x$, where the sign '$\propto$' means 'is proportional to'. Another way of stating the relationship between y and x is $y = kx$, where k is a constant called the **coefficient of proportionality.** In this example, $k = 2$. Similarly, when $c = 2\pi r$ then c is directly proportional to r, i.e. $c \propto r$ and the coefficient of proportionality in this case is $k = 2\pi$.

Inverse proportion

When an increase in an independent variable leads to a decrease of the same proportion in the dependent variable (or vice-versa) then this is termed **inverse proportion.** If y is inversely proportional to x then $y \propto \frac{1}{x}$ or $y = \frac{k}{x}$.

Rearranging gives $k = xy$, that is, **for inverse proportionality the product of the variables is constant.**

Some practical examples of laws involving direct and inverse proportion in science include:

(*a*) **Hooke's law,** which states that within the elastic limit of a material, the strain ϵ produced is directly proportional to the stress σ producing it, i.e. $\epsilon \propto \sigma$ or $\epsilon = k\sigma$.

(*b*) **Boyle's law,** which states that for a gas at constant temperature, the volume V of a fixed mass of gas is inversely proportional to its absolute pressure p, i.e. $p \propto \frac{1}{V}$ or $p = \frac{k}{V}$ or $k = pV$.

(*b*) **Charles' law,** which states that for a given mass of gas at constant pressure the volume V is directly proportional to its thermodynamic temperature T, i.e. $V \propto T$ or $V = kT$.

(*d*) **Ohm's law,** which states that the current I flowing through a fixed resistor R is directly proportional to the applied voltage V, i.e. $I \propto V$ or $I = kV$.

Alternatively, for a fixed voltage V, $I \propto \frac{1}{R}$ or $I = \frac{k}{R}$.

Worked problems on direct and inverse proportionality

Problem 1. If y is directly proportional to x and $y = 3.54$ when $x = 0.6$, determine (*a*) the coefficient of proportionality, (*b*) the value of y when x is 0.7, and (*c*) the value of x when y is 8.85.

(*a*) $y \propto x$ or $y = kx$

When $y = 3.54$ and $x = 0.6$ then $3.54 = k\ (0.6)$

Hence the coefficient of proportionality, $k = \frac{3.54}{0.6} = \mathbf{5.9}$

(*b*) $y = kx$, hence when $x = 0.7$, $y = (5.9)\ (0.7) = \mathbf{4.13}$

(*c*) $y = kx$, from which, $x = \frac{y}{k}$

Hence when $y = 8.85$ and $k = 5.9$, $x = \frac{8.85}{5.9} = \mathbf{1.5}$

Problem 2. Hooke's law states that stress σ is directly proportional to strain ϵ within the elastic limit of a material. For a mild steel specimen when the stress is 20×10^6 pascals the strain is 0.000 1. Determine

(*a*) the coefficient of proportionality and (*b*) the value of strain when the stress is 16×10^6 pascals.

(*a*) $\sigma \propto \epsilon$ or $\sigma = k\epsilon$, from which, $k = \dfrac{\sigma}{\epsilon}$

Hence the coefficient of proportionality, $k = \dfrac{20 \times 10^6}{0.000\,1} = \mathbf{200 \times 10^9}$ **pascals**

The coefficient of proportionality k in this case is called the Young modulus.

(*b*) Since $\sigma = k\epsilon$ then $\epsilon = \dfrac{\sigma}{k}$

Hence when $\sigma = 16 \times 10^6$, strain $\epsilon = \dfrac{16 \times 10^6}{200 \times 10^9} = \mathbf{0.000\,08}$

Problem 3. Boyle's law states that at a given constant temperature the volume V of a fixed mass of gas is inversely proportional to its absolute pressure p. If a gas occupies a volume of 0.060 m^3 at a pressure of 1.80×10^6 pascals and at constant temperature, determine (*a*) the coefficient of proportionality, (*b*) the volume if the pressure is changed to 4.0×10^6 pascals and (*c*) the pressure when the volume is 0.045 m^3.

(*a*) $V \propto \dfrac{1}{p}$, i.e. $V = \dfrac{k}{p}$ or $k = pV$

Hence the coefficient of proportionality, $k = (1.80 \times 10^6)(0.060)$

$$= \mathbf{0.108 \times 10^6\ Pa\ m^3}$$

(*b*) Volume $V = \dfrac{k}{p} = \dfrac{0.108 \times 10^6}{4.0 \times 10^6} = \mathbf{0.027\ m^3}$

(*c*) Pressure $p = \dfrac{k}{V} = \dfrac{0.108 \times 10^6}{0.045} = \mathbf{2.4 \times 10^6}$ **pascals.**

Problem 4. The electrical resistance R of a piece of wire is inversely proportional to its cross-sectional area A. When $A = 8$ mm^2, $R = 4.2$ ohms. Determine (*a*) the coefficient of proportionality, (*b*) the cross-sectional area corresponding to a resistance of 3 ohms, and (*c*) the resistance when the cross-sectional area is 10 mm^2.

(*a*) $R \propto \dfrac{1}{A}$ or $R = \dfrac{k}{A}$ or $k = RA$

Hence when $R = 4.2$ and $A = 8$, the coefficient of proportionality, $k = (4.2)(8) = \mathbf{33.6}$

(*b*) Since $k = RA$ then $A = \frac{k}{R}$

When $R = 3$ ohms, the cross-sectional area $A = \frac{33.6}{3} = \mathbf{11.2\ mm^2}$

(*c*) $R = \frac{k}{A}$, hence when $A = 10$, $R = \frac{33.6}{10} = \mathbf{3.36\ ohms.}$

Further problems on direct and inverse proportionality may be found in Section 8.3 (Problems 1–8), page 193.

8.2 Straight line graphs

The usual method of drawing graphs is to use two **axes,** these being two lines drawn mutually at right angles. One line is drawn horizontally and is called the **horizontal axis,** and the other is drawn vertically and is called the **vertical axis.** The point where these axes cross is called the **origin.** The information to be depicted is given in two sets of data, say, set A and set B.

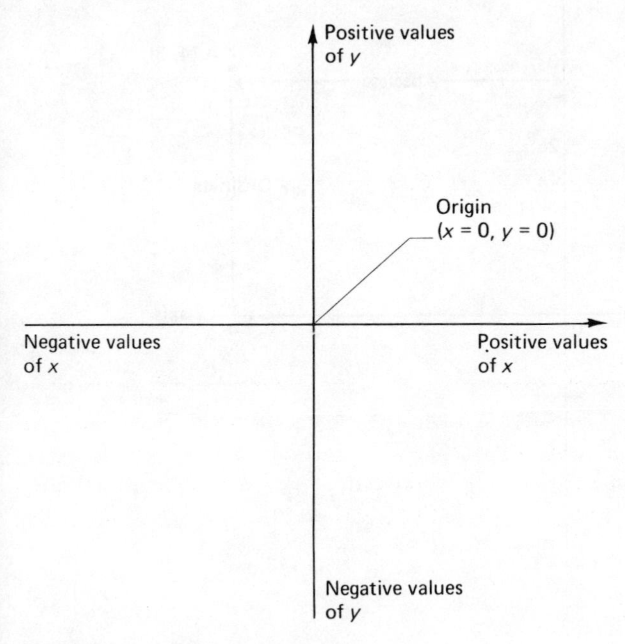

Fig. 8.1 Cartesian or rectangular axes

Figure 8.1 shows the conventions which have been adopted. The values from set A are called x-values and those from set B y-values. The **ordered pairs** (an x-value with its assigned value of y) are called **cartesian coordinates** and the system uses **cartesian** or **rectangular axes.**

A graph drawn on rectangular axes is very much a pictorial representation of information. It is usually drawn on squared paper showing how a quantity (denoted by y) depends on, or varies with a related quantity (denoted by x). Often the corresponding values of x and y are given in tabulated form and the graph is the best straight line joining the system of points plotted on the squared paper.

To analyse two sets of related data (or figures) meaningfully can often take a lot of time, but if the data can be transferred on to a graph an immediate visual analysis is provided.

A **scale** can take many forms. One of these is a set of marks graded at equal distances as on a ruler; that is, the distance from 0 to 1 is the same as the distance from 1 to 2 and so on. Such a scale is called **linear.**

The scales on each of the rectangular axes are chosen so that each space is equal to 1, 2, 4, 5, 10 or any other easily divisible unit. Scales where each space is equal to 3, 7, 11, 13 or similar units usually prove extremely awkward

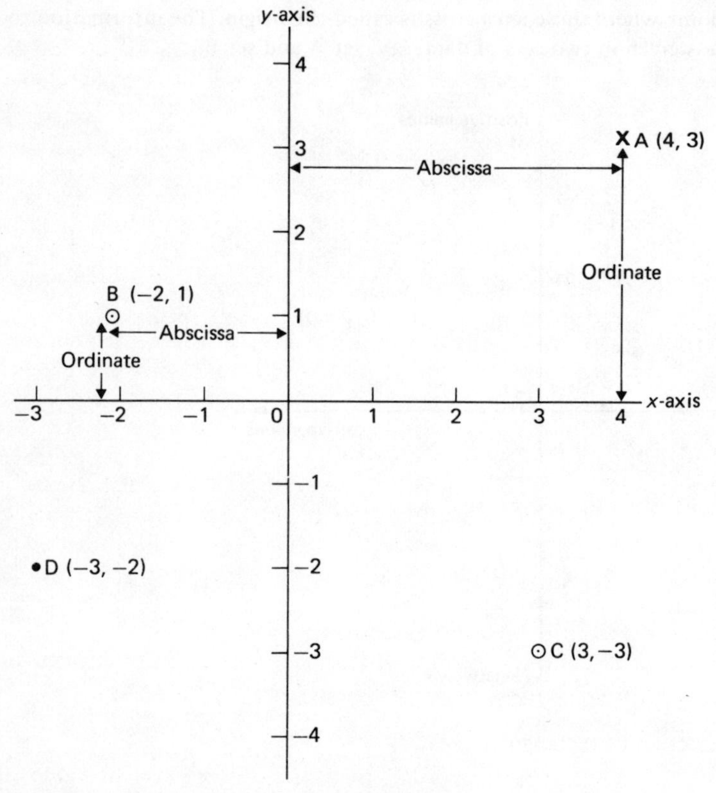

Fig. 8.2 Coordinates on rectangular axes

and should be avoided since these numbers cannot be readily divided into smaller parts. The scale chosen for the x-axis need not be the same as the scale chosen for the y-axis. Scales are usually chosen such that the graph occupies as much of the squared paper as possible, since the larger the graph the greater the accuracy.

The points on a graph are called **coordinates.** To specify a point on a graph two pieces of information are needed; the value of x, the horizontal component, and the value of y, the vertical component. If x is 4 units and the corresponding value of y is 3 units then the point or coordinate is termed (4,3). Note that the x value is always specified first. In Fig. 8.2 rectangular axes x and y are shown. The coordinate A is shown at (4,3) and marked by a cross. The coordinates B, shown at (−2, 1), and C, shown at (3, −3), are marked by a dot and circle. Coordinate D, shown at (−3, −2), is marked by a dot only. All three methods may be used to denote coordinates on graphs.

The x distance of a point is called the **abscissa** and the y value the **ordinate.** Thus for the point A, the abscissa is 4 and the ordinate 3; for the point D, the abscissa is −3 and the ordinate −2.

Problem 1. The table below shows the temperature in degrees Celsius and the corresponding values in degrees Fahrenheit. Draw rectangular axes, choose a suitable scale and plot the graph of degrees Celsius (on the horizontal axis) against degrees Fahrenheit (on the vertical axis).

°C	10	25	40	65	80	100
°F	50	77	104	149	176	212

The coordinates (10,50), (25,77), (40,104), and so on, are marked on the squared paper. If the coordinates are now joined, a straight line is produced as shown in Fig. 8.3.

Since the resulting graph is a straight line, this shows that a linear relationship exists; that is, degrees Celsius are directly proportional to degrees Fahrenheit.

Interpolation and extrapolation

However, there is more information that can be obtained from the graph in Problem 1 above. For example, if the temperature in degrees Fahrenheit is required at 50°C this can be found from the graph. If the table given above was the only information available then all that could be said was that 50°C lies somewhere between 104°F and 149°F. However, the actual value may be found from the graph as follows:

1. The vertical broken line AB at 50°C cuts the graph at B.
2. The point where the horizontal line BD cuts the vertical axis indicates the corresponding temperature in degrees Fahrenheit, that is 122°F.

Hence 50°C ≙ 122°F. (The sign, ≙, means ‘corresponds to’.)

Similarly the reverse process, that of finding an equivalent value of temperature in degrees Celsius to a given Fahrenheit temperature, can be read

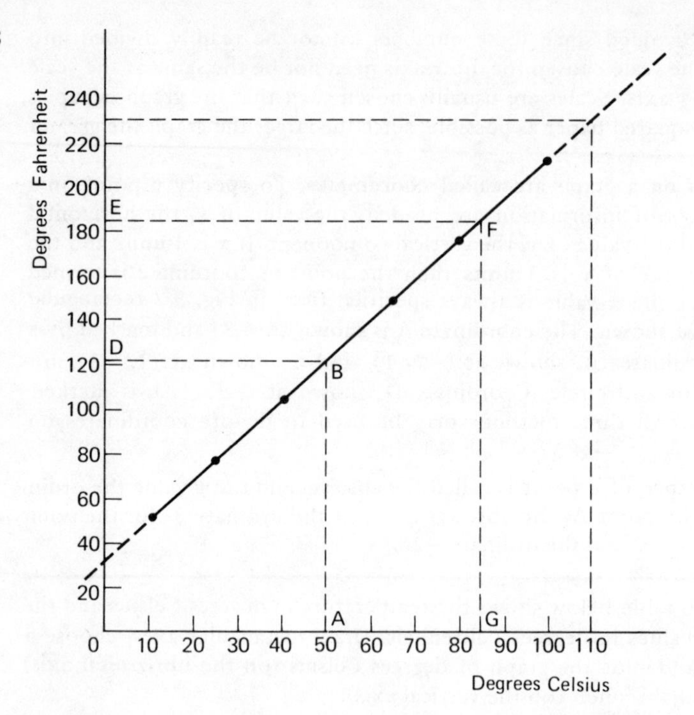

Fig. 8.3 Graph of degrees Celsius against degrees Fahrenheit

straight from the graph. Thus, to find the Celsius equivalent to 185° F, the horizontal line EF is drawn at 185°F which cuts the graph at F. Then the point where the vertical line FG cuts the horizontal axis represents the equivalent temperature in Celsius, that is, 85° C.

This process of finding equivalent values in between the given information is called **interpolation.**

If the temperature in Fahrenheit is required at, say, 0°C this can be found even though it lies outside the given information. If the graph is assumed to be linear even outside the given data then the graph may be extended at both ends, and this is shown by the broken lines in Fig. 8.3. (Such an assumption cannot always be made, since although a graph may be linear over the part being depicted, it could change its shape outside the range shown.) Thus at 0°C the equivalent value in Fahrenheit is 32°F. Similarly, at 230° F for example, the equivalent value of Celsius temperature is 110°C.

This process of finding equivalent values which are outside a given range of values is called **extrapolation.**

Axes where zero values are not at the origin

Whenever possible scales are chosen so that the zero value corresponds with the origin. Sometimes, however, this results in an accumulation of points in one corner of the graph paper. Whenever this happens it is best not to start the scales at zero. This is illustrated by Problem 2.

Problem 2. The resistance R ohms of a copper winding is measured at various temperatures t°C and the results are recorded in the following table:

R ohms	127.5	129.4	131.7	134.1	136
t°C	50	54	59	64	68

Plot a graph of resistance against temperature and find from it
(*a*) the temperature when the resistance is 130 ohms, and
(*b*) the resistance when the temperature is 66°C.

If a graph is plotted with both the scales starting at zero at the origin, then the most appropriate scales would be as shown in Fig. 8.4.

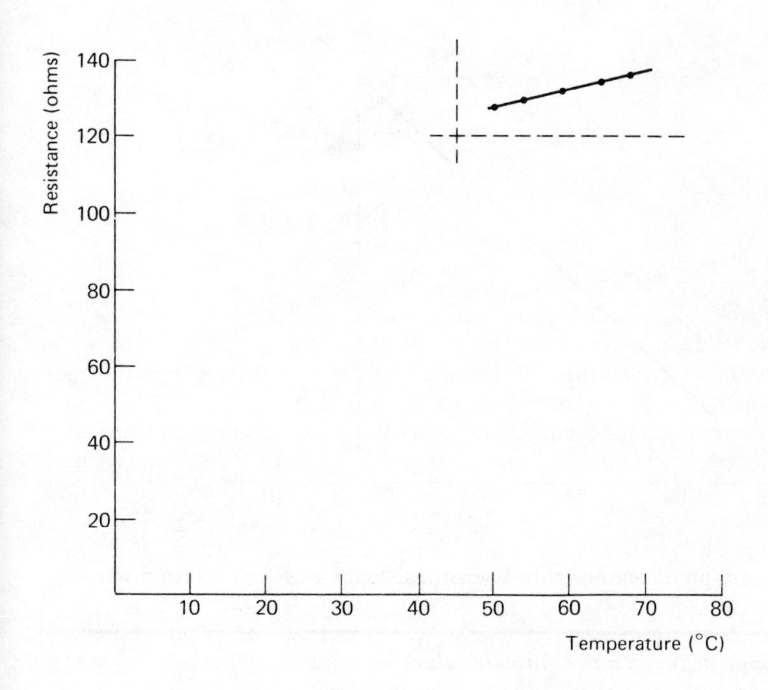

Fig. 8.4 Graph of temperature against resistance with cramped scale

It will be seen that all of the points fall in the top right-hand side of the graph paper, making interpolation difficult. A much more accurate graph is obtained if the x-axis (temperature) starts at 45°C and the y-axis (resistance) starts at 125 ohms. The axes corresponding to these values are shown by a broken line in Fig. 8.4 and are called **false axes** since the origin is not now at zero.

Choosing more appropriate scales the graph shown in Fig. 8.5 is produced. This is, in fact, a magnified version of the relevant part of the graph shown in Fig. 8.4.

(*a*) When the resistance is 130 ohms the equivalent value of temperature is 55° C.

(*b*) When the temperature is 66°C the equivalent value of resistance is 135 ohms.

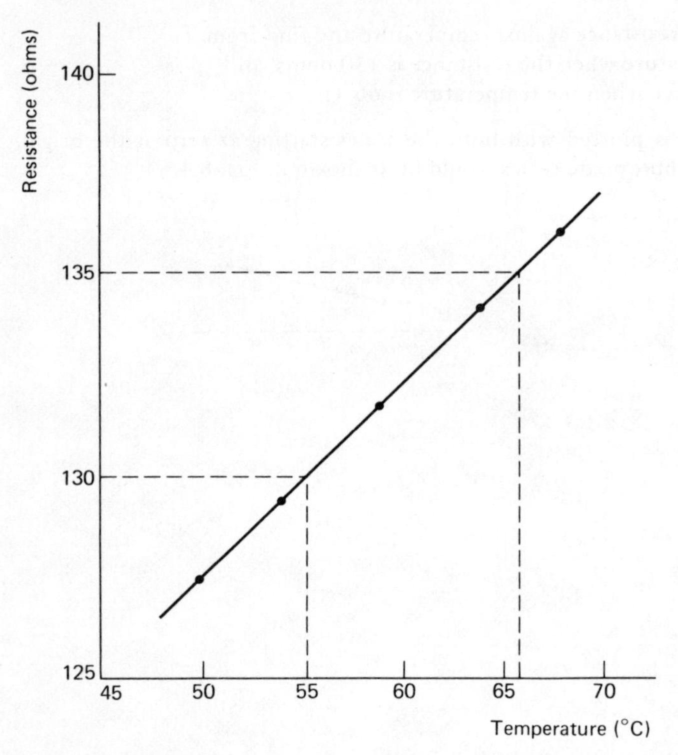

Fig. 8.5 Graph of temperature against resistance with a non-zero origin

Graphs involving negative coordinate values

In Problems 1 and 2 it was necessary to draw only the part of the rectangular axes which had positive values of x and y. When negative values of x and y are involved it is necessary to draw the four quadrants representing the complete rectangular axes. This is illustrated in Problem 3.

Problem 3. Draw a graph of $y = 3x + 2$ between the range of values $x = -3$ to $x = +5$. Hence find
(*a*) the value of y when $x = 2.5$ (*b*) the value of x when $y = -5.5$.

Now $y = 3x + 2$ represents a simple equation and so for every value of x there is a corresponding value of y. For example, when $x = 0$, $y = 3(0) + 2 = 2$. Similarly, when $x = 1$, $y = 3(1) + 2 = 5$ and when $x = -2$, $y = 3(-2) + 2 = -6 + 2 = -4$, and so on. In this way a table of values may be drawn up as follows:

x	−3	−2	−1	0	1	2	3	4	5
y	−7	−4	−1	2	5	8	11	14	17

When only the equation is given, such a table, giving corresponding values of the two variables, is necessary.

The coordinates (−3,−7), (−2,−4), (−1,−1), and so on, are plotted and joined together to produce the straight line as shown in Fig. 8.6.

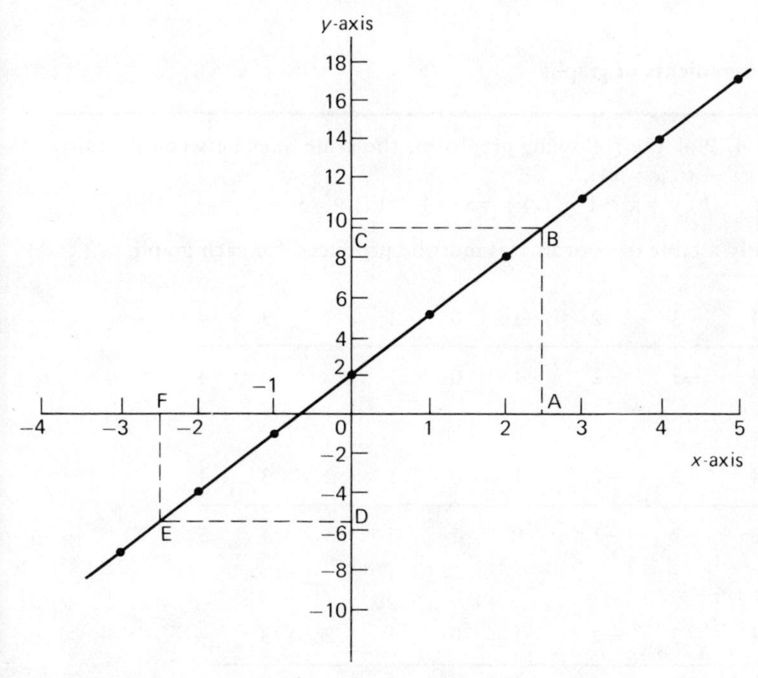

Fig. 8.6 Graph of $y = 3x + 2$

(*a*) **When $x = 2.5$, the value of y is 9.5.** This is obtained by drawing the vertical AB at $x = 2.5$ and reading off directly from the graph the y ordinate of BC, that is 9.5.

(*b*) **When $y = -5.5$, the value of x is -2.5.** This is obtained by drawing the horizontal ED at $y = -5.5$ and reading off directly from the graph the x value corresponding to EF, that is, −2.5.

The answers to (*a*) and (*b*) may be checked by calculation.

(*a*) $y = 3x + 2$

Therefore, when $x = 2.5$

$y = 3(2.5) + 2$

$y = 7.5 + 2$

$y = 9.5.$

(*b*) If $y = 3x + 2$

then $3x = y - 2$

and $x = \frac{y - 2}{3}$

Therefore, when $y = -5.5$

$x = \frac{-5.5 - 2}{3}$

$x = \frac{-7.5}{3}$

$x = -2.5$

Slopes or gradients of graphs

Problem 4. Plot the following graphs on the same axes between the range of values $x = -4$ to $x = +4$:
(*a*) $y = x$ (*b*) $y = x + 1$ (*c*) $y = x + 4$ (*d*) $y = x - 2$

Firstly a table of coordinates must be produced for each graph.

(*a*) $y = x$

x	−4	−3	−2	−1	0	1	2	3	4
y	−4	−3	−2	−1	0	1	2	3	4

(*b*) $y = x + 1$

x	−4	−3	−2	−1	0	1	2	3	4
y	−3	−2	−1	0	1	2	3	4	5

(*c*) $y = x + 4$

x	−4	−3	−2	−1	0	1	2	3	4
y	0	1	2	3	4	5	6	7	8

(*d*) $y = x - 2$

x	−4	−3	−2	−1	0	1	2	3	4
y	−6	−5	−4	−3	−2	−1	0	1	2

The coordinates are plotted and joined for each graph. The result is shown in Fig. 8.7.

It will be noticed that the straight lines shown are parallel to each other, that is, the **slope or gradient** is the same for each. The gradient of a straight line may be calculated as shown by the following example.

Let OA represent a horizontal surface and BC a flat piece of metal hinged at C. This is illustrated in Fig. 8.8.
As BC lies flat on OA, BC is parallel to OA. The slope or gradient of BC in this position is zero, that is, BC has no gradient.

In Fig. 8.9, B has been raised.

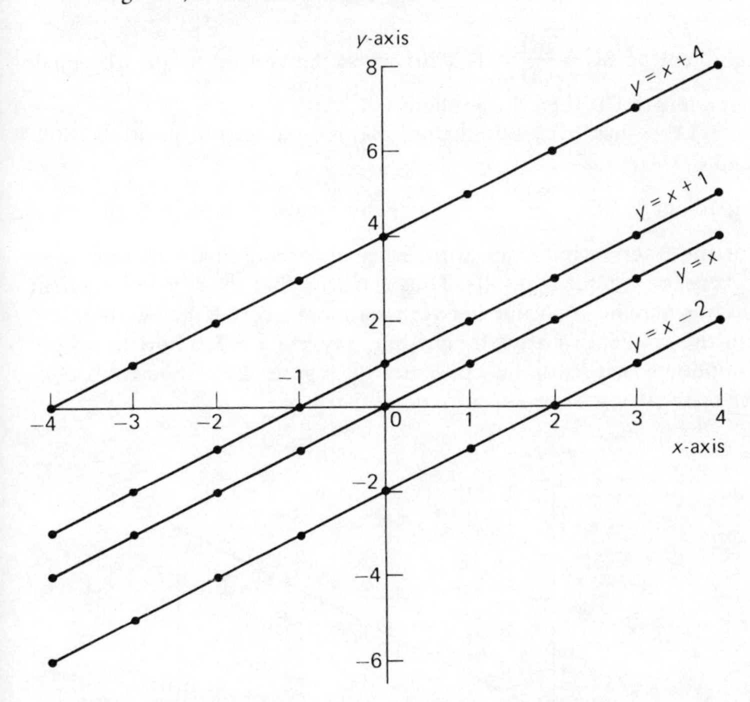

Fig. 8.7 Graph of $y = x$, $y = x + 1$, $y = x + 4$ and $y = x - 2$

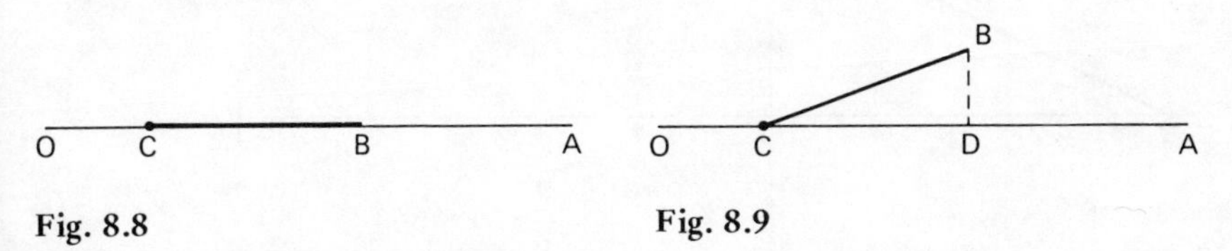

Fig. 8.8 **Fig. 8.9**

If the broken line BD is vertical to OA then the **slope or gradient of BC is defined as the ratio** $\mathbf{\frac{BD}{CD}}$. Hence if BD is 1 cm in length and CD is 10 cm in length then the gradient would be $\frac{1}{10}$. In other words BC rises 1 unit vertically in 10 units horizontally.

In Fig. 8.10, B is raised until BD = CD.

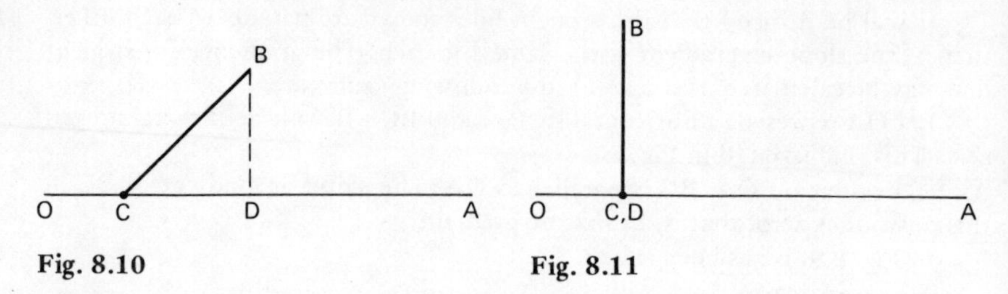

Fig. 8.10 **Fig. 8.11**

Then the gradient of BC $= \frac{BD}{CD} = 1$. Thus when the vertical height BD equals the horizontal length CD then the gradient is 1.

In Fig. 8.11, B has been raised until BC is in a vertical position. Now BC = BD and CD = 0.

Hence the gradient of BC $= \frac{BD}{CD} = \frac{BD}{0}$. As a denominator of a fraction decreases the value of the fraction increases until, when the denominator is zero, as in this case, a very large number results. Thus it is seen that the slope or gradient of a straight line may be any value between zero and a very large number.

To find the gradient of any straight line, say $y = x + 3$, a horizontal and vertical component needs to be constructed. Figure 8.12 shows AB constructed vertically at $x = 4$ and BC horizontally at $y = 3$.

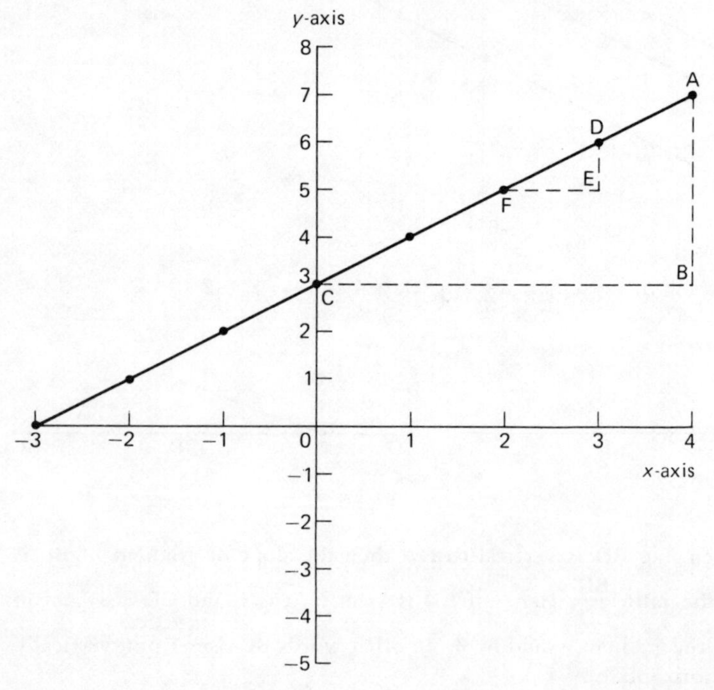

Fig. 8.12 Graph of $y = x + 3$

Thus the gradient of the straight line, that is, the gradient of AC is $\frac{AB}{BC}$.

Therefore its gradient is

$$\frac{AB}{BC} = \frac{7-3}{4-0} = \frac{4}{4} = 1$$

Thus the gradient of the straight line $y = x + 3$ is 1.

The positioning of AB and BC is unimportant. It would be equally correct in Fig. 8.12 to make the vertical component at DE and the horizontal component at EF. However, the greater the length involved the greater the accuracy. Then the gradient is

$$\frac{DE}{EF} = \frac{6-5}{3-2} = \frac{1}{1} = 1$$

The position of these components generally is best determined by trying to make the denominator an integer.

In Problem 4 the gradient of each of the straight lines can be shown to be unity.

Problem 5.

(*a*) Plot the following graphs on the same axes between the range of values $x = -3$ to $x = +3$ and find the gradient of each.

(*i*) $y = 2x$ (*ii*) $y = 2x + 3$ (*iii*) $y = 2x - 5$

(*b*) Plot the following graphs on the same axes between the range of values $x = -3$ to $x = +3$ and find the gradient of each.

(*i*) $y = 5x$ (*ii*) $y = 5x + 2$ (*iii*) $y = 5x - 6$

(*a*) A table of coordinates is firstly drawn up for each equation:

(*i*) $y = 2x$

x	−3	−2	−1	0	1	2	3
y	−6	−4	−2	0	2	4	6

(*ii*) $y = 2x + 3$

x	−3	−2	−1	0	1	2	3
y	−3	−1	1	3	5	7	9

(*iii*) $y = 2x - 5$

x	−3	−2	−1	0	1	2	3
y	−11	−9	−7	−5	−3	−1	1

Plotting and joining the coordinates, the graphs shown in Fig. 8.13 are produced.

Each of the graphs are straight lines and are parallel to each other.

If the gradient of the graph $y = 2x$ is found, this will also be the gradient of $y = 2x + 3$, and $y = 2x - 5$, since parallel lines have the same gradient.

Hence the gradient of AC $= \frac{AB}{BC} = \frac{6}{3} = 2$

 Therefore **for each of the straight lines $y = 2x$, $y = 2x + 3$ and $y = 2x - 5$ the gradient is 2.**

(*b*) A table of coordinates is again drawn up for each equation:

(*i*) $y = 5x$

x	−3	−2	−1	0	1	2	3
y	−15	−10	−5	0	5	10	15

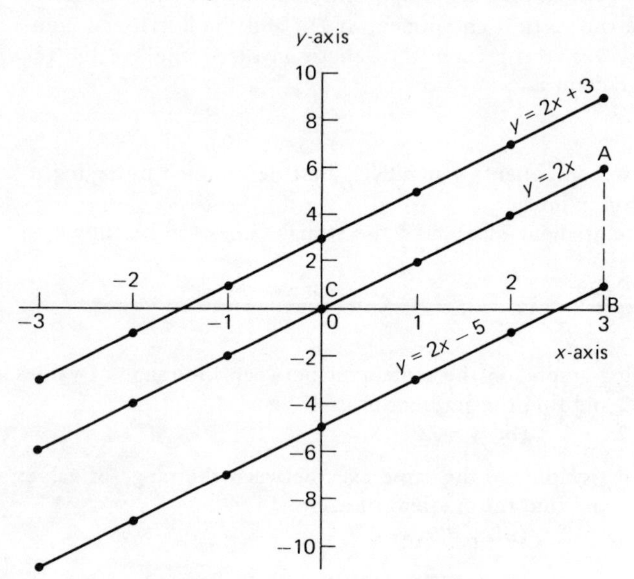

Fig. 8.13 Graphs of $y = 2x$, $y = 2x + 3$ and $y = 2x - 5$

(*ii*) $y = 5x + 2$

x	−3	−2	−1	0	1	2	3
y	−13	−8	−3	2	7	12	17

(*iii*) $y = 5x - 6$

x	−3	−2	−1	0	1	2	3
y	−21	−16	−11	−6	−1	4	9

Plotting and joining the coordinates, the graphs shown in Fig. 8.14 are produced.

The graphs are straight lines and are parallel to each other.

If the gradient of the graph $y = 5x$ is found, this will also be the gradient of $y = 5x + 2$ and $y = 5x - 6$.

Hence the gradient of $DF = \frac{DE}{EF} = \frac{15}{3} = 5$

Therefore **for each of the straight lines $y = 5x$, $y = 5x + 2$, and $y = 5x - 6$ the gradient is 5.**

Important properties relating to straight-line graphs can be deduced from problems 4 and 5.

When an equation is of the form $y = mx + c$ (where m and c are constants) then

(*i*) a graph of y against x produces a straight line;
(*ii*) m represents the slope or gradient of the line with the x-axis; and

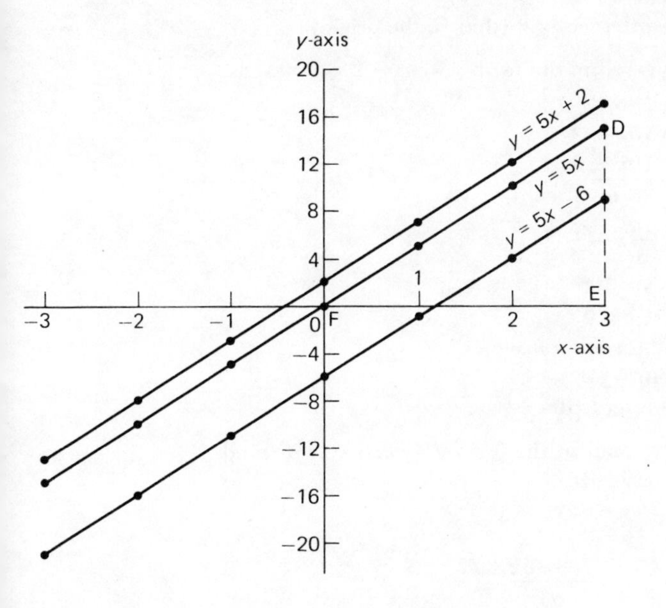

Fig. 8.14 Graphs of $y = 5x$, $y = 5x + 2$, $y = 5x - 6$

(*iii*) c represents the value of y where the straight line cuts the y-axis. This is known as the intercept. (This is only true when the x scale is zero at the origin.)

When plotting a graph of the form $y = mx + c$, only two coordinates need be determined. When the coordinates are plotted a straight line is drawn between the two points. Normally, however, three coordinates are determined, the third one acting as a check.

If an equation with two variables can be arranged into the straight-line form $y = mx + c$ then the slope and the intercept with the y-axis can be readily found. An equation such as $y = 4$ represents a horizontal straight line intercepting the y-axis at 4. (That is, the line has no gradient with the x-axis.) An equation such as $x = 3$ represents a vertical straight line intercepting the x-axis at 3.

Problem 6. The following equations represent straight lines. Find the gradient to the x-axis and the y-axis intercept for each.
(*a*) $y = 7x + 2$ (*b*) $y = -4x$ (*c*) $3 = 10x - 6y$ (*d*) $7 - 63y = 36x$

(*a*) $y = 7x + 2$ is of the form $y = mx + c$. Hence
m = **the gradient = 7**
c = **the y-axis intercept = 2**

(*b*) $y = -4x$ is of the form $y = mx + c$ where c is zero.
Hence: **the gradient = −4**
the y-axis intercept = 0 (that is, the origin)

(*c*) $3 = 10x - 6y$ is not in the form $y = mx + c$ as it stands.
Therefore if $3 = 10x - 6y$
then $6y = 10x - 3$

$$y = \frac{10x - 3}{6}$$

$$y = \frac{10}{6}x - \frac{3}{6}$$

$$y = \frac{5}{3}x - \frac{1}{2}$$

This is now of the form $y = mx + c$.
Hence: **the gradient = $\frac{5}{3}$**
the y-axis intercept = $-\frac{1}{2}$

(*d*) $7 - 63y = 36x$ is not of the form $y = mx + c$ as it stands.
Therefore if $7 - 63y = 36x$
then $7 - 36x = 63y$
or $63y = 7 - 36x$

Therefore $$y = \frac{7 - 36x}{63}$$

$$y = \frac{7}{63} - \frac{36}{63}x$$

$$y = \frac{1}{9} - \frac{4}{7}x$$

$$y = -\frac{4}{7}x + \frac{1}{9}$$

This is now of the form $y = mx + c$

Hence: **the gradient = $-\frac{4}{7}$**

the y-axis intercept = $\frac{1}{9}$

Problem 7. Plot a graph of $2y - 5x = 14$ between the range $x = -3$ and $x = +3$. Find from the graph:
(*a*) the slope to the x-axis (*b*) the value of y when x is $1\frac{1}{2}$
(*c*) the value of x when y is $-3\frac{1}{2}$
Check the answers by calculations.

If $2y - 5x = 14$
then $2y = 5x + 14$

Therefore $y = \frac{5}{2}x + 7$

$y = 2.5x + 7$

A table of coordinates can now be produced:

x	-3	-2	-1	0	1	2	3
y	-0.5	2	4.5	7	9.5	12	14.5

Plotting and joining the coordinates the straight-line graph shown in Fig. 8.15 is produced.

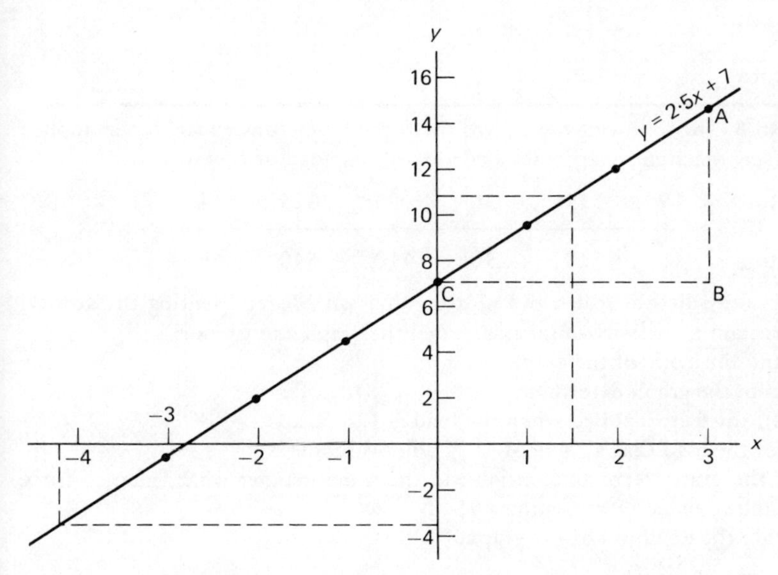

Fig. 8.15 Graph of $2y - 5x = 14$ (or $y = 2.5x + 7$)

(*a*) From the graph, the gradient or slope of AC is $\frac{AB}{BC}$.

Hence **the slope of graph** $= \frac{14\frac{1}{2} - 7}{3 - 0} = \frac{7.5}{3} = \mathbf{2.5}$

This may be checked from the fact that $y = 2.5x + 7$ is of the form $y = mx + c$. Therefore the coefficient of x represents the gradient, that is, 2.5.

(*b*) From the graph, when x is 1.5, the value of y is **10.8**. (The decimal part can only be an estimation.)

By calculation: $y = 2.5x + 7$

therefore $y = (2.5)(1.5) + 7$

$= 3.75 + 7$

$y = 10.75$

 (*c*) From the graph, when y is -3.5, the value of x is **-4.2**. (This may be read from the graph by extrapolating.)

By calculation: $y = 2.5x + 7$

therefore $y - 7 = 2.5x$

$$2y - 14 = 5x$$
$$5x = 2y - 14$$
$$x = \frac{2y - 14}{5}$$

When $y = -3.5$ $\quad x = \frac{2(-3.5) - 14}{5}$

$$= \frac{-7 - 14}{5}$$
$$= \frac{-21}{5}$$

Therefore $\quad x = -4.2$

Problem 8. The following table gives the force F newtons which, when applied to a lifting machine, overcomes a corresponding load of L newtons.

F newtons	19	35	50	93	125	147
L newtons	40	120	230	410	540	680

(*a*) Choose suitable scales and plot a graph with F representing the vertical axis and L the horizontal axis. Label the graph and its axes.
(*b*) Find the slope of the graph.
(*c*) From the graph determine
  (*i*) the force applied when the load is 325 N, and
  (*ii*) the load that a force of 45 N will overcome.
(*d*) If the graph were to continue in the same manner what value of force will be needed to overcome a 750 N load.
(*e*) State the equation of the graph.

(*a*) Plotting the coordinates (40,19), (120,35), (230,50), and so on, the graph produced is the best straight line which can be drawn corresponding to these points, as shown in Fig. 8.16. The graph and each of its axes are labelled.
(*b*) To find the slope of the graph, horizontal and vertical components are constructed. Such components may be positioned anywhere along the length of the straight line. The horizontal component AB is chosen so that it is a number which makes any calculations easier, in this case 500. Thus the slope of graph = the slope of AC = $\frac{BC}{AB}$. Therefore the slope of the graph is

$$\frac{130 - 30}{600 - 100} = \frac{100}{500} = \frac{1}{5} \text{ or } 0.2$$

Hence the slope of the graph is **0.2**

(*c*) (*i*) When the load is 325 N the corresponding force is **75 N.**
  (*ii*) When the force is 45 N the corresponding load is **170 N.**
Each of these can be seen from the broken lines shown in Fig. 8.16.

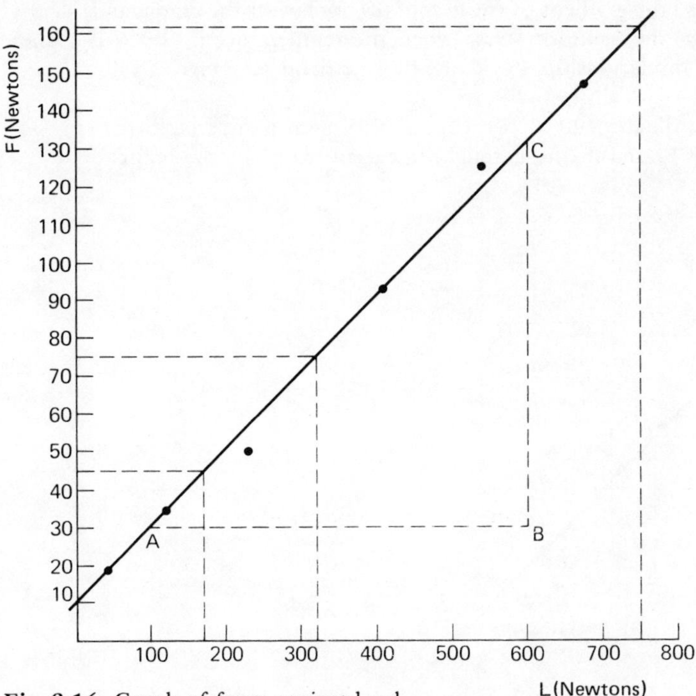

Fig. 8.16 Graph of force against load

(*d*) The graph is extended at its upper end and a load of 750 N corresponds to a force of **161 N.**

(*e*) The equation of a straight line is $y = mx + c$ where y is plotted on the vertical axis and x on the horizontal axis. Also m represents the slope and c the y-axis intercept. Thus, in this case:
Force F corresponds to y,
Load L corresponds to x,
m, the slope is 0.2,
c, the y-axis intercept is 11 (by extrapolation).
Hence the equation of the graph is $\mathbf{F = 0.2L + 11}$.

Problem 9. Results obtained from determining the breaking stress σ of rolled copper at various temperatures, t, are as follows:

Stress, σ ($\times 10^4$ Pa)	8.28	7.99	7.70	7.48	7.30	7.03
temperature, t (°C)	90	205	320	410	480	590

(*a*) Plot a graph of stress (vertical) against temperature (horizontal), drawing the best straight line through the plotted coordinates.

 Determine (*b*) the gradient of the graph, (*c*) the law of the graph, and (*d*) the value of stress when the temperature is 700°C if a linear relationship is assumed outside the given range of values.

(*a*) The coordinates (90, 8.28), (205, 7.99), and so on, are plotted as shown in Fig. 8.17 and the best straight line drawn through the points.

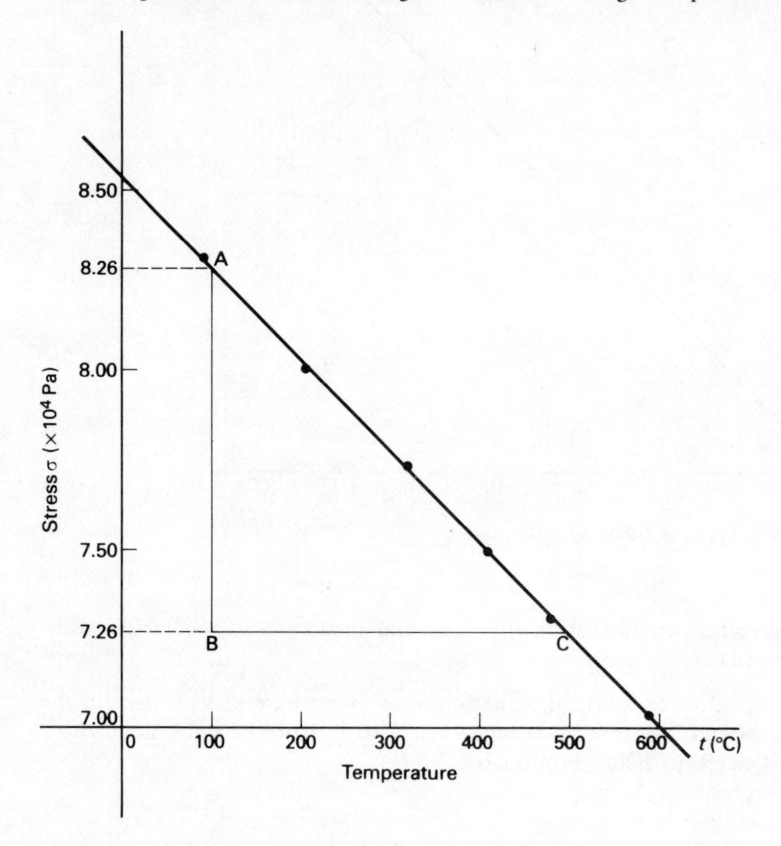

Figure 8.17

(*b*) Gradient of the straight line $= \dfrac{AB}{BC} = \dfrac{(8.26 - 7.26)10^4}{100 - 500} = \mathbf{(-0.002\ 5) \times 10^4}$

(*c*) Intercept on the vertical axis $= 8.52 \times 10^4$

Hence the equation of the graph is $\sigma = \mathbf{(-0.002\ 5\ t + 8.52)10^4}$

(*d*) When $t = 700$°C, the stress $\sigma = (-0.002\ 5 \times 700 + 8.52)10^4$

$= (-1.75 + 8.52)10^4 = \mathbf{6.77 \times 10^4}$ **Pa.**

Further problems on graphs may be found in Section 8.3 (Problems 9–26), page 194.

Summary

The following general rules can be applied to any graph:

1. Give the graph a title, clearly explaining what is being illustrated.
2. The scales are chosen such that the graph occupies as much space as possible on the graph paper being used, consistent with simplicity of axes.
3. Scales are chosen so that interpolation is made as easy as possible. Usually, scales such as 1 cm = 1 unit, or 1 cm = 2 units or 1 cm = 10 units are used. Awkward scales such as 1 cm = 3 units or 1 cm = 7 units are not used.
4. The scales should not start at zero if starting at zero produces an accumulation of points within a small area of the graph paper.
5. The coordinates, or points, should be clearly marked. This may be done either by a cross, by a dot and circle or by a dot only.
6. A statement should be made along each axis explaining the numbers represented with their appropriate units.
7. Sufficient numbers should be written along each axis without cramping.
8. If the equation of a graph is of the form $y = mx + c$ the graph will always be a straight line, where m represents the slope or gradient and c the intercept with the y-axis, provided false zeros have not been introduced. Do not join each point in turn but draw the best straight line through the coordinates.
9. To find the gradient of a straight line graph a horizontal and a vertical component is constructed for any two points on the straight line. The gradient is then given by the value of the vertical component divided by the horizontal component.

8.3 Further problems

Direct and inverse proportionality

1. If y is directly proportional to x and $y = 45$ when $x = 7.2$, determine (*a*) the coefficient of proportionality and (*b*) the value of y when x is 10. (*a*) [6.25] (*b*) [62.5]
2. Charles' law states that for a given mass of gas at constant pressure the volume is directly proportional to its thermodynamic temperature. A gas occupies a volume of 4.5 dm^3 at 300 K. Determine, for constant pressure, (*a*) the coefficient of proportionality, (*b*) the volume at 380 K and (*c*) the temperature when the volume is 4.8 dm^3. (*a*) [0.015] (*b*) [5.7 dm^3] (*c*) [320 K]
3. Hooke's law states that stress is directly proportional to strain within the elastic limit of a material. A spring has a coefficient of proportionality k of 150 N m^{-2}. (*a*) Determine the load required to extend it by

20 mm. (*b*) What extension corresponds to 13.5 N? (*a*) [3 N] (*b*) [90 mm]

4. 0.002 m^3 of aluminium has a mass of 5.3 kg. If mass is directly proportional to volume determine (*a*) the mass of 0.001 2 m^3 of aluminium and (*b*) the volume of 8.48 kg of aluminium. (*a*) [3.18 kg] (*b*) [0.003 2 m^3]
5. Ohm's law states that the current flowing in a resistor is directly proportional to the applied voltage. When 40 volts is applied across a resistor the current flowing through the resistor is 3.2×10^{-3} amperes. Determine (*a*) the coefficient of proportionality, (*b*) the current when the voltage is 56 volts, and (*c*) the voltage when the current is 5.4×10^{-3} amperes. (*a*) [8×10^{-5}] (*b*) [4.48×10^{-3} A] (*c*) [67.5 V]
6. The distillation rate, R (in cm^3/min), of a given gas is inversely proportional to the temperature, T(K). If $R = 3$ when $T = 5$, (*a*) what is the rate when $T = 4$, and (*b*) at what temperature is $R = 5$? (*a*) [3¾ cm^3/min] (*b*) [3K]
7. Boyle's law states that for a gas at constant temperature, the volume of a fixed mass of gas is inversely proportional to its absolute pressure. If a gas occupies a volume of 2.25 m^3 at a pressure of 2×10^5 pascals determine for constant temperature (*a*) the coefficient of proportionality, (*b*) the volume when the pressure is 7.5×10^5 pascals, and (*c*) the pressure when the volume is 1.5 m^3. (*a*) [4.5×10^5] (*b*) [0.6 m^3] (*c*) [3×10^5 Pa]
8. The resistor R of a length of cable is inversely proportional to its cross-sectional area, a. When $a = 6$ mm^2, $R = 5.5$ ohms. Determine (*a*) the coefficient of proportionality, (*b*) the cross-sectional area corresponding to a resistance of 2 ohms, and (*c*) the resistance when the cross-sectional area is 12 mm^2. (*a*) [33] (*b*) [16.5 mm^2] (*c*) [2.75 ohms]

Straight line graphs

9. Assuming graph paper measuring 20 cm by 20 cm is available, suggest suitable scales for the following ranges of values:
   (*a*) Horizontal axis: 3 volts to 54 volts. [1 cm rep. 4 volts]
   Vertical axis: 10 ohms to 180 ohms. [1 cm rep. 10 ohms]
   (*b*) Horizontal axis: 7 cm to 85 cm. [1 cm rep. 5 cm]
   Vertical axis: 0.30 volts to 1.68 volts. [1 cm rep. 0.1 volt]
   (*c*) Horizontal axis: 210 N to 340 N.
   [1 cm rep. 10 N (Note: not starting at zero)]
   Vertical axis: 0.6 mm to 3.3 mm. [1 cm rep. 0.2 mm]
10. Corresponding values obtained experimentally for two quantities are:

x	−2	−1	0	1	2.5	4	5
y	−12	−7	−2	3	10.5	18	23

Using a horizontal scale for x of 1 cm = $\frac{1}{2}$ unit and a vertical scale for y of 1 cm = 2 units, draw a graph of x against y. Label the graph and each of its axes. By interpolation, find from the graph the value of y when x is 1.5 [5.5]

11. The equation of a line is $y = 3x + 2$. A table of corresponding values is produced and is shown below. Complete the table and plot a graph of y against x. Find the gradient of the graph.

x	-4	-3	-2	-1	0	1	2	3	4
y		-7			2		8		

[Gradient = 3]

12. Find the gradient and intercept on the y-axis for each of the following equations:
(*a*) $y = 3x + 4$ (*b*) $y = -2x + 3$ (*c*) $y = -5x - 1$ (*d*) $y = 6x$
(*a*) [3, 4] (*b*) [−2, 3] (*c*) [−5, −1] (*d*) [6, 0]

13. Find the gradient and the y-axis intercept for each of the following equations:

(*a*) $3y + 1 = 2x$ (*b*) $5x + 2y = -4$ (*c*) $4(2y - 3) = \frac{x}{2}$

(*d*) $3x - \frac{y}{2} = 5\frac{1}{2}$ (*e*) $y = 5$

Sketch the graph in each case showing the slope and y-axis intercept.
(*a*) $[\frac{2}{3}, -\frac{1}{3}]$ (*b*) $[-2\frac{1}{2}, -2]$ (*c*) $[\frac{1}{16}, 1\frac{1}{2}]$ (*d*) [6, −11] (*e*) [0, 5]

14. Draw on the same axes the graphs of $y = 3x - 1$ and $y + 2x = 4$. Find the coordinates of the point of intersection. [1, 2]

15. In an experiment on Charles' Law, the value of the volume of gas $V\text{m}^3$ was measured for various temperatures T°C. The results are tabulated below:

$V\text{m}^3$	26.68	27.16	27.56	28.08	28.52	29.00
T°C	10	20	30	40	50	60

Choose suitable scales and plot a graph of volume (vertically) against temperature (horizontally). Interpolate to find the volume at 35°C. Extrapolate to find the temperature when the volume is 30.0 m^3, assuming the relationship holds. [27.85 m^3, 81.7°C]

16. Plot the graphs of $y - 6.5 = 2x$ and $3x + 1 = -y$ and find the value of x and y at the point of intersection. Check the result obtained by solving the two simultaneous equations algebraically. [−1.5, 3.5]

17. In an experiment demonstrating Hooke's Law, the strain in a wire was measured for various stresses. The results are shown below:

Stress N/mm²	10.8	21.6	33.3	37.8	45.9
Strain	0.000 12	0.000 24	0.000 37	0.000 42	0.000 51

Using scales of 1 cm = 5 N/mm² stress and 2 cm = 0.000 1 strain and plotting stress on the vertical axis, draw a graph of stress against strain.

Find Young's Modulus of elasticity which is given by the gradient of the graph. [90 kN/mm^2]

18. The speed n rev/min of a motor changes when the voltage V across the armature is varied. The results are shown in the table below:

n rev/min	560	720	900	1 010	1 240	1 440
V volts	80	100	120	140	160	180

It is suspected that one of the readings taken of the speed is inaccurate. Plot a graph of speed (horizontally) against voltage (vertically) and find this value. Find also (*a*) the speed at a voltage of 132 V and (*b*) the voltage at a speed of 1 300 rev/min.
[The 1 010 rev/min reading should be 1 088 rev/min (*a*) 1 018 rev/min (*b*) 164 V]

19. A test on a metal filament lamp gave the following values of resistance R ohms at various voltages V volts.

V volts	131		195	221	255	296	355
R ohms	78	95	121		162	190	230

Draw a graph of voltage (horizontally) against resistance (vertically) and find the missing values in the table.
[V = 156 V at 95 ohms; R = 139 ohms at 221 V]

20. Tests carried out to determine the breaking stress S (Ncm^{-2}) of rolled copper at different temperatures T(°C) gave the following results:

S Ncm^{-2}	8.90	8.45	8.20	7.70	7.50	7.10
T °C	60	210	300	410	500	600

For the graph paper you have available, choose as big a scale as possible with stress on the vertical axis and temperature on the horizontal axis. Draw the best straight line through the plotted coordinates. Find the slope of the graph and the intercept with the y-axis. [$-\frac{1}{300}$, 9.1]

21. The velocity v ms^{-1} of a body at the end of a time interval t seconds was found by a series of experiments to be as tabulated below:

t seconds	1	3	6	8	11	12
v ms^{-1}	13.8	15.4	17.8	19.4	21.8	22.6

By plotting v vertically and t horizontally draw the graph of velocity against time. Find the equation of the graph. [$v = 0.8t + 13$]

22. The solubility of potassium bromide in g per 100 g of water varies with temperature in degrees Celsius as shown below.

Temperature (t)	10	20	30	40	50	60	70	80
Mass of potassium bromide (x)	2.97	3.23	3.50	3.77	4.04	4.30	4.57	4.8

Show that these results obey a law of the form $t = mx + c$ where m and c are constants. Determine m and c. [$t = 38x - 103$]

23. The crushing strength of mortar in tonnes varies with the percentage of water used in its preparation as shown below.

Crushing strength (F)	1.51	1.33	1.15	0.97	0.79	0.61
% of water used (x)	8.0	10	12	14	16	18

Prove that these values obey the law $F = ax + b$ where a and b are constants. Determine a and b. [$F = -0.090x + 2.23$]

24. The mass, m, of a steel joist varies with length, l, as follows:

mass (kg)	95.0	100	110	120
length (m)	3.480	3.670	4.030	4.400

Find the relationship between them. [$m = 27.40l - 0.550\,0$]

25. The amount of light (L) absorbed by a coloured sulphide solution varies with the concentration (C in μg per 100 g of solution) as follows:

Amount of light (L)	0	0.09	0.18	0.27	0.36	0.45
Concentration (C)	0	2	4	6	8	10

Find the relationship between L and C. [$L = 0.045C$]

26. The molar heat capacity (c in J deg^{-1} mole^{-1}) varies with absolute temperature (T in K) as shown. (The results are for silicon carbide.)

Molar heat capacity (c)	38.77	41.29	43.81	46.33
Temperature (T)	100	300	500	700

How is c related to T? [$c = 0.012\,6\,T + 37.51$]

Section III

Geometry and trigonometry

Chapter 9

Geometry

9.1 Definitions

A **line** is that which has length without breadth or thickness.

A **point** is that which has position but no magnitude.

Geometry is a part of mathematics in which the properties of points, lines, surfaces and solids are investigated.

9.2 Angles and their measurement

An **angle** is formed by the intersection of two **straight lines.** Let AOX, in Fig. 9.1 represent an angle. Consider the line OX to be fixed and the other line OA free to rotate about the point O. The point O is called the **vertex** of the angle, while OA and OX are called the **arms** of the angle.

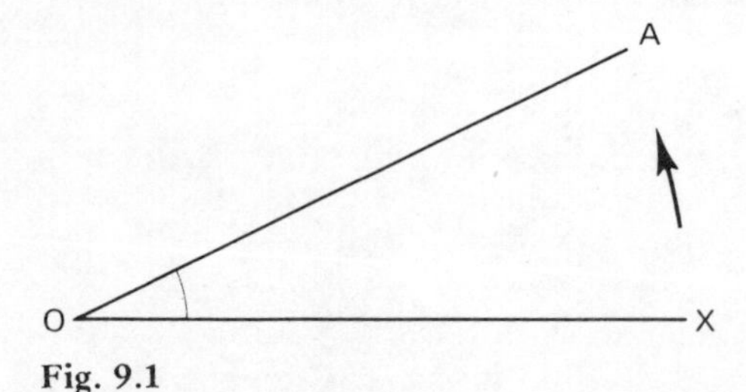

Fig. 9.1

By definition, an **angle is the amount of rotation between two straight lines.** Thus, in Fig. 9.1, OA has rotated from OX. The amount of rotation is known as angle AOX and is written as ∠ AOX.

Angles may be measured in either **degrees** or **radians** (radian measure is dealt with in Section 9.10).

If the line OA lies along OX and is rotated until it lies along OX again, it is said to have made one **revolution.** If one revolution is divided into 360 equal parts then each part is called 1 degree. Thus

1 revolution = 360 degrees (or 1 rev. = 360°)

A degree is subdivided into 60 minutes (written as $60'$) and a minute is subdivided into 60 seconds (written as $60''$). Hence, summarising:

1 degree = $\frac{1}{360}$th of one revolution
1 minute = $\frac{1}{60}$th of a degree
1 second = $\frac{1}{60}$th of a minute

Worked problems on the measurement of angles

Problem 1. Add (*a*) $13^\circ\ 52'$ and $26^\circ\ 29'$ (*b*) $25^\circ\ 37'\ 51''$ and $41^\circ\ 29'\ 16''$.

$$\begin{array}{r} 13^\circ\ 52' \\ 26^\circ\ 29' \\ \hline 40^\circ\ 21' \\ \hline 1^\circ \end{array}$$

(*a*) $52' + 29'$ gives $81'$. Since $60' = 1^\circ$, $81' = 1^\circ\ 21'$. Thus the $21'$ is placed in the minutes column and 1° is carried in the degrees column. Then $13^\circ + 26^\circ + 1^\circ$ (carried) $= 40^\circ$. This is placed in the degrees column.

Thus $13^\circ\ 52' + 26^\circ\ 29' = \mathbf{40^\circ\ 21'}$

$$\begin{array}{r} 25^\circ\ 37'\ 51'' \\ 41^\circ\ 29'\ 16'' \\ \hline 67^\circ\ \ 7'\ \ 7'' \\ \hline 1^\circ\ \ 1'\ \ \ \ \end{array}$$

(*b*) $51'' + 16''$ gives $67''$. Since $60'' = 1'$, $67'' = 1'\ 7''$. Thus the $7''$ is placed in the seconds column and $1'$ is carried in the minutes column. Then $37' + 29' + 1'$ (carried) $= 67'$. Since $60' = 1^\circ$, $67' = 1^\circ\ 7'$. Thus the $7'$ is placed in the minute column and 1° is carried in the degrees column. Then $25^\circ + 41^\circ + 1^\circ = 67^\circ$. This is placed in the degrees column.

Thus $25^\circ\ 37'\ 51'' + 41^\circ\ 29'\ 16'' = \mathbf{67^\circ\ 7'\ 7''}$

Problem 2. Subtract (*a*) $9^\circ\ 43'\ 52''$ from $12^\circ\ 12'\ 12''$
(*b*) $22^\circ\ 45'\ 39''$ from $49^\circ\ 28'\ 15''$

$$\begin{array}{r} 11^\circ\ 11'\ \ \ \ \ \ \\ \not{12}^\circ\ \not{12}'\ 12'' \\ 9^\circ\ 43'\ 52'' \\ \hline 2^\circ\ 28'\ 20'' \\ \hline \end{array}$$

(*a*) $12'' - 52''$ cannot be done. Hence $1'$ or $60''$ is 'borrowed' from the minutes column which leaves $11'$ in that column. Now $(60'' + 12'') - 52'' = 20''$. This is placed in the seconds column. $11' - 43'$ cannot be done. Hence 1° or $60'$ is 'borrowed' from the degrees column which leaves 11° in that column. Now $(60' + 11') - 43' = 28'$. This is placed in the minutes column. $11^\circ - 9^\circ = 2^\circ$. This is placed in the degrees column.

Thus $12^\circ\ 12'\ 12'' - 9^\circ\ 43'\ 52'' = \mathbf{2^\circ\ 28'\ 20''}$
(This answer can be checked. Adding $2^\circ\ 28'\ 20''$ to $9^\circ\ 43'\ 52''$ should give $12^\circ\ 12'\ 12''$.)

$$\begin{array}{r} 48° \ 27' \\ (b) \ \cancel{49}° \ \cancel{28}' \ 15'' \\ 22° \ 45' \ 39'' \\ \hline 26° \ 42' \ 36'' \\ \hline \end{array}$$

15″ − 39″ cannot be done. Hence 1′ or 60″ is 'borrowed' from the minutes column which leaves 27′ in that column. Now (60″ + 15″) − 39″ = 36″. This is placed in the seconds column. 27′ − 45′ cannot be done. Hence 1° or 60′ is 'borrowed' from the degrees column which leaves 48° in that column. Now (60′ + 27′) − 45′ = 42′. This is placed in the minutes column 48° − 22° = 26°. This is placed in the degrees column.

Thus 49° 28′ 15″ − 22° 45′ 39″ = **26° 42′ 36″**

(This may be checked by adding the bottom two lines of the subtraction sum. That is, 22° 45′ 39″ plus 26° 42′ 36″ should be equal to the top line, 49° 28′ 15″.)

Problem 3. Convert the following angles to degrees and decimals of a degree: (*a*) 23° 42′ (*b*) 84° 33′ 17″

(*a*) Since 1 minute = $\frac{1}{60}$th of a degree

42 minutes = $\frac{42}{60}$ of a degree

= 0.70 degrees

Therefore 23° 42′ = **23.70°**

(*b*) Since 1 second = $\frac{1}{60}$th of a minute

17 seconds = $\frac{17}{60}$ minutes

= 0.283 3 minutes

Therefore 84° 33′ 17″ = 84° 33.283 3′

Since 1 minute = $\frac{1}{60}$th of a degree

$$33.283\ 3 \text{ minutes} = \frac{33.283\ 3}{60} \text{ degrees}$$

= 0.554 7 degrees correct to 4 decimal places.

Hence 84° 33′ 17″ = **84.554 7° correct to 4 decimal places.**

Problem 4. Convert 37.468° into degrees, minutes and seconds.

Since 1 degree = 60 minutes

0.468 degrees = 0.468 × 60 minutes

= 28.08 minutes

Since 1 minute = 60 seconds

0.08 minutes = 0.08 × 60 seconds

= 4.8 seconds

Hence 37.468° = **37° 28′ 5″, to the nearest second.**

Further problems on angular measurement may be found in Section 9.11 (Problems 1–4), page 251.

9.3 Acute, obtuse, reflex and right angles

There are four names given to angles, according to their size.

1. Any angle between 0° and 90° is called an **acute angle.** For example, 30° is an acute angle and is shown in Fig. 9.2.

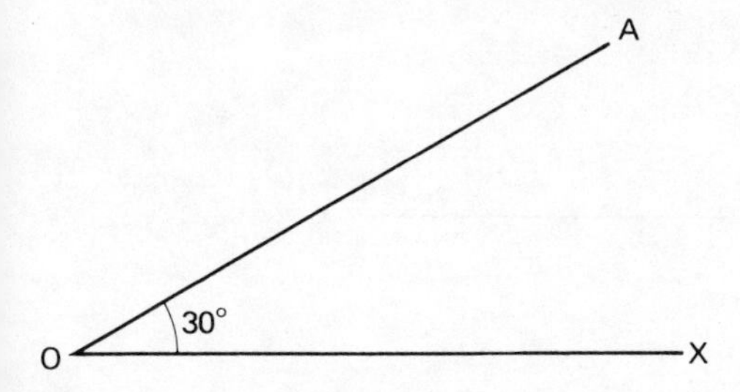

Fig. 9.2

2. An angle exactly equal to 90° is called a **right angle.** OA and OX in Fig. 9.3 are 90° apart and are said to be **perpendicular** to one another.

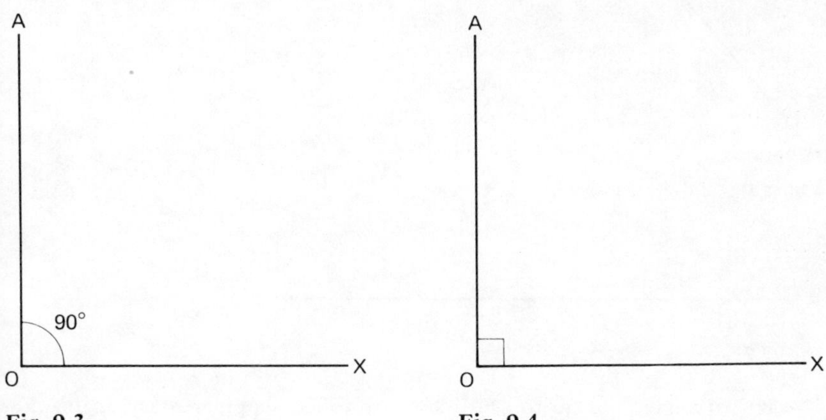

Fig. 9.3 **Fig. 9.4**

There is a special sign used to denote a right angle and this is shown in Fig. 9.4. When this sign is used it is unnecessary to write the value of the angle as 90°, it is assumed to mean this.

Note that there are four right angles in one complete revolution.

3. Any angle between 90° and 180° is called an **obtuse angle.** For example 145° is an obtuse angle and is shown in Fig. 9.5.
4. Any angle greater than 180° and less than 360° is called a **reflex angle.** For example 230° is a reflex angle and is shown in Fig. 9.6.

9.4 Properties of angles and straight lines

1. An angle of 180° lies on a straight line. See Fig. 9.7.
2. If two angles add up to 90° they are called **complementary angles.** For example, 15° and 75° are complementary angles and each angle is called the complement of the other. Similarly, the complement of 27° is 63° and the complement of $12\frac{1}{2}°$ is $77\frac{1}{2}°$.

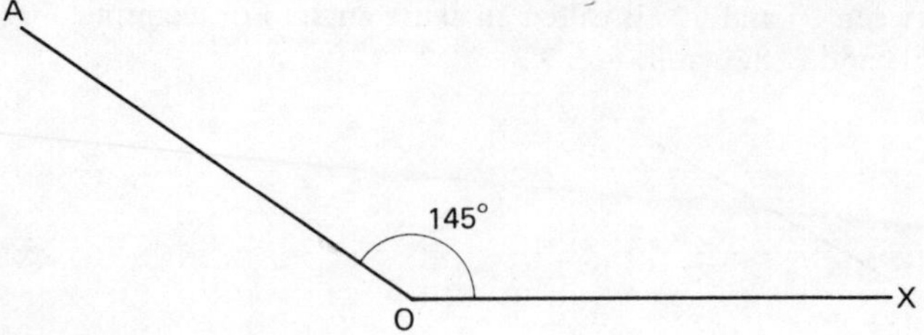

Fig. 9.5

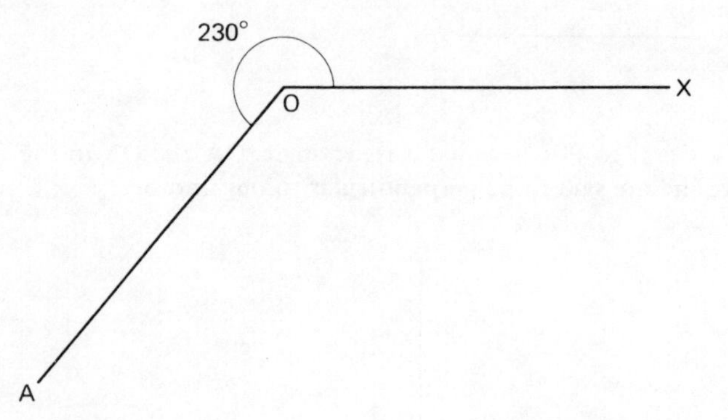

Fig. 9.6

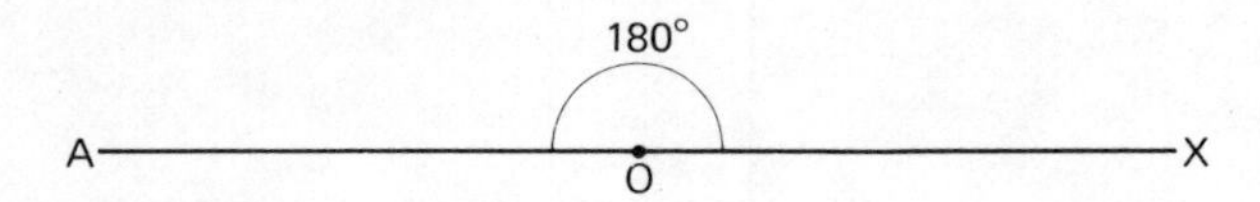

Fig. 9.7

3. If two angles add up to 180° they are called **supplementary angles.** For example 76° and 104° are supplementary angles and each angle is called the supplement of the other. Similarly the supplement of 129° is 51° and the supplement of $14^\circ\ 23'$ is $165^\circ\ 37'$.
4. When two straight lines intersect, the vertically opposite angles are equal.

Proof: Let two straight lines RT and PQ intersect at O as shown in Fig. 9.8. Let the four angles produced be denoted by a, b, c and d, where $\angle a = \angle \text{POR}$, $\angle b = \angle \text{POT}$, $\angle c = \angle \text{TOQ}$ and $\angle d = \angle \text{ROQ}$

Now $\angle a$ and $\angle b$ are supplementary, that is

$$\angle a + \angle b = 180 \qquad (1)$$

Also $\angle b$ and $\angle c$ are supplementary, that is

$$\angle b + \angle c = 180 \qquad (2)$$

Subtracting equation (2) from equation (1) gives

$$\angle a - \angle c = 0$$

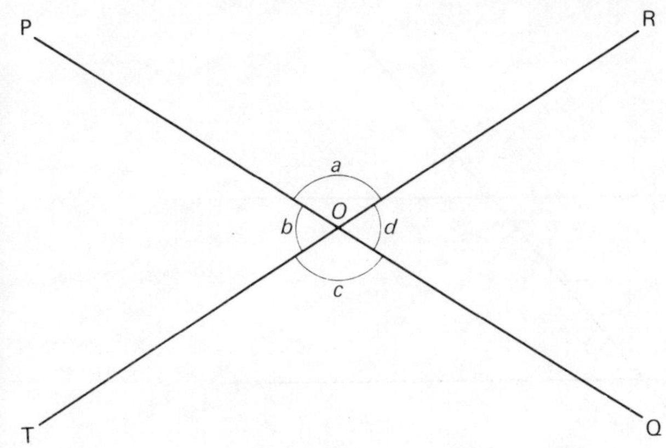

Fig. 9.8

Therefore $\angle a = \angle c$

Similarly $\angle a$ and $\angle d$ are supplementary, that is

$\angle a + \angle d = 180$ (3)

Subtracting equation (3) from equation (1) gives

$$\angle b - \angle d = 0$$

Therefore $\angle b = \angle d$

The pairs of angles, *a* and *c*, and *b* and *d* are called vertically opposite angles.

Thus when two straight lines intersect at a point the vertically opposite angles are equal.

---

5. **Parallel lines** may be defined as straight lines which are in the same plane (flat or level surface) and never meet, however far they may be produced in either direction. Therefore there can never be any angle between two parallel lines.

Let two parallel lines be represented by AB and CD in the plane of the paper as shown in Fig. 9.9. The arrows drawn on AB and CD are a way of denoting parallel lines.

A straight line which crosses two parallel lines is called a **transversal.** Let a transversal be represented by EF as shown in Fig. 9.9. Let *a, b, c, d, e, f, g* and *h* represent the angles formed at the intersections of the transversal with the parallel lines.

Since the lines AB and CD are parallel and the transversal EF is equally inclined to both lines, the following facts emerge and may be stated:

(*i*) $\angle a = \angle e, \angle b = \angle f$
$\angle c = \angle g, \angle d = \angle h$

Such pairs of angles are called **corresponding angles.**

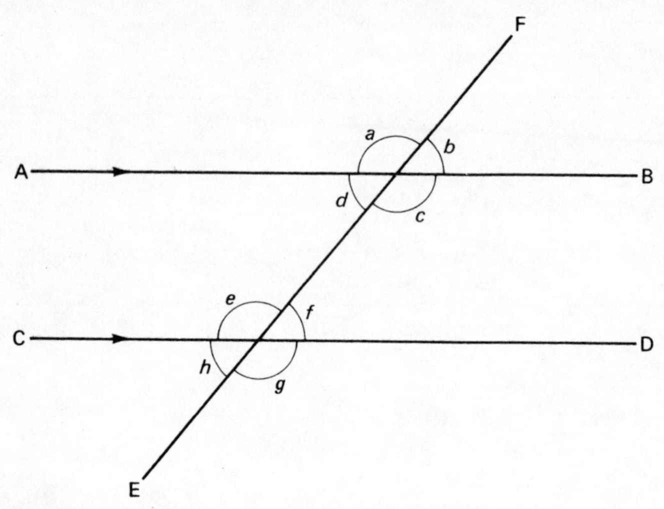

Fig. 9.9

(*ii*) $\angle d = \angle f, \angle c = \angle e$

Such pairs of angles are called **alternate angles.**

(*iii*) $\angle d + \angle e = 180^\circ$
$\angle c + \angle f = 180^\circ$

Such pairs of angles are called **interior angles.**

Thus the sum of the interior angles on the same side of the transversal is equal to 180°. That is, $\angle d$ and $\angle e$, and also $\angle c$ and $\angle f$, are supplementary.

If a transversal crosses two straight lines, and any one of the three conditions, (*i*), (*ii*) or (*iii*), holds good, then the two straight lines must be parallel.

Worked problems on types and properties of angles

Problem 1. State the general name given to the following angles:
(*a*) 163° (*b*) 90° (*c*) 47° (*d*) 215°

(*a*) 163° lies between 90° and 180° and is therefore called an **obtuse angle.**
(*b*) 90° is called a **right angle.**
(*c*) 47° lies between 0° and 90° and is therefore called an **acute angle.**
(*d*) 215° is greater than 180° and less than 360° and is therefore called a **reflex angle.**

Problem 2. Find the angles complementary to (*a*) 37° (*b*) $49^\circ\ 51'$ (*c*) $78^\circ\ 33'\ 19''$

(*a*) The complement of 37° is $(90^\circ - 37^\circ)$ i.e. 53°

(*b*) The complement of 49° 51′ is (90° − 49° 51′) i.e. **40° 9′**
(*c*) The complement of 78° 33′ 19″ is (90° − 78° 33′ 19″) i.e. **11° 26′ 41″**

Problem 3. Find the angles supplementary to
(*a*) 16° (*b*) 91° 10′ (*c*) 138° 23′ 5″

(*a*) The supplement of 16° is (180° − 16°) i.e. **164°**
(*b*) The supplement of 91° 10′ is (180° − 91° 10′) i.e. **88° 50′**
(*c*) The supplement of 138° 23′ 5″ is (180° − 138° 23′ 5″) i.e. **41° 36′ 55″**

Problem 4. Two straight lines MN and PQ intersect at O. If ∠ MOP is 32° find ∠ MOQ, ∠ QON and ∠ PON

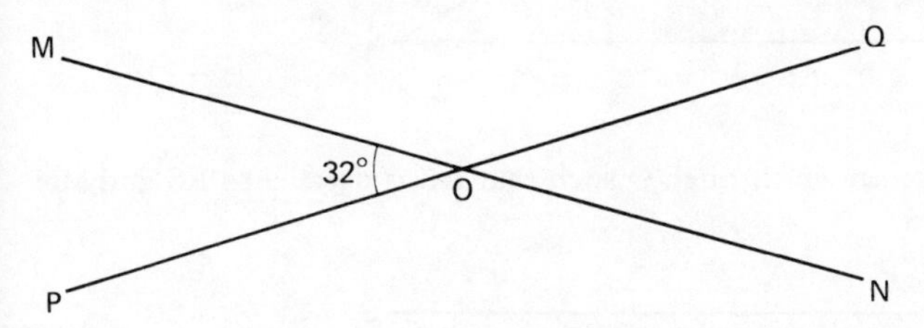

Fig. 9.10

In Fig. 9.10 ∠ MOQ is supplementary to ∠ MOP
Therefore ∠ MOQ = (180° − 32°) = **148°**
When two straight lines intersect the vertically opposite angles are equal.
Therefore ∠ QON = **32°**
∠ PON = **148°**

Problem 5. Find the angle x in Fig. 9.11.

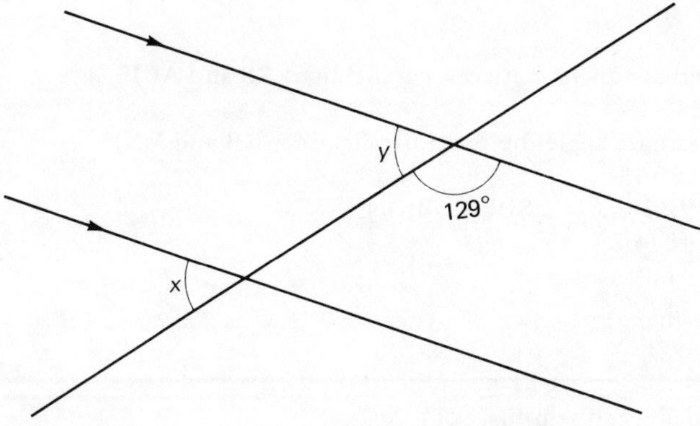

Fig. 9.11

Let ∠ y be the angle shown in Fig. 9.11, the supplement of 129°.
Therefore ∠ y = 180° − 129° = 51°

But $\angle x = \angle y$ (corresponding angles between parallel lines)
Thus $\angle x = \mathbf{51^\circ}$

Problem 6. Find the angle θ in Fig. 9.12.

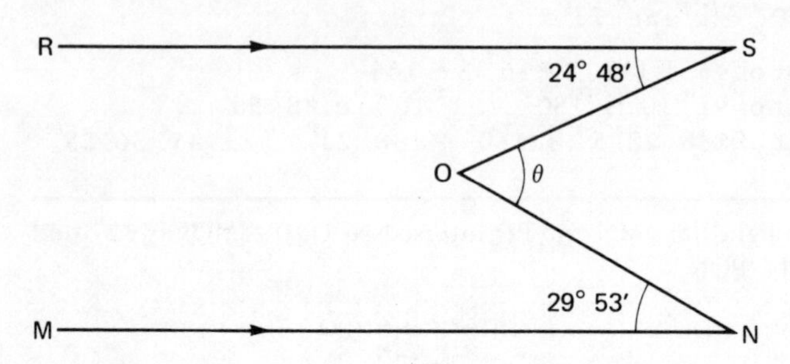

Fig. 9.12

Let a line AB be drawn through O such that AB is parallel to RS and MN as shown in Fig. 9.13.

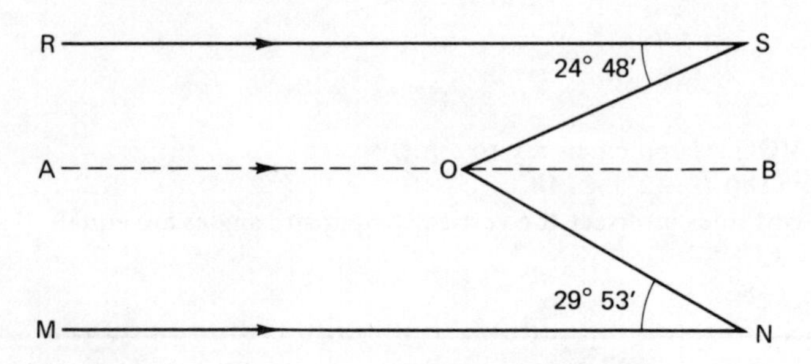

Fig. 9.13

$\angle$ RSO = $\angle$ SOB (alternate angles between parallel lines RS and AB)
Therefore $\angle$ SOB = $24^\circ\ 48'$
$\angle$ ONM = $\angle$ BON (alternate angles between parallel lines AB and MN)
Therefore $\angle$ BON = $29^\circ\ 53'$
From Figs 9.12 and 9.13, $\angle \theta = \angle$ SOB + $\angle$ BON
Therefore $\angle \theta = 24^\circ\ 48' + 29^\circ\ 53'$
$= 54^\circ\ 41'$
Hence $\angle \theta = \mathbf{54^\circ\ 41'}$

Problem 7. Find angles x and y in Fig. 9.14.

Let a and b be the angles shown in Fig. 9.14.
Therefore $\angle a = 52^\circ$ (corresponding angles between parallel lines)
Also $\angle a + \angle x + 90^\circ = 180^\circ$ (angles on a straight line)

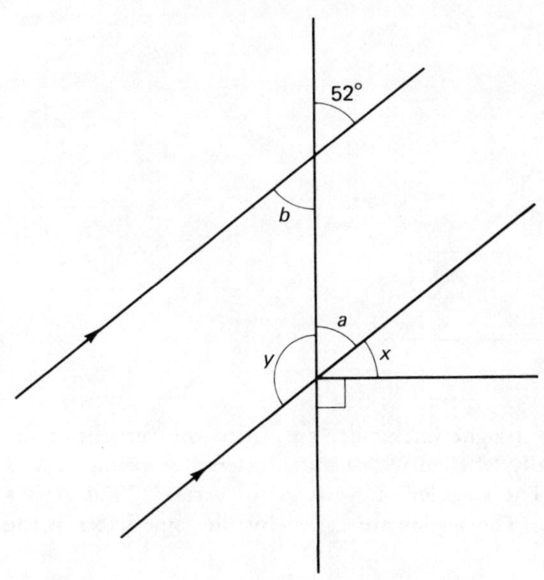

Fig. 9.14

Therefore $52^\circ + \angle x + 90^\circ = 180^\circ$

$\angle x = 180^\circ - 90^\circ - 52^\circ$

$\angle x = \mathbf{38^\circ}$

Similarly $\angle a + \angle y = 180^\circ$

Therefore $52^\circ + \angle y = 180^\circ$

$\angle y = 180^\circ - 52^\circ$

$\angle y = \mathbf{128^\circ}$

There are often alternative methods of finding unknown angles. For example:

$90^\circ + \angle x = \angle y$ (vertically opposite angles)

Therefore $\angle y = 90^\circ + 38^\circ$

$\angle y = \mathbf{128^\circ}$

or $\angle b = 52^\circ$ (vertically opposite angles)

Therefore $\angle y + \angle b = 180^\circ$ (interior angles on the same side of a transversal)

$\angle y = 180^\circ - \angle b$

$= 180^\circ - 52^\circ$

$\angle y = \mathbf{128^\circ}$

Further problems on types and properties of angles may be found in Section 9.11 (Problems 5–14), page 251.

9.5 Properties of triangles

A triangle is a figure enclosed by three straight lines.

1. The sum of the three angles of a triangle is equal to 180°.

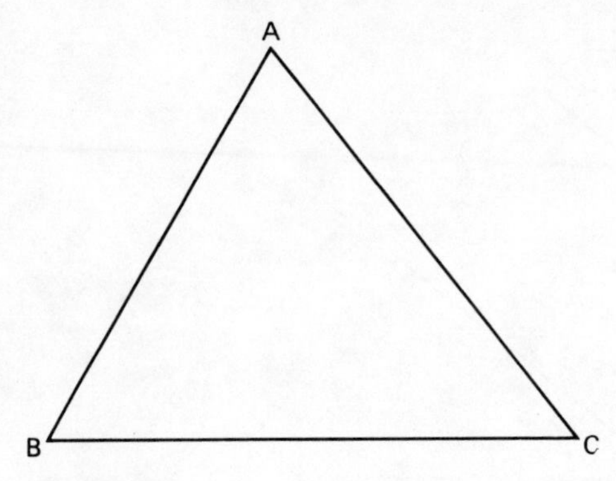

Fig. 9.15

The points where the straight lines meet are called the **vertices** of the triangle and are normally labelled with capital letters, for example, A, B and C of Fig. 9.15. (The singular of 'vertices' is '**vertex**'. Thus A is a vertex of the triangle.) The angles are called by the same letter as the vertices. Thus

$$\angle A + \angle B + \angle C = 180^\circ$$

2. An **acute-angled triangle** is one in which all the angles are acute, that is, all the angles are less than 90°. For example, the triangle shown in Fig. 9.16 is a typical acute-angled triangle.

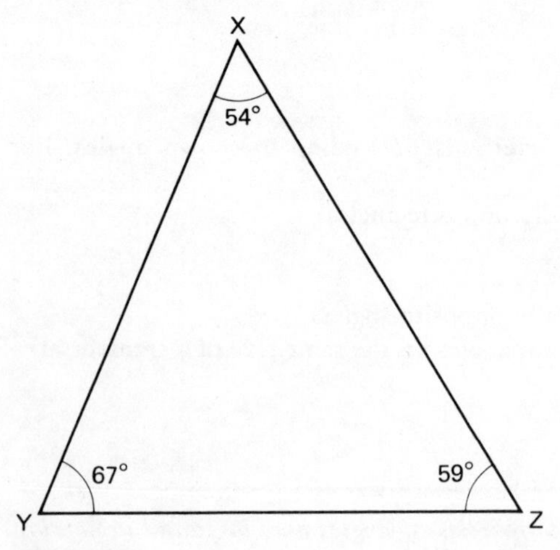

Fig. 9.16

3. A **right-angled triangle** is one which contains a right angle. There can be only one right angle in any one triangle as the three angles add up to 180°. (Therefore the other two angles must together add up to 90°.) The triangle shown in Fig. 9.17 is a typical right-angled triangle.

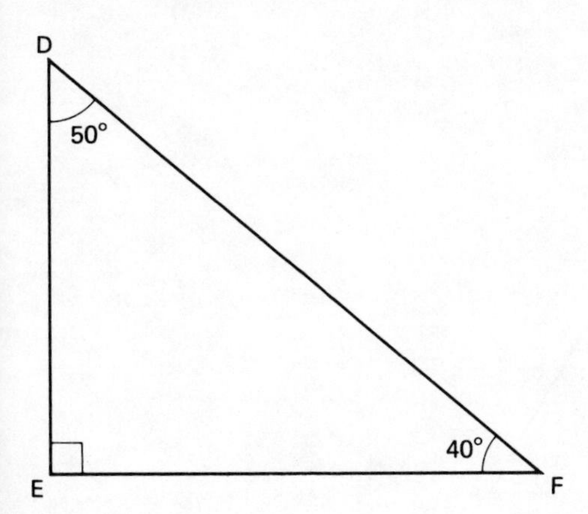

Fig. 9.17

The side DF, that is, the side opposite the right angle, is called the **hypotenuse.**

4. An **obtuse-angled triangle** is one which contains one obtuse angle, that is, one angle which lies between 90° and 180°. There can be only one such angle in any one triangle as the three angles add up to 180°. (Therefore the other two angles must together add up to less than 90°.) Figure 9.18 shows a typical obtuse-angled triangle.

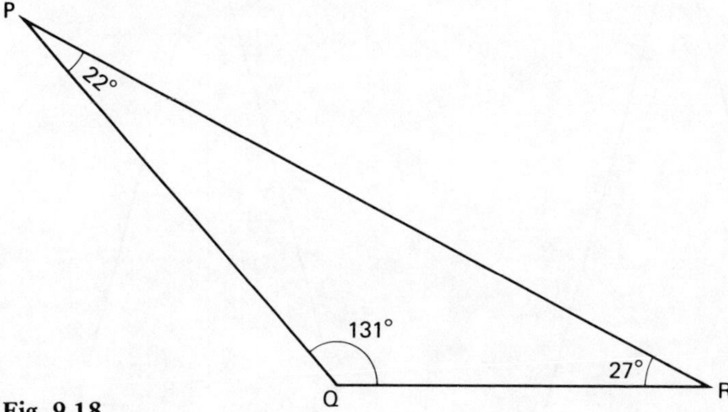

Fig. 9.18

5. An **equilateral triangle** is one in which all the sides and all the angles are equal. Since the sum of the angles is 180° then each angle must be 60°. An equilateral triangle is shown in Fig. 9.19.
   It is normal convention to label the side lying opposite an angle with the small letter corresponding to that of the angle. For example, the side opposite angle A (i.e. BC) is labelled '*a*', the side opposite angle B (i.e. AC) is labelled '*b*' and the side opposite angle C (i.e. AB) is labelled '*c*'. Thus, in an equilateral triangle as shown in Fig. 9.19, $a = b = c$.

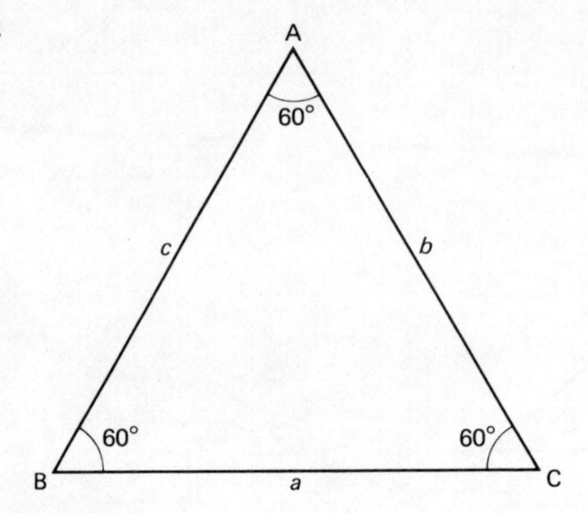

Fig. 9.19

6. An **isosceles triangle** is one in which two angles and two sides are equal, the equal sides lying opposite the equal angles. Figure 9.20 shows a typical triangle where ∠ E = ∠ G. Hence side e = side g.

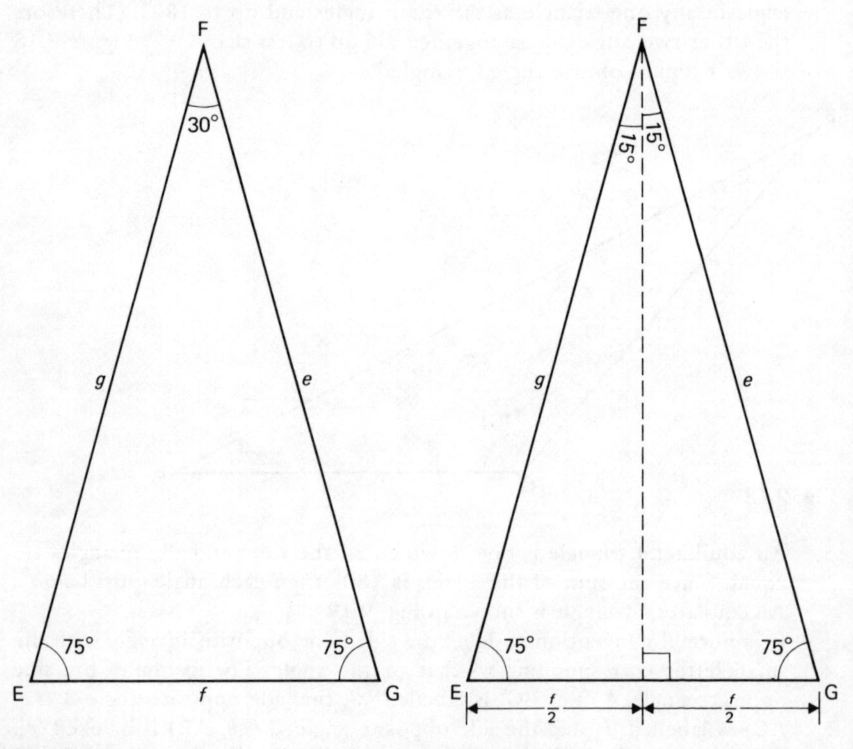

Fig. 9.20 **Fig. 9.21**

In the isosceles triangle of Fig. 9.20, if a perpendicular to EG is drawn from F, then the angle F will be bisected (that is, cut equally into two parts). The base of the triangle EG will also be bisected. This is shown in Fig. 9.21.

7. A **scalene triangle** is one with unequal angles and therefore unequal sides. Two typical scalene triangles are shown in Fig. 9.22 where (*a*) is an acute-angled scalene triangle and (*b*) is an obtuse angled scalene triangle.

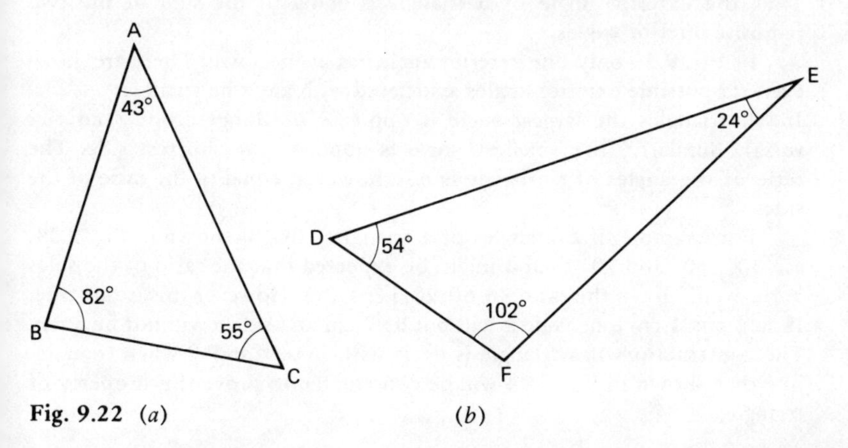

Fig. 9.22 (*a*) (*b*)

8. The three angles contained within a triangle are called interior angles. In Fig. 9.23, $\angle A$, $\angle B$ and $\angle C$ are called **interior** angles. If the side BC is produced (i.e. extended) as shown, then the angle X is known as an

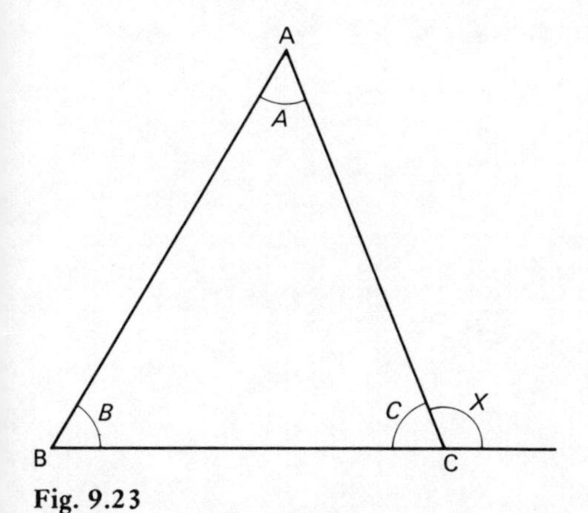

Fig. 9.23

exterior angle of the triangle.

If $\angle A + \angle B + \angle C = 180^\circ$

then $\angle C = 180^\circ - \angle A - \angle B$

Also, $\angle C$ and $\angle X$ are supplementary

i.e. $\angle C + \angle X = 180^\circ$

$$\angle X = 180^\circ - \angle C$$
$$= 180^\circ - (180^\circ - \angle A - \angle B)$$
$$= 180^\circ - 180^\circ + \angle A + \angle B$$

Therefore $\angle X = \angle A + \angle B.$

Thus the exterior angle of a triangle is equal to the sum of the two opposite interior angles.

In Fig. 9.23 only one exterior angle has been shown. There are, however, six possible exterior angles associated with any one triangle.

9. In all triangles the largest angle is opposite the longest side (and vice versa). Similarly, the smallest angle is opposite the shortest side. **The ratio of the angles of a triangle is not, however, equal to the ratio of the sides.**

For example, if the angles of a triangle ABC, as shown in Fig. 9.24, are 30°, 60° and 90° then it might be expected that the ratio of the sides a:b:c would be in the ratio 30:60:90, i.e. 1:2:3. However this is **not** true. If side a is 1 cm long, side b will **not** be 2 cm and side c will **not** be 3 cm. The construction of a triangle is dealt with in section 9.9 when triangles like that shown in Fig. 9.24 will be constructed to prove this property of triangles.

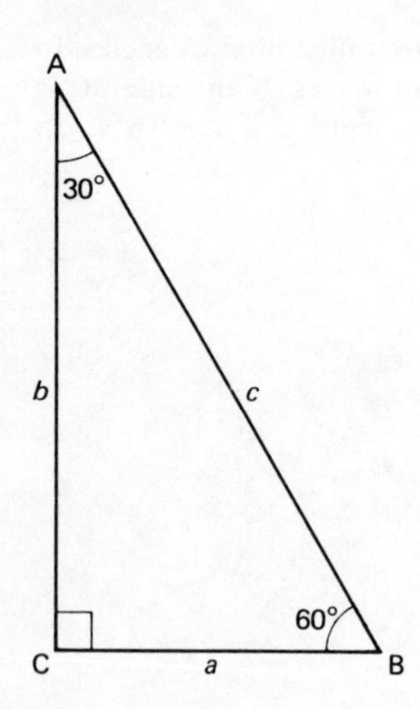

Fig. 9.24

10. In any triangle the sum of the lengths of any two of the sides is always greater than the length of the third side. In Fig. 9.25, y is the longest side since it lies opposite the largest angle.

Always $x + z > y$ (The sign > means 'is greater than')

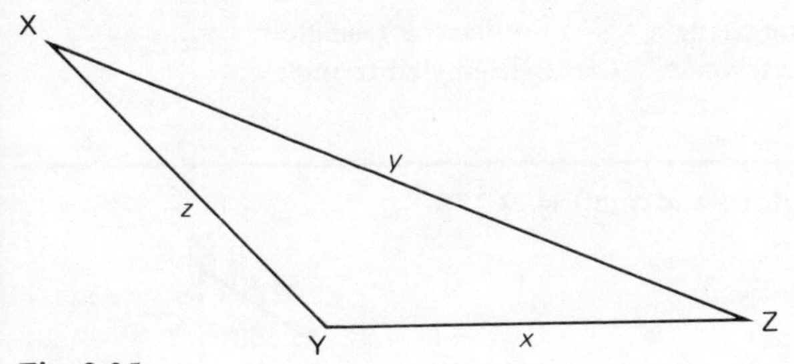

Fig. 9.25

Note that $x + y + z$ is called the **perimeter** of the triangle.

Worked problems on the properties of triangles

Problem 1. Name the types of triangles shown in Fig. 9.26

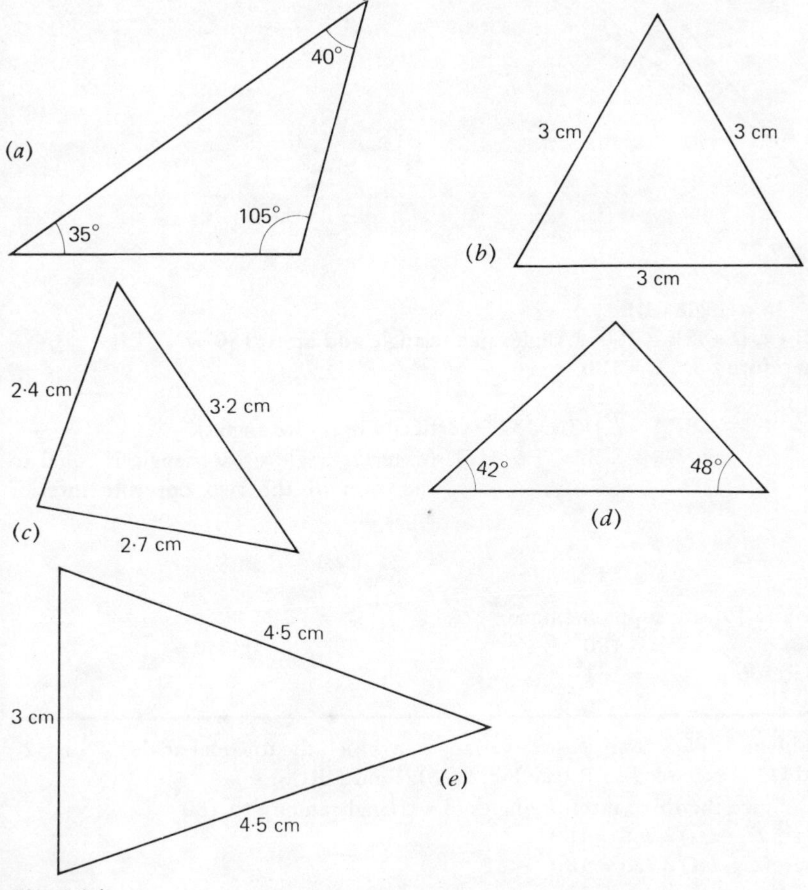

Fig. 9.26

(*a*) Obtuse-angled scalene triangle (*b*) Equilateral triangle
(*c*) Acute-angled scalene triangle (*d*) Right-angled triangle
(*e*) Isosceles triangle

Problem 2. Find the angles ϕ and x in Fig. 9.27.

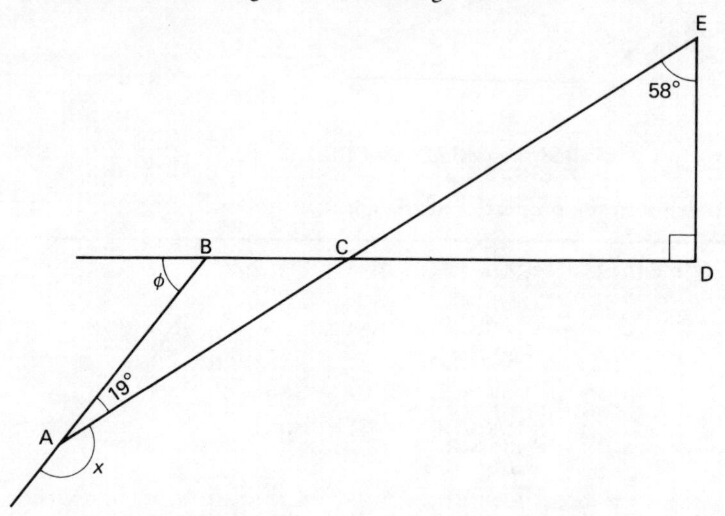

Fig. 9.27

In triangle CDE

$\angle C + \angle D + \angle E = 180^\circ$ (Angles in a triangle add up to 180°.)

Therefore $\angle DCE = 180^\circ - 90^\circ - 58^\circ$

$= 32^\circ$

$\angle BCA = \angle DCE = 32^\circ$ (vertically opposite angles).

$\angle \phi = \angle BAC + \angle ACB$ (exterior angle of a triangle is equal to the sum of the two opposite interior angles).

$\angle \phi = 19^\circ + 32^\circ$

$\angle \phi = \mathbf{51^\circ}$

$\angle x$ and 19° are supplementary.

Thus $\angle x = 180^\circ - 19^\circ$

Therefore $\angle x = \mathbf{161^\circ}$

Problem 3. PQS is an isosceles triangle in which the unequal angle QPS is 52° and PQ is extended to R (see Fig. 9.28). Find $\angle$ RQS.

Since the three interior angles of a triangle add up to 180°

$52^\circ + \angle Q + \angle S = 180^\circ$

Therefore $\angle Q + \angle S = 180^\circ - 52^\circ$

$= 128^\circ$

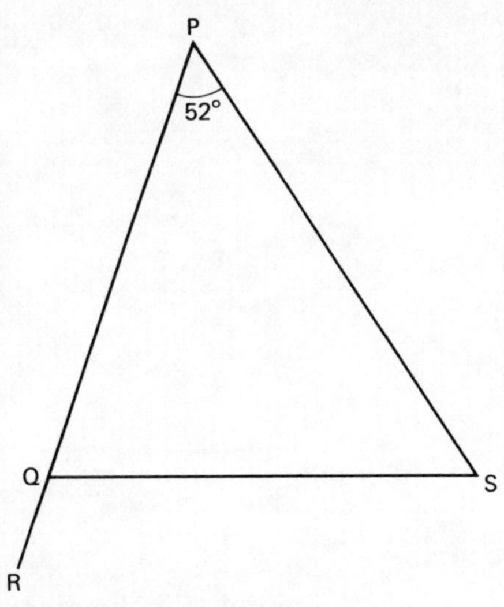

Fig. 9.28

But triangle PQS is isosceles, therefore

$$\angle \mathbf{Q} = \angle \mathbf{S} = \frac{128^\circ}{2} = \mathbf{64^\circ}$$

The exterior angle RQS is equal to the sum of the two interior opposite angles, i.e. $\angle$ QPS + $\angle$ QSP.

Therefore $\angle \text{RQS} = 52^\circ + 64^\circ$
$= \mathbf{116^\circ}$

(This may be checked since $\angle$ PQS and $\angle$ RQS are supplementary.

Therefore $\angle \text{PQS} + \angle \text{RQS} = 180^\circ$
$(64^\circ + 116^\circ = 180^\circ)$

Problem 4. Find the angles α and β shown in Fig. 9.29.

In triangle ABC:

$41^\circ\ 37' + 12^\circ\ 49' + \angle\alpha = 180^\circ$ (angles in a triangle add up to 180°)

Therefore $\angle\alpha = 180^\circ - 41^\circ\ 37' - 12^\circ\ 49'$

$\angle\alpha = \mathbf{125^\circ\ 34'}$

In triangle ABD:

$90^\circ + 41^\circ\ 37' + (12^\circ\ 49' + \angle\beta) = 180^\circ$ (angles in a triangle add up to 180°)

Therefore $\angle\beta = 180^\circ - 90^\circ - 41^\circ\ 37' - 12^\circ\ 49'$

$\angle\beta = \mathbf{35^\circ\ 34'}$

An alternative method to find $\angle\beta$ is:

$\angle\alpha = \angle\beta + 90^\circ$ since the exterior angle of a triangle is equal to the sum of the two opposite interior angles.

Therefore $\angle\beta = \angle\alpha - 90^\circ$
$= 125^\circ\ 34' - 90^\circ$
$\angle\beta = \mathbf{35^\circ\ 34'}$

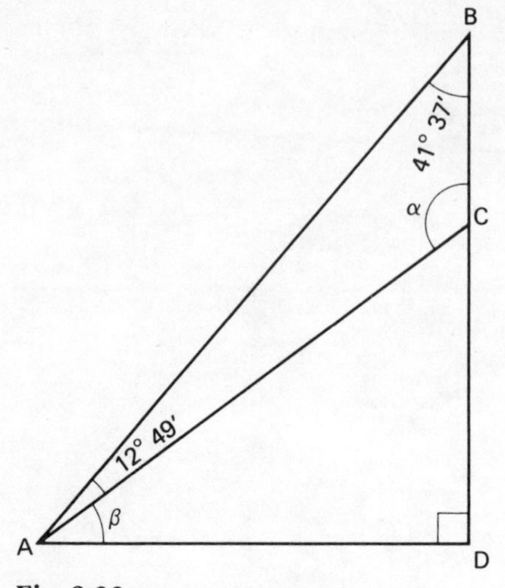

Fig. 9.29

Problem 5. Find the angles a, b, c, d and e in Fig. 9.30.

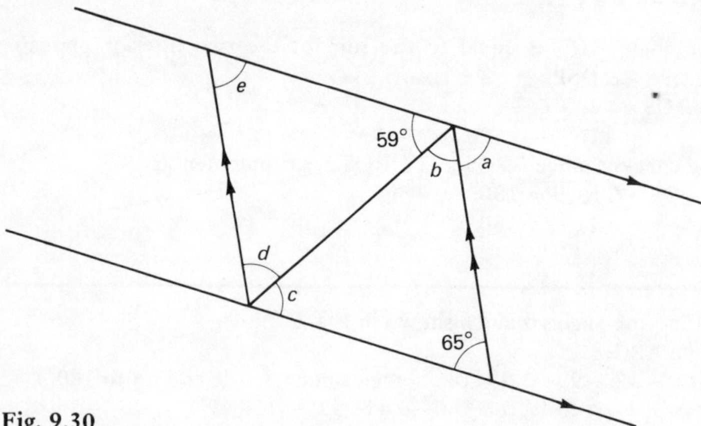

Fig. 9.30

When two pairs of parallel lines are depicted on the same figure, single arrows can be used to denote one pair of parallel lines and double arrows used to denote the other pair of parallel lines, as shown in Fig. 9.30.

$\angle c$ = **59°** (alternate angles between parallel lines)
$\angle a$ = **65°** (alternate angles between parallel lines)
$\angle b + 59° + 65° = 180°$ (angles in a triangle add up to 180° or angles on a straight line add up to 180°.)

Therefore $\angle b = 180° - 59° - 65°$
$\angle$ **b = 56°**

$\angle b = \angle d =$ **56°** (alternate angles between parallel lines marked by double arrows)

$\angle e + 59° + 56° = 180°$ (angles in a triangle add up to 180°)

Therefore $\angle e = 180° - 59° - 56°$

$\angle e =$ **65°**

Check:

$\angle e = \angle a = 65°$ (corresponding angles between parallel lines)

Problem 6. In Fig. 9.31 find all the unknown angles.

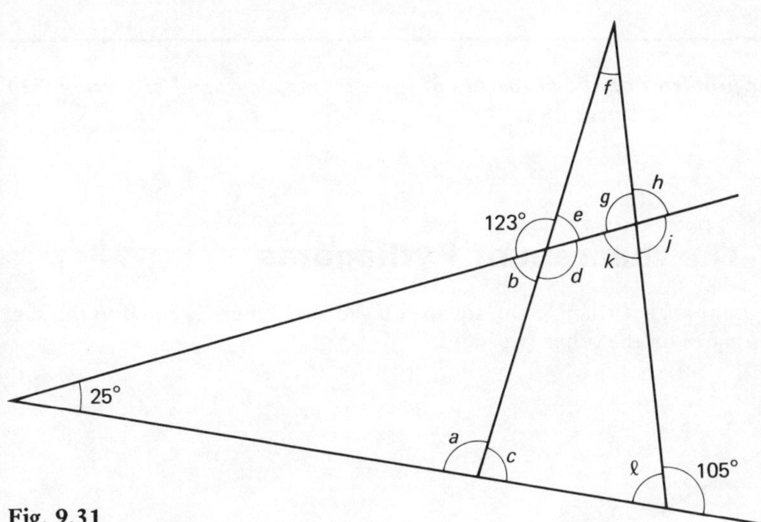

Fig. 9.31

$\angle d =$ **123°** (vertically opposite angles)

$\angle e$ is the supplement of 123°

Therefore $\angle e = 180° - 123°$

$\angle e =$ **57°**

$\angle e = \angle b =$ **57°** (vertically opposite angles)

$\angle a + \angle b + 25° = 180°$ (angles in a triangle add up to 180°)

Therefore $\angle a = 180° - 25° - 57°$

$\angle a =$ **98°**

$\angle a$ and $\angle c$ are supplementary.

Therefore $\angle c = 180° - 98°$

$\angle c =$ **82°**

$\angle l$ and 105° are supplementary, so

Therefore $\angle l = 180° - 105°$

$\angle l =$ **75°**

$\angle c + \angle f + \angle l = 180°$ (angles in a triangle add up to 180°)

Therefore $\angle f = 180° - 82° - 75°$

$\angle f =$ **23°**

$\angle e + \angle f + \angle g = 180°$ (angles in a triangle add up to 180°)

Therefore $\angle g = 180^\circ - 57^\circ - 23^\circ$

$\angle g = 100^\circ$

$\angle g = \angle j = 100^\circ$ (vertically opposite angles)

$\angle h$ is the supplement of 100°, therefore

$\angle h = 180^\circ - 100^\circ$

$\angle h = 80^\circ$

$\angle h = \angle k = 80^\circ$ (vertically opposite angles)

Check:

$25^\circ + \angle k + \angle l = 180^\circ$ (angles in a triangle add up to 180°)

$25^\circ + 80^\circ + 75^\circ = 180^\circ$

Further problems on the properties of triangles may be found in Section 9.11 (Problems 15–24), page 254.

9.6 The theorem of Pythagoras

In any right-angled triangle the square on the hypotenuse is equal to the sum of the squares on the other two sides.

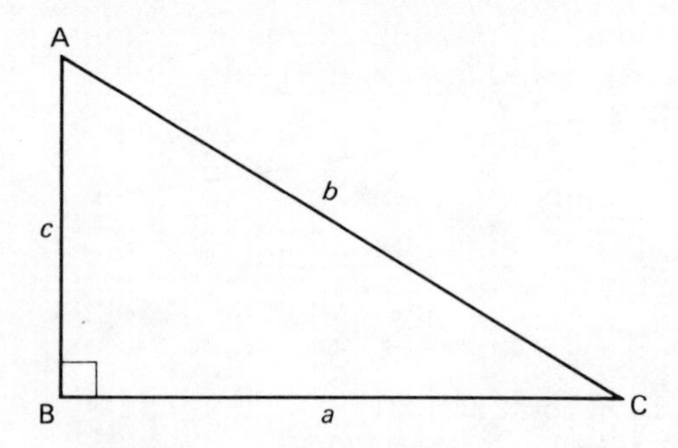

Fig. 9.32

For the triangle ABC, shown in Fig. 9.32, b is the hypotenuse. (The hypotenuse is always the side opposite the right angle.) Therefore, from above $(AC)^2 = (BC)^2 + (AB)^2$

that is $b^2 = a^2 + c^2$

Worked problems on the theorem of Pythagoras

Problem 1. In a triangle PQR, $\angle Q = 90^\circ$, PQ = 3 cm, and RQ = 4 cm. Find the length of PR.

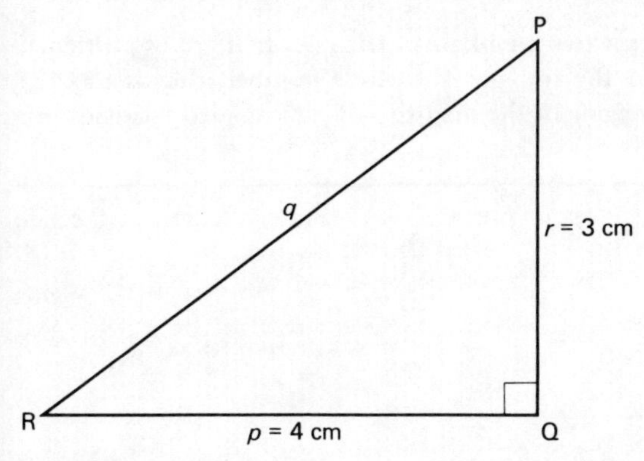

Fig. 9.33

The triangle PQR is shown in Fig. 9.33. By Pythagoras's theorem:

$$q^2 = p^2 + r^2$$

Therefore $q^2 = (4)^2 + (3)^2$

$$= 16 + 9$$

$$q^2 = 25$$

$$q = \sqrt{25} = \pm 5$$

The value -5 is meaningless in the context and is thus ignored.

Therefore PR = **5 cm**

Triangle PQR is known as the 3:4:5 triangle.

Problem 2. In a triangle XYZ, ∠ Y is a right angle, the hypotenuse is 13 cm long and the side XY is 5 cm long. Find the length of the side YZ.

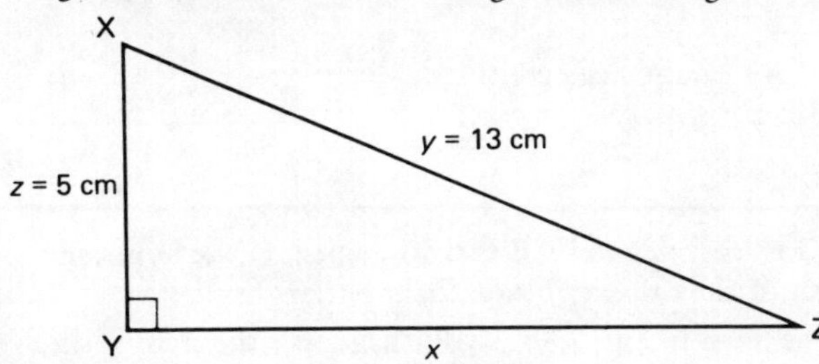

Fig. 9.34

The triangle XYZ is shown in Fig. 9.34. By Pythagoras's theorem:

$$y^2 = x^2 + z^2$$

Therefore $(13)^2 = x^2 + (5)^2$

$$169 = x^2 + 25$$

$$x^2 = 169 - 25$$

$$= 144$$

$$x = \sqrt{144}$$

$$= \pm 12$$

YZ = **12 cm** (-12 can be neglected as it has no meaning in this context.)

Triangle XYZ is known as the 5:12:13 triangle.

Problems 1 and 2 have shown two well-known triangles that are of particular interest because each side of the triangle is a whole number, that is a 3, 4, 5 triangle and a 5, 12, 13 triangle. In the majority of right angled triangles this is not found to be so.

Problem 3. In the triangle ABC shown in Fig. 9.35 find the length of the side b.

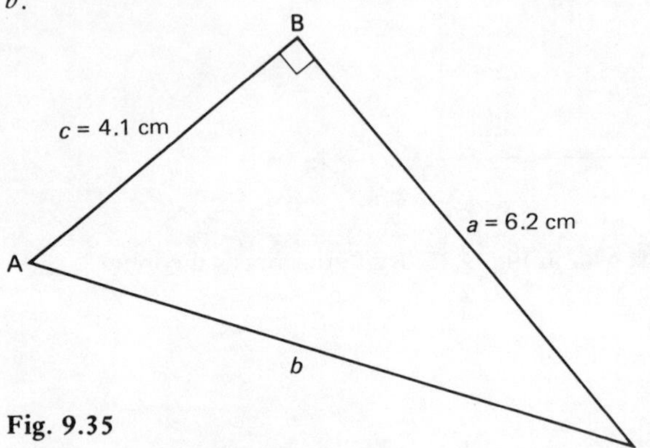

Fig. 9.35

By Pythagoras's theorem:

$$b^2 = a^2 + c^2$$
$$= (6.2)^2 + (4.1)^2$$
$$= 38.44 + 16.81$$
$$= 55.25$$

Therefore $b = \sqrt{55.25}$

$= \pm 7.43$ (-7.43 can be neglected)

Hence the side b is **7.43 cm**

Problem 4. Triangle EFG is isosceles and $\angle$ E is right angled. If the hypotenuse is 7.5 mm find the length of the other two sides of the triangle.

The triangle EFG is shown in Fig. 9.36. By Pythagoras's theorem:
$e^2 = f^2 + g^2$

However if triangle EFG is isosceles then $\angle G = \angle F = 45^\circ$ and side f = side g.

Therefore $e^2 = f^2 + f^2$

$$e^2 = 2f^2$$
$$(7.5)^2 = 2f^2$$
$$56.25 = 2f^2$$
$$f^2 = \frac{56.25}{2}$$
$$= 28.125$$
$$f = \sqrt{28.125}$$
$$= \pm 5.303$$

$= 5.30$ correct to 3 significant figures (-5.30 can be neglected).

Hence **side f = side g = 5.30 mm**

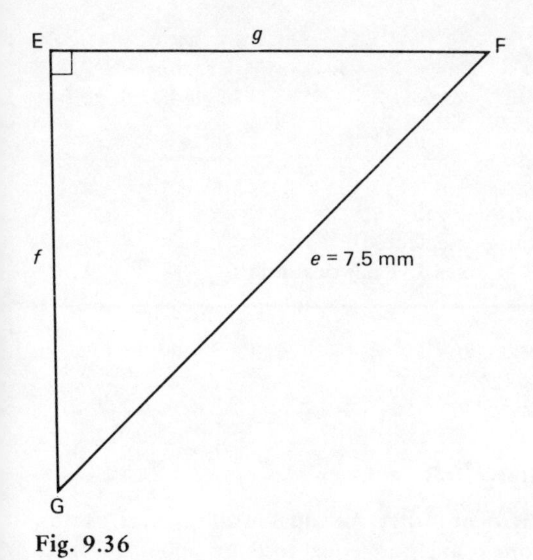

Fig. 9.36

Problem 5. Two ships leave a port together. One travels due south at 26 knots and the other due west at 35 knots. If 1 knot = 1 nautical mile per hour, find how far apart the two ships are after 3 hours.

After 3 hours the first ship would have travelled a distance of $3 \times 26 =$ 78 nautical miles due south.

Similarly, after 3 hours, the second ship would have travelled a distance of $3 \times 35 = 105$ nautical miles due west.

If, in Fig. 9.37, PR represents 78 nautical miles and PQ represents 105 nautical miles, then the distance apart after 3 hours is given by the distance QR.

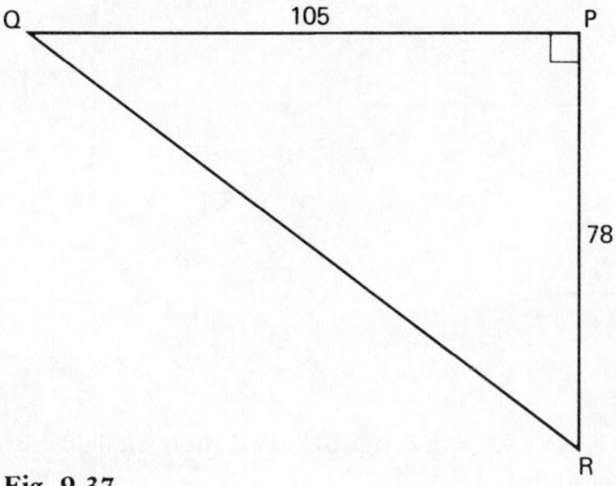

Fig. 9.37

By Pythagoras's theorem:

$$(QR)^2 = (PQ)^2 + (PR)^2$$

Therefore

$$QR^2 = (105)^2 + (78)^2$$
$$= 11\,025 + 6\,084$$
$$= 17\,109$$
$$QR = \sqrt{17\,109}$$
$$= \pm 131$$
$$= 131 \text{ } (-131 \text{ can be neglected})$$

Hence the distance apart after 3 hours is **131 nautical miles.**

Further problems on the theorem of Pythagoras may be found in Section 9.11 (Problems 25–35), page 257.

9.7 Congruent triangles

Two triangles are said to be **congruent** if they are equal in all respects, that is, three angles and three sides in one triangle are equal to three angles and three sides in the other triangle. Their areas must consequently be the same. Six facts determine a triangle completely – the three sides and the three angles.

However, to show that two triangles are congruent it is only necessary to show that three particular facts are equal in both triangles. There are only four possible combinations of facts and these are shown below:

1. Two triangles are congruent if the three sides of one are equal to the three sides of the other. (That is, side, side, side, which is sometimes written as *S.S.S.* for short.)

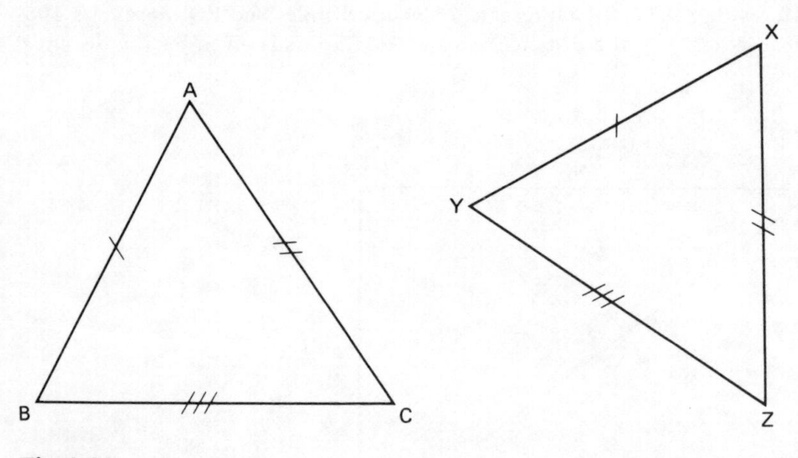

Fig. 9.38

Thus, in Fig. 9.38, if AB = XY, AC = XZ and BC = YZ then triangle ABC and triangle XYZ are congruent.

2. Two triangles are congruent if they have two sides of the one equal to two sides of the other, and if the angles included by these sides are equal. (That is, side, angle, side, which is sometimes written as *S.A.S.* for short.)

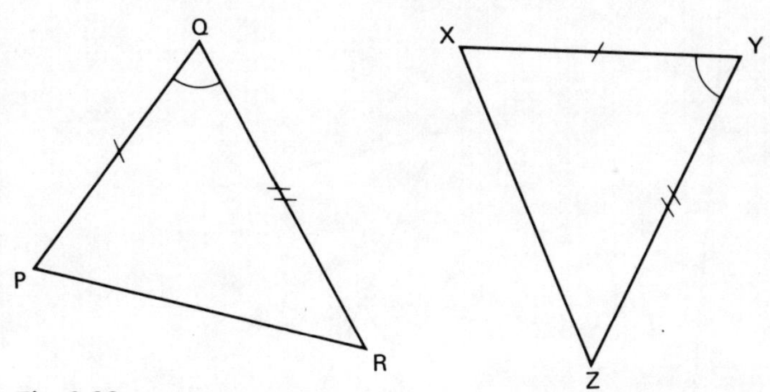

Fig. 9.39

If, in Fig. 9.39, PQ = XY, QR = YZ and ∠ Q = ∠ Y then triangle PQR and triangle XYZ are congruent.

3. Two triangles are congruent if two angles of the one are equal to two angles of the other, and any side of the first is equal to the corresponding side of the other. (That is, angle, side, angle, which is sometimes written as *A.S.A.* for short.)

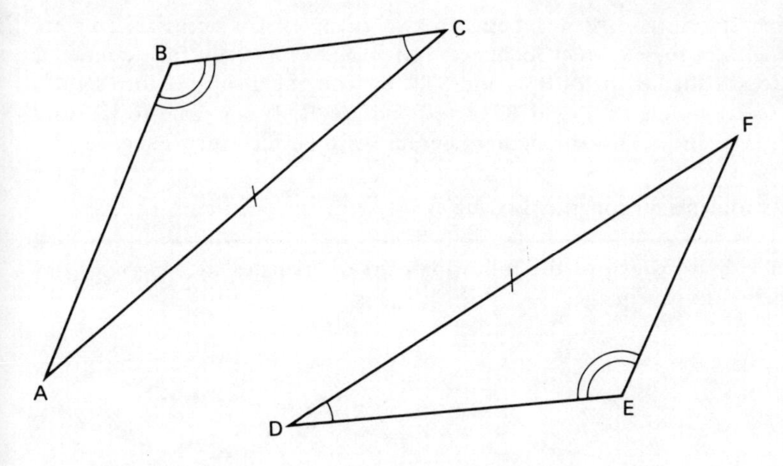

Fig. 9.40

If, in Fig. 9.40, AC = DF, ∠ C = ∠ D and ∠ B = ∠ E then triangle ABC and triangle FED are congruent.

Note that if two angles of one triangle are equal to two angles of the other triangle then the third angle in the two triangles are also equal since the sum of the interior angles of a triangle is 180°.

4. Two right-angled triangles are congruent if their hypotenuses are equal, and if one other side of one is equal to the corresponding side of the other. (That is, right-angle, hypotenuse, side, which is sometimes written as *R.H.S.* for short.)

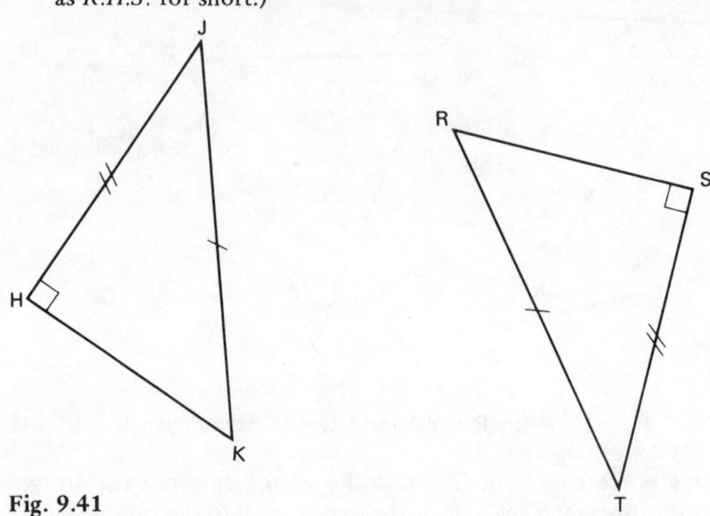

Fig. 9.41

If, in Fig. 9.41, JK = RT, HJ = ST and ∠ H = ∠ S = 90°, then triangle HJK and triangle STR are congruent.

When stating congruency between two triangles it is necessary to state the triangles in their correct sequence. Referring to Fig. 9.41, for example, it is true to say that triangles HJK and STR are congruent since H corresponds to S, J corresponds to T and K corresponds to R. It is untrue to say that triangles HJK and RST are congruent because of their incorrect sequence.

Worked problems on congruent triangles

Problem 1. State which of the following pairs of triangles are congruent and name their correct sequence.

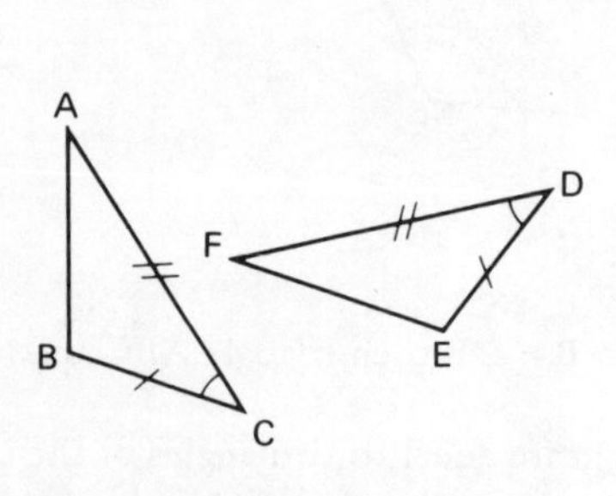

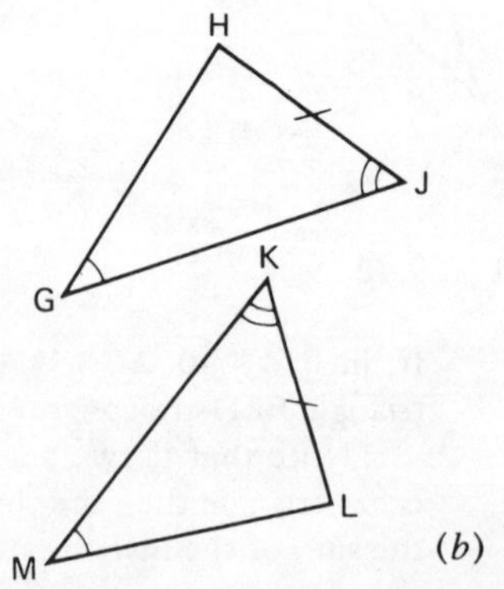

Fig. 9.42 (*a*) (*b*)

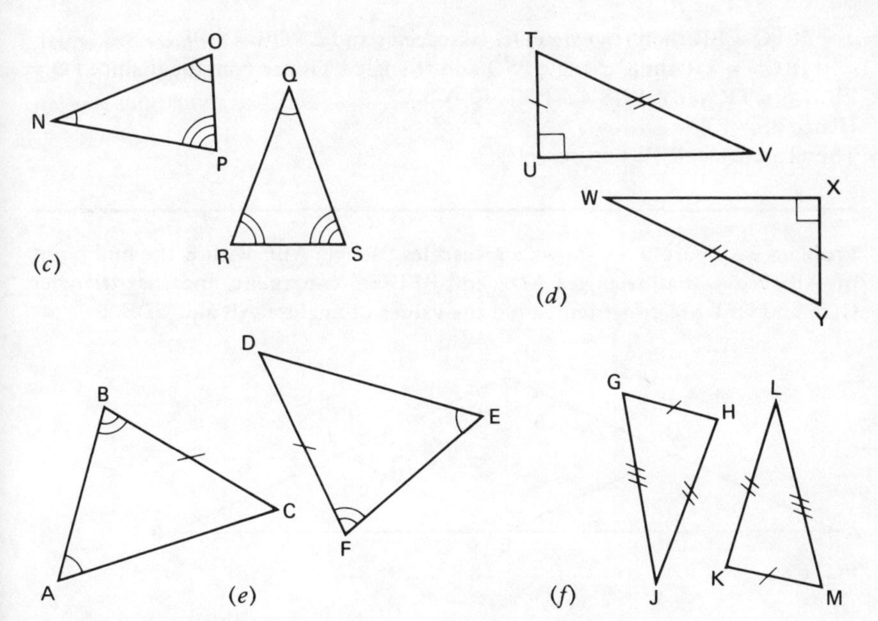

Fig. 9.42 (cont.)

(*a*) Congruent. ACB, FDE (*S.A.S.*)
(*b*) Congruent. GHJ, MLK (*A.S.A.*)
(*c*) Not necessarily congruent.
(*d*) Congruent. TUV, YXW (*R.H.S.*)
(*e*) Congruent. ABC, EFD (*A.S.A.*)
(*f*) Congruent. GHJ, MKL (*S.S.S.*)

Problem 2. If, in Fig. 9.43, PQ = PR and QS = TR show that triangle PST is isosceles.

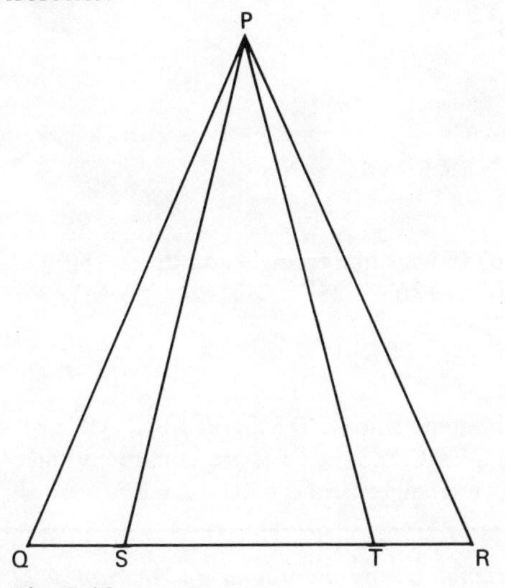

Fig. 9.43

If $PQ = PR$ then triangle PQR is isosceles and $\angle PQR = \angle PRQ$.

If $QS = TR$ then triangle PSQ and triangle PTR are congruent since $PQ = PR$, $QS = TR$ and $\angle PQS = \angle PRT$. (S.A.S.).
Hence $PS = PT$
Therefore triangle PST is isosceles.

Problem 3. Figure 9.44 shows an isosceles triangle ABC with E the mid point of AB. Prove that triangles ADE and BFE are congruent, and that triangles CDE and CFE are congruent. Find the values of angles CAE and CDE.

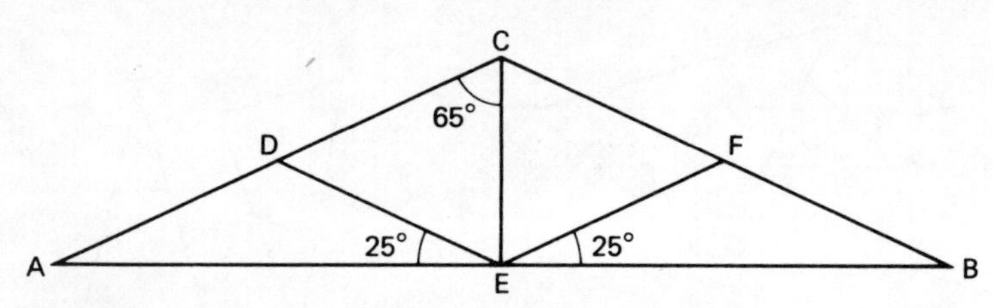

Fig. 9.44

Since triangle ABC is isosceles, $AC = CB$.

If E is the mid point of AB then triangles ACE and CEB are congruent (S.S.S.) and CE bisects the angle ACB and CE is perpendicular to AB, that is, $\angle AEC = 90^\circ$.
Therefore $\angle CED = 90^\circ - 25^\circ = 65^\circ$
Similarly $\angle CEF = 65^\circ$

In triangle CDE, $\angle CDE + 65^\circ + 65^\circ = 180^\circ$ (angles in a triangle add up to 180°)
Therefore $\angle CDE = 180^\circ - 65^\circ - 65^\circ$
$= 50^\circ$
Similarly $\angle CFE = 50^\circ$
$\angle ADE$ is supplementary to $\angle EDC$, hence
$\angle ADE = 180^\circ - 50^\circ$
$= 130^\circ$
Similarly $\angle BFE = 130^\circ$
Therefore $\angle DAE + 130^\circ + 25^\circ = 180^\circ$ (angles in a triangle add up to 180°)
$\angle DAE = 180^\circ - 130^\circ - 25^\circ$
$= 25^\circ$
Similarly $\angle FBE = 25^\circ$
Hence $\angle$ **CAE** $= \mathbf{25^\circ}$ and $\angle$ **CDE** $= \mathbf{50^\circ}$.

Triangles ADE and BFE are congruent since $\angle DAE = \angle FBE$, $AE = BE$ and $\angle DEA = \angle FEB$ (A.S.A.). Triangles CDE and CFE are congruent since $\angle DCE = \angle FCE$, CE is common to both triangles and $\angle CED = \angle CEF$.

Further problems on congruent triangles may be found in Section 9.11 (Problems 36–37), page 259.

9.8 Similar triangles

Two triangles are said to be **similar** if the angles of one triangle are equal to the angles of the other triangle.

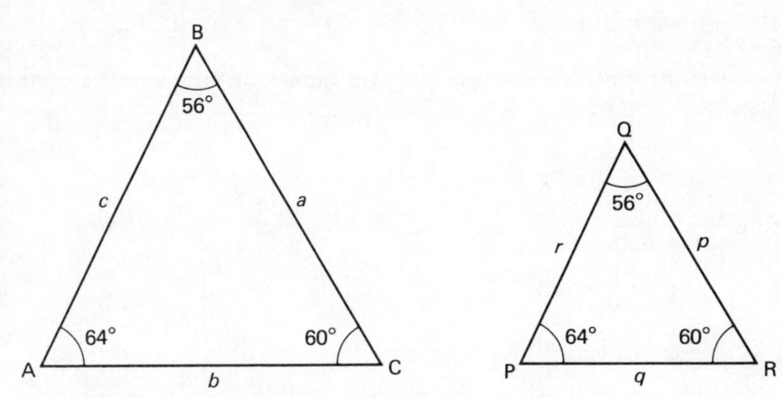

Fig. 9.45

Triangles ABC and PQR of Fig. 9.45 are similar triangles. However, unlike congruent triangles, the corresponding sides of the triangles are not the same, but are in proportion to each other. For example, if AB = 6 cm and PQ = 3 cm then by proportion the lengths of all the sides in triangle PQR will each be one half of the sides of triangle ABC.

Since the sides of the similar triangles are in proportion:

$$\frac{p}{a} = \frac{q}{b} = \frac{r}{c}$$

Worked problems on similar triangles

Problem 1. In Fig. 9.46 find the length of side b.

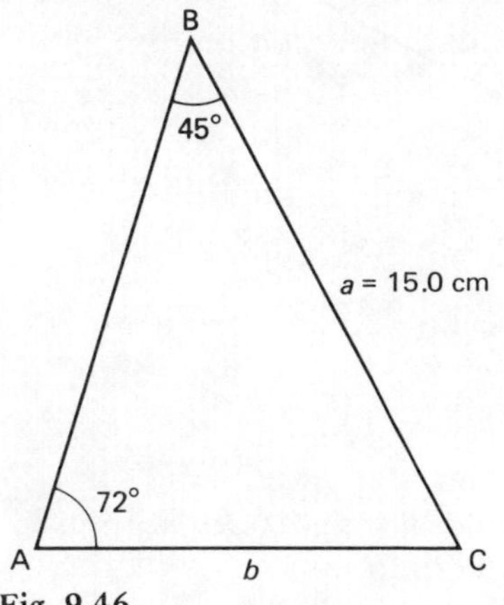

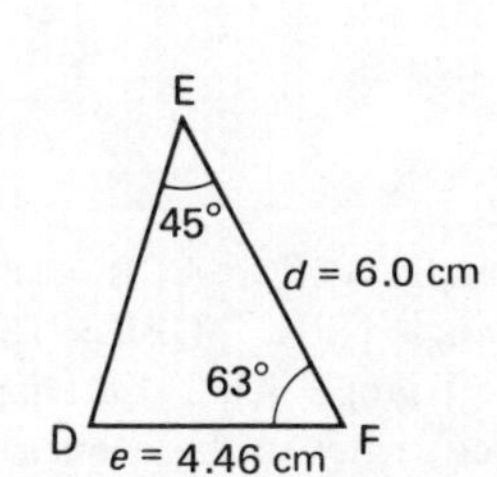

Fig. 9.46

In triangle ABC, $72^\circ + 45^\circ + \angle C = 180^\circ$ (angles in a triangle add up to 180°)

Therefore $\angle C = 180^\circ - 72^\circ - 45^\circ$

$\angle C = 63^\circ$

Similarly, in triangle DEF, $\angle D = 180^\circ - 63^\circ - 45^\circ$

Therefore $\angle D = 72^\circ$

Therefore triangle ABC and triangle DEF are similar, as their angles are the same. Hence by proportion:

$$\frac{a}{d} = \frac{b}{e}$$

$$\text{Therefore } \frac{15.0}{6.00} = \frac{b}{4.46}$$

$$b = \frac{15.0}{6.00}(4.46)$$

$$b = 11.15$$

Hence the length of side b = **11.15 cm**

Problem 2. In Fig. 9.47 find the dimensions a, b and q.

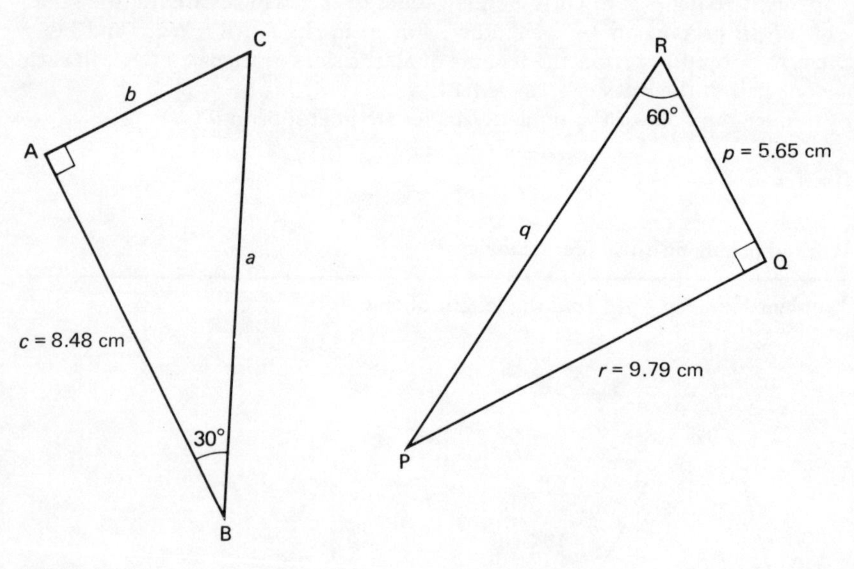

Fig. 9.47

In triangle ABC, $\angle ACB = 180^\circ - 90^\circ - 30^\circ = 60^\circ$

In triangle QPR, $\angle QPR = 180^\circ - 90^\circ - 60^\circ = 30^\circ$

Hence triangle ABC and triangle QPR are similar since three angles in one equals three angles in the other.

Redrawing the triangles gives Fig. 9.48.

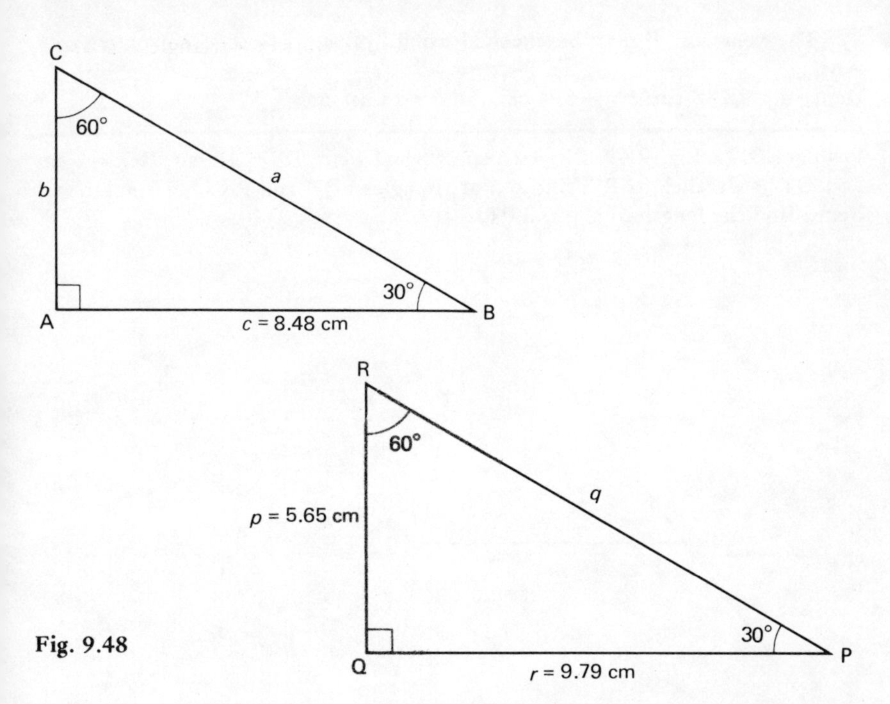

Fig. 9.48

Since the two triangles are similar their sides are in proportion:

Therefore $\dfrac{q}{a} = \dfrac{p}{b} = \dfrac{r}{c}$

$$\frac{q}{a} = \frac{5.65}{b} = \frac{9.79}{8.48}$$

$$b = 8.48 \left[\frac{5.65}{9.79}\right]$$

$\boldsymbol{b}$ **= 4.894 cm**

In triangle QPR, q can be found using the theorem of Pythagoras.

That is, $q^2 = p^2 + r^2$

$= (5.65)^2 + (9.79)^2$

$= 31.92 + 95.84$

$= 127.76 = 127.8$ (to 4 significant figures)

Therefore $q = \sqrt{127.8} = \pm 11.30$

$\boldsymbol{q}$ **= 11.30 cm** (-11.30 can be neglected).

By proportion: $\dfrac{a}{q} = \dfrac{c}{r}$

Therefore $\dfrac{a}{11.30} = \dfrac{8.48}{9.79}$

$$a = 11.30 \left[\frac{8.48}{9.79}\right]$$

$\boldsymbol{a}$ **= 9.788 cm**

The value of 'a' may be checked using Pythagoras's theorem on triangle ABC.

Hence a = **9.788 cm**, b = **4.894 cm** and q = **11.30 cm**

Problem 3. In Fig. 9.49, PR = 16 cm, PS = 13 cm, RS = 12 cm, RQ = 7 cm and QT is parallel to RS. Show that triangles PQT and PRS are similar and hence find the length of QT and PT.

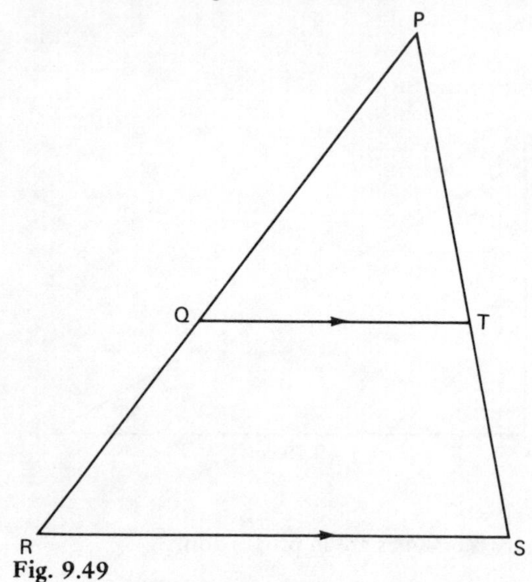

Fig. 9.49

Since QT is parallel to RS then:

∠PQT = ∠PRS (corresponding angles between parallel lines QT and RS) and ∠PTQ = ∠PSR (corresponding angles between parallel lines QT and RS)
Also, ∠P is common to triangle PQT and triangle PRS.

Since the three angles in triangle PQT are equal to the three angles in triangle PRS then the triangles are similar.

Hence the sides of triangles PQT and PRS are in proportion.

Therefore $\dfrac{PQ}{PR} = \dfrac{PT}{PS} = \dfrac{QT}{RS}$

$$\frac{16-7}{16} = \frac{PT}{13}$$

$$PT = 13\left(\frac{9}{16}\right)$$

$$\mathbf{PT = 7\tfrac{5}{16}\ cm\ or\ 7.31\ cm}$$

Also $\dfrac{9}{16} = \dfrac{QT}{12}$

Therefore $QT = 12\left(\dfrac{9}{16}\right)$

$$\mathbf{QT = 6\tfrac{3}{4}\ cm\ or\ 6.75\ cm}$$

Problem 4. In Fig. 9.50, prove that triangle QST is similar to triangle QRS. If RT = 10 cm, RS = 6 cm and ST = 8 cm find the length of SQ.

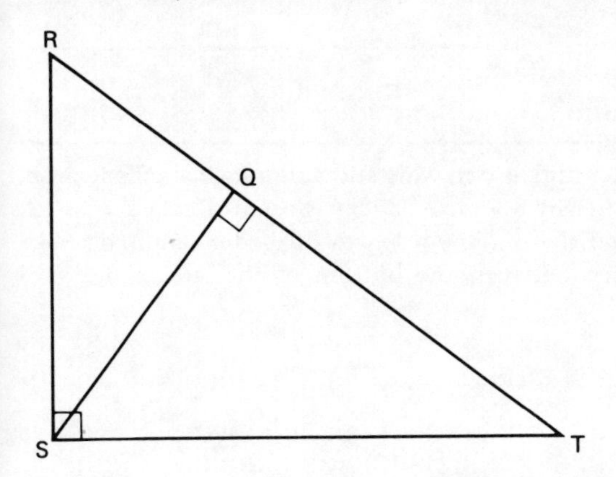

Fig. 9.50

Now, we are given

$\angle$ TSQ + $\angle$ QTS = 90° (1)

Also

$\angle$ SRT + $\angle$ QTS = 90° (The two other angles in a right-angled triangle add up to 90°) (2)

Therefore $\angle$ SRT = $\angle$ TSQ.

Hence triangles QST and SRT each contain a right angle and another pair of corresponding angles. Therefore they are similar triangles. Also, $\angle$ QRS and $\angle$ SRT are equal.

Therefore, triangles QRS and STQ are similar.

Redrawing two of the triangles, RST and SQT separately gives Fig. 9.51.

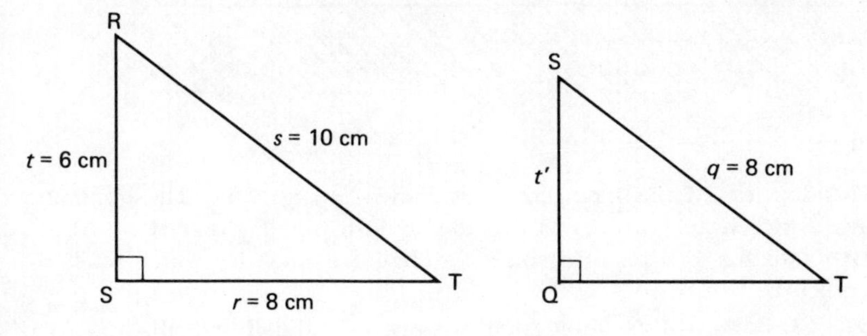

Fig. 9.51

Hence by proportion:

$$\frac{q}{s} = \frac{t'}{t}$$

$$\text{Therefore } \frac{8}{10} = \frac{t'}{6}$$

$$t' = 6\left(\frac{8}{10}\right)$$

$$= 4.8$$

Hence the length SQ = **4.8 cm**

Problem 5. A rectangular water tank 4.0 m wide and 5.0 m high stands against a perpendicular building of height 8.5 m. A ladder is used to gain access to the roof of the building. Find the minimum length of ladder required, and, using this ladder the distance between the bottom of the ladder and the building.

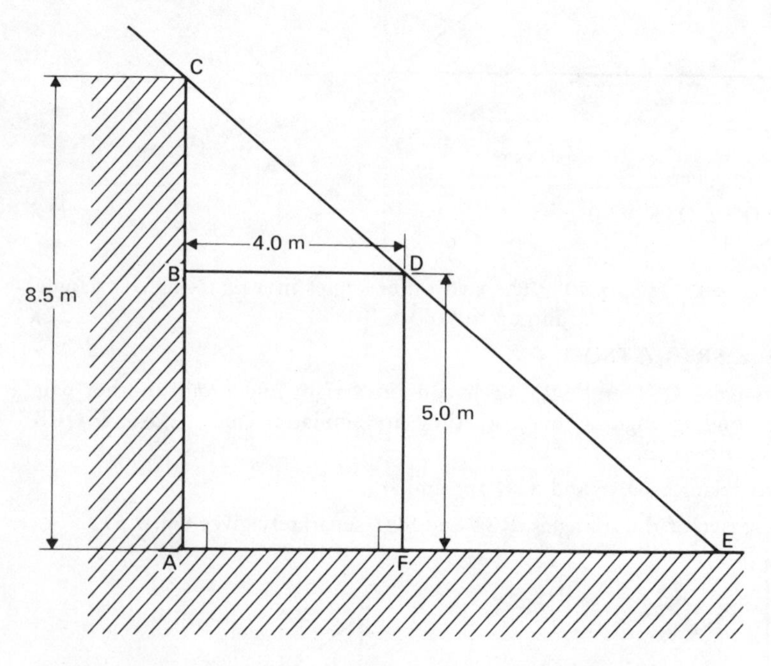

Fig. 9.52

A side view of the arrangement is shown in Fig. 9.52. The minimum distance between the bottom of the ladder and the building is given by AE.

BD and AE are parallel since the building and the water tank are perpendicular.

$\angle$ CDB = $\angle$ DEF (corresponding angles between parallel lines BD and AE). Similarly, $\angle$ BCD = $\angle$ FDE.

Therefore, triangle BCD is similar to triangle FDE since three angles in one equals three angles in the other.

If AC = 8.5 m and AB (= DF) = 5.0 m

then BC = (8.5 − 5.0)m = 3.5 m

By proportion: $\dfrac{BC}{FD} = \dfrac{BD}{FE}$

Therefore $\dfrac{3.5}{5.0} = \dfrac{4.0}{FE}$

$$FE = 4.0\left[\frac{5.0}{3.5}\right]$$

$$= 5.71 \text{ m}$$

Therefore **the distance of the bottom of the ladder from the building is 5.71 m + 4.0 m = 9.71 m.**

If AC = 8.5 m and AE = 9.71 m then the length CE (i.e. the minimum length of the ladder) may be found using Pythagoras's theorem.

Hence $(CE)^2 = (AC)^2 + (AE)^2$

$= (8.5)^2 + (9.71)^2$

$= 72.25 + 94.28$

$= 166.53 = 167$ (to 3 significant figures)

Therefore $CE = \sqrt{167} = \pm 12.9$

$= 12.9$ m (-12.9 can be neglected)

Hence a ladder of minimum length 12.9 m is required.

Further problems on similar triangles may be found in Section 9.11 (Problems 38–47) page 260.

9.9 Construction of triangles

To construct any triangle the following drawing instruments are needed:

(*i*) ruler and/or straight edge;
(*ii*) compass;
(*iii*) protractor;
(*iv*) pencil; and
(*v*) set square (possibly)

The method of constructing triangles depends on the available information. It was mentioned in Section 9.7 that six facts determine a triangle completely – the three sides and the three angles. When constructing triangles only three particular facts are required and the only four possible combinations are shown, by examples, in cases 1–4 below.

Case 1. Given the lengths of the three sides

Example: To construct a triangle whose sides are 7 cm, 5 cm and 4 cm.

(*i*) Let one side, say that 7 cm long be the base. Draw a straight line of any length, and, with a pair of compasses, mark out a 7 cm length and label it AB as shown in Fig. 9.53(*i*)

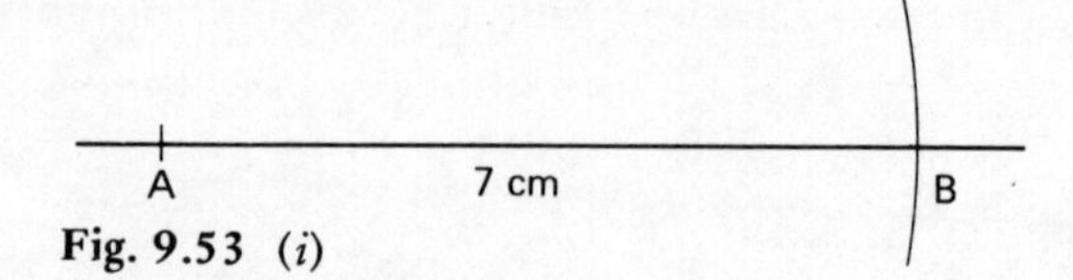

Fig. 9.53 (*i*)

(*ii*) Using a compass set to 5 cm and, with centre at A describe the arc MN as shown in Fig. 9.53(*ii*)

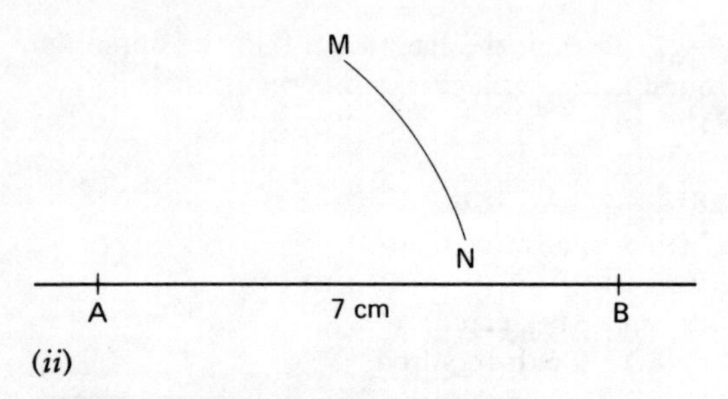

(*ii*)

(*iii*) Set the compass to 4 cm and, with centre at B, describe the arc OP as shown in Fig. 9.53(*iii*)

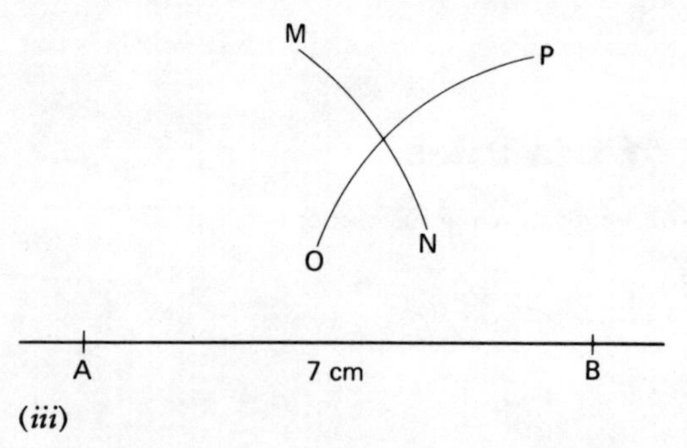

(*iii*)

(*iv*) The intersection of the two arcs (produced if necessary) at C is a vertex of the required triangle. Thus join AC and BC by straight lines as shown in Fig. 9.53(*iv*)

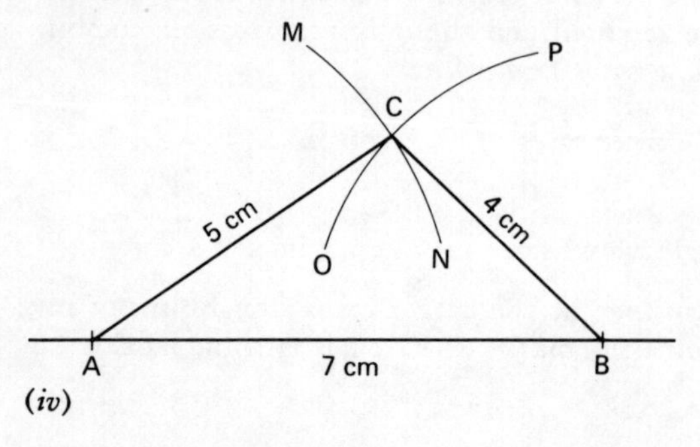

(*iv*)

Thus triangle ABC with sides 7 cm, 5 cm and 4 cm has been constructed.

It was stated in Section 9.5 that the ratio of the angles of a triangle is not equal to the ratio of the sides. This can be proved by referring to Fig. 9.53(*iv*).

BC = 4 cm. As a proportion of the perimeter of the triangle ABC

$$BC = \frac{4}{4+5+7} = \frac{4}{16} = \frac{1}{4}$$

Similarly, AC = $\frac{5}{16}$ of the perimeter and

AB = $\frac{7}{16}$ of the perimeter.

If it is assumed that the ratio of the angles is equal to the ratio of the sides then:

$\angle$ A (the angle opposite side BC) should be $\frac{1}{4}$ of 180° that is, 45°;
$\angle$ B (the angle opposite AC) should be $\frac{5}{16}$ of 180°, that is, 56°; and
$\angle$ C (the angle opposite AB) should be $\frac{7}{16}$ of 180°, that is, 79°.

Using a protractor, the angles of triangle ABC can be measured, with the following values resulting:

$\angle A = 34°; \angle B = 44\frac{1}{2}°; \angle C = 101\frac{1}{2}°$

It can therefore be clearly seen that **the ratio of the angles of a triangle is not equal to the ratio of the sides.**

Case 2. Given the lengths of two sides and the included angle

Example: To construct a triangle given two sides, of 6 cm and 3 cm, and the angle between these two sides as 50°.

(*i*) Let one side, say that 6 cm long, be the base. Draw a line 6 cm long and label it PQ as shown in Fig. 9.54(*i*)

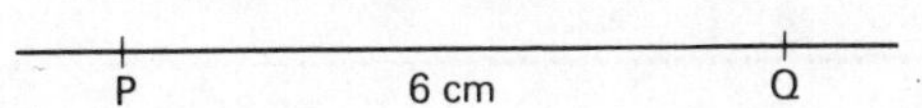

Fig. 9.54 (*i*)

(*ii*) Using a protractor centred at P make an angle of 50° to PQ as shown in Fig. 9.54(*ii*)

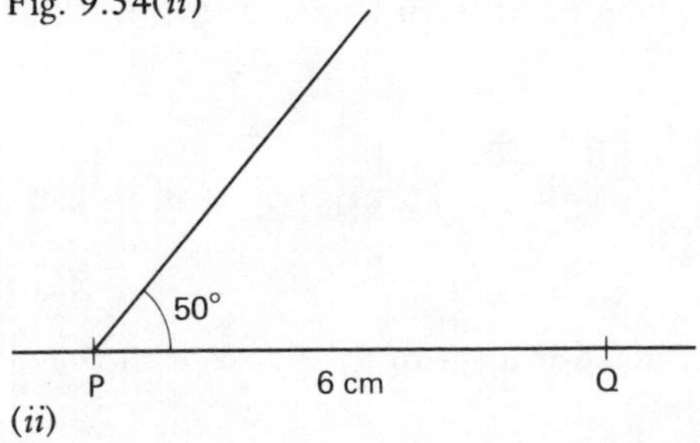

(*ii*)

 (*iii*) From P, measure a length of 3 cm and label the length PR, as shown in Fig. 9.54(*iii*)

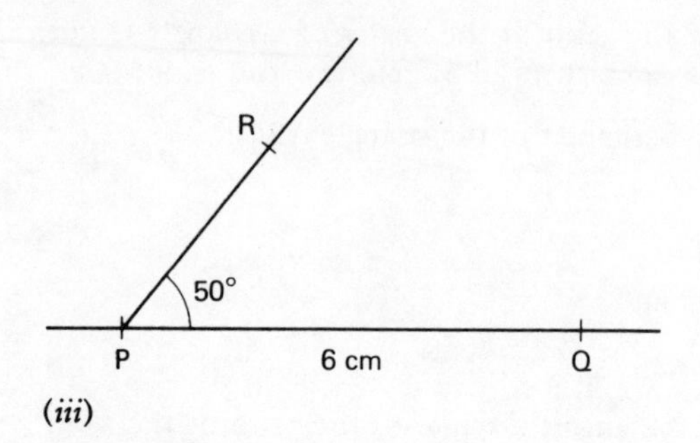

(*iii*)

(*iv*) Join the point Q to R by a straight line as shown in Fig. 9.54(*iv*)

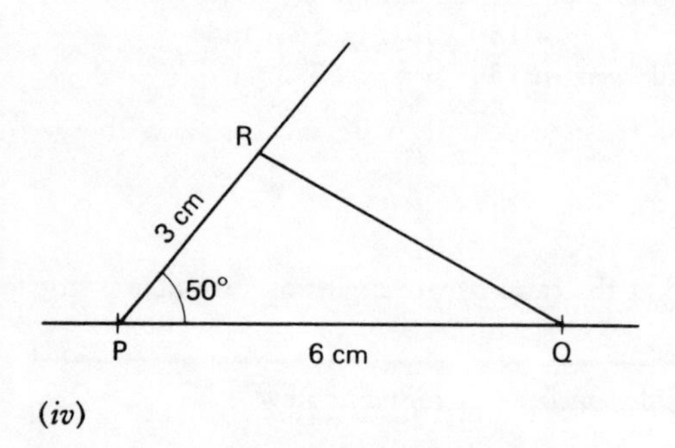

(*iv*)

Thus, given two sides, of 6 cm and 3 cm, and an included angle of 50°, a triangle PQR has been constructed.

Case 3. Given the length of one side and the value of two angles

Example (a): To construct a triangle XYZ given that XY = 8 cm, $\angle$ X = 63° and $\angle$ Y = 42°

(*i*) Let the 8 cm side be the base. Draw a line 8 cm long and label it XY as shown in Fig. 9.55(*i*)

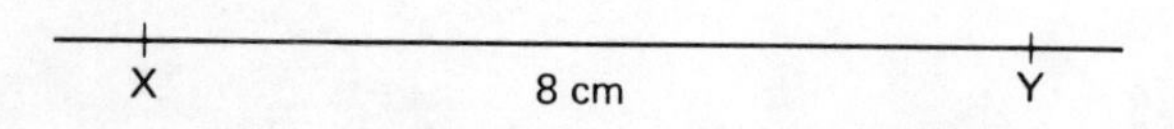

Fig. 9.55 (*i*)

(*ii*) Using a protractor centred at X make an angle of 63° to XY as shown in Fig. 9.55(*ii*)

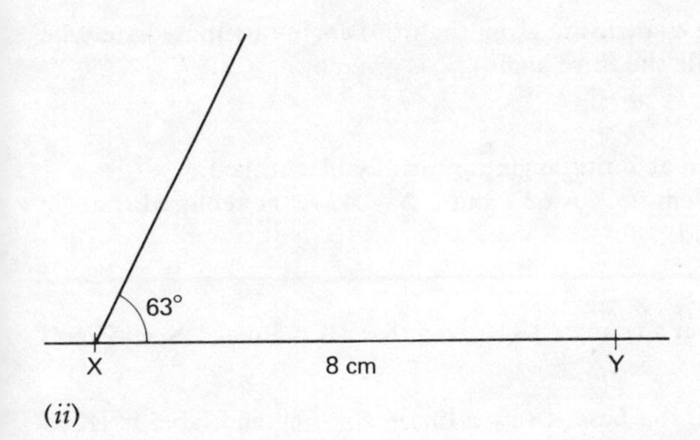

(*ii*)

(*iii*) Using a protractor centred at Y make an angle of 42° to XY as shown in Fig. 9.55(*iii*)

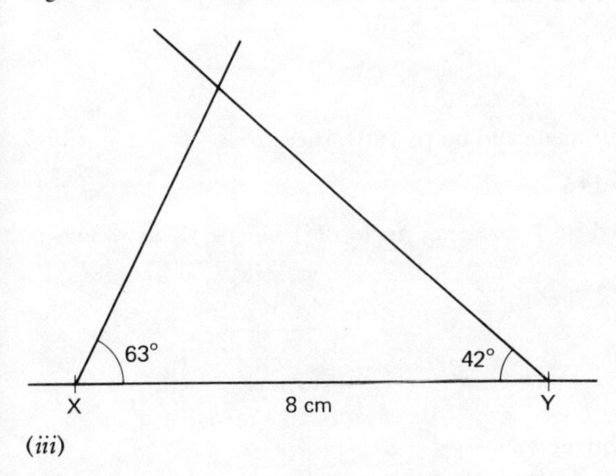

(*iii*)

(*iv*) The intersection of the arms (produced if necessary) form the vertex, Z, of the triangle as shown in Fig. 9.55(*iv*)

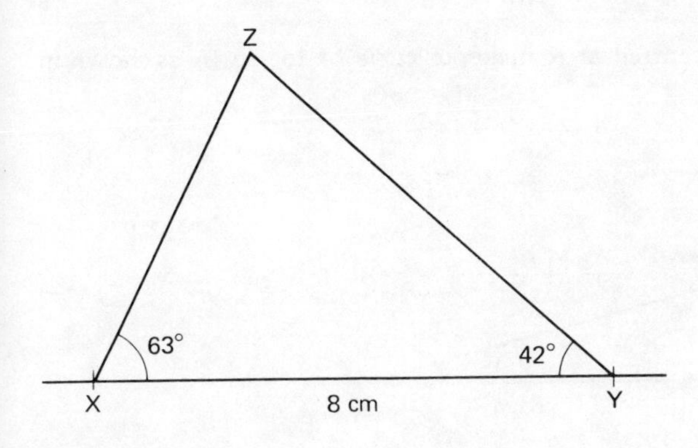

(*iv*)

Wherever two angles of a triangle are given the third angle can immediately be calculated. In this example the third angle $\angle Z$ is given by:

$\angle Z = 180° - 63° - 42° = 75°$

Using a protractor centred at Z this angle can readily be checked.

Thus, given XY = 8 cm, $\angle X = 63°$ and $\angle Y = 42°$, the required triangle XYZ has been constructed.

Example (b): To construct a triangle JKL given that JK = 5 cm, $\angle K = 15°$ and $\angle L = 19°$

(*i*) Let the 5 cm side be the base. Draw a line 5 cm long and label it JK as shown in Fig. 9.56(*i*)

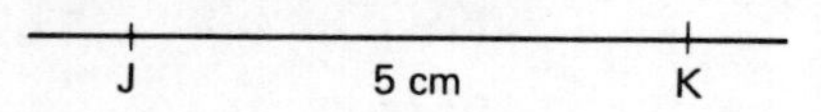

Fig. 9.56 (*i*)

(*ii*) The three angles of a triangle add up to 180°, therefore

$\angle J = 180° - 15° - 19° = 146°$

Using a protractor centred at J make an angle of 146° to JK as shown in Fig. 9.56(*ii*)

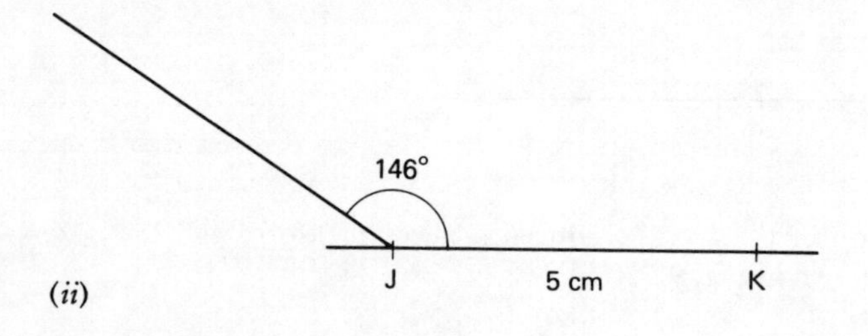

(*ii*)

(*iii*) Using a protractor centred at K make an angle of 15° to JK as shown in Fig. 9.56(*iii*)

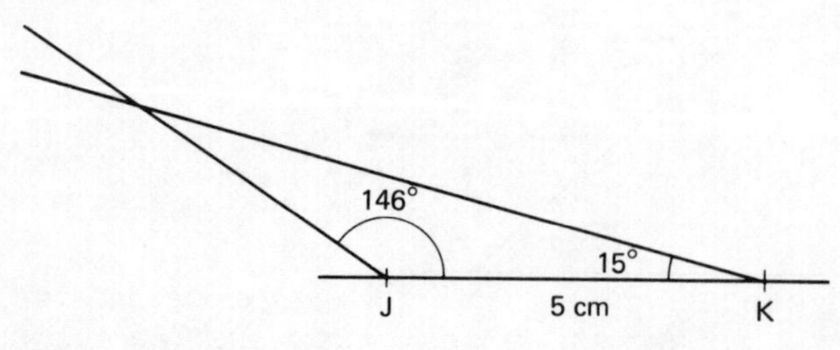

(*iii*)

(*iv*) The intersection of the arms (produced if necessary) form the vertex, L, of the triangle as shown in Fig. 9.56(*iv*). (The angle L must be equal to 19° and this should be checked using a protractor centred at L.)

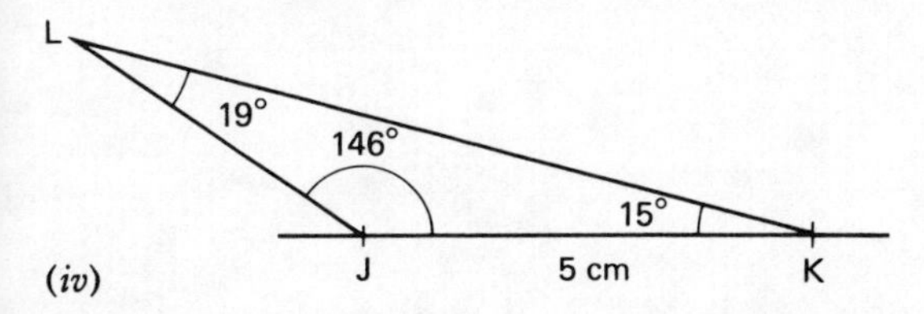

(*iv*)

Thus, given JK = 5 cm, ∠ K = 15° and ∠ L = 19°, the required triangle JKL has been constructed.

Case 4. To construct a triangle given one side, the hypotenuse and a right angle

Example: To construct a triangle MNO given that MN = 6 cm, the hypotenuse, NO = 9 cm, and ∠ M = 90°

(*i*) Let the 6 cm side be the base. Draw a line 6 cm long and label it MN as shown in Fig. 9.57(*i*)

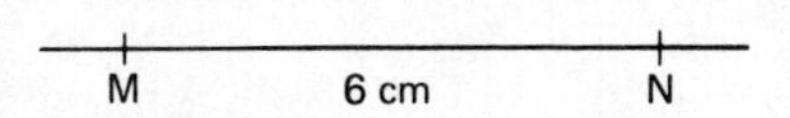

Fig. 9.57 (*i*)

(*ii*) Produce the line MN any distance to P. With compass centred at M make an arc at A and A′. (The length AM is arbitrary.) With compass centred at A, draw an arc BC greater in length than AM. With the same compass setting, and centred at A′, draw the arc DE. This is shown in Fig. 9.57(*ii*).

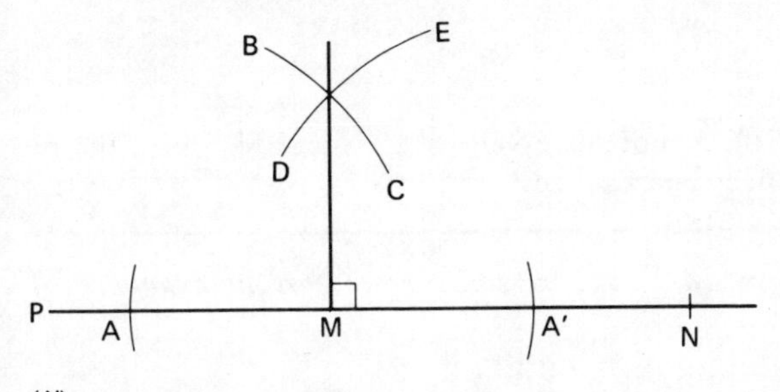

(*ii*)

Join the intersection of the arcs to M and a right angle to MN is produced at M, as shown in Fig. 9.57(*iii*). (Alternatively, a protractor or set square can be used to construct a 90° angle.)

 (*iii*) The hypotenuse is always opposite the right angle. Thus the side NO is opposite ∠ M. Using a compass, centred at N and set to 9 cm describe the arc EF as shown in Fig. 9.57(*iii*)

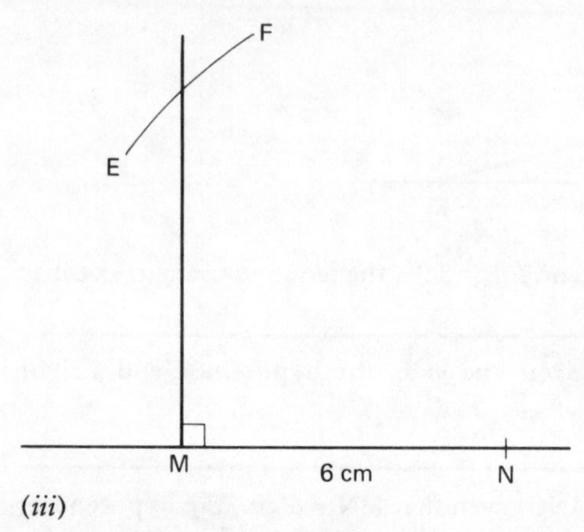

(*iii*)

(*iv*) The intersection of the arms (produced if necessary) form the vertex, O, of the required triangle. Join ON by a straight line as shown in Fig. 9.57(*iv*)

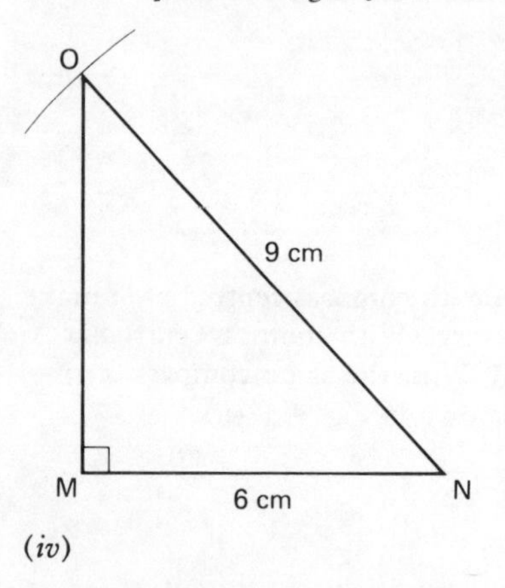

(*iv*)

Thus, given MN = 6 cm, the hypotenuse NO = 9 cm and ∠ M = 90°, the required triangle MNO has been constructed.

In Section 9.11 (Problem 48), page 263, *there are further examples of triangles for construction.*

9.10 Terminology and properties of the circle

1. A circle is a plane figure enclosed by a curved line, every point on which is equidistant from a point within, called the **centre**. In Fig. 9.58, O is the centre of the circle.

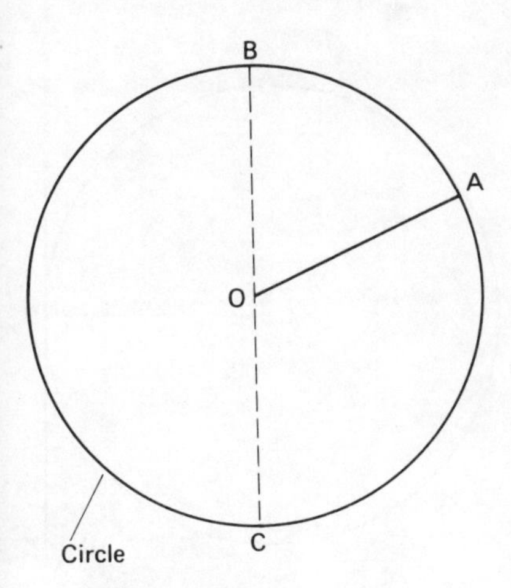

Fig. 9.58

2. The distance from the centre to the curve, OA, is called the **radius,** r, of the circle.
3. Any straight line passing through the centre and touching the circumference at each end as BC does in Fig. 9.58 is called the **diameter,** d. The diameter is twice the length of the radius. Thus $d = 2r$.
4. The boundary of a circle, that is, the perimeter, is called the **circumference,** c.
   It is found that if the length of the circumference, c, of any circle is divided by the length of the diameter, d, the answer is always the same, that is, $\frac{c}{d}$ is a constant.

   The constant is denoted by a Greek letter π (pronounced as 'pie'), where $\pi = 3.141\,59$ to 5 decimal places.

   Thus $\quad \frac{c}{d} = \pi$

   or $\quad c = \pi d$

   But $\quad d = 2r$

   Therefore $c = 2\pi r$

   Often the value of π is taken as 3.142 or 3.14 and sometimes $\frac{22}{7}$, depending on the accuracy required.
5. A **semicircle** is one half of the whole circle.
6. A **quadrant** is one quarter of a whole circle.
7. A **tangent** to a circle is a straight line which meets the circle in one point only and does not cut the circle when produced. In Fig. 9.59, FG is a tangent to the circle, since it touches the circle at the point H only. If radius OH is drawn, as in Fig. 9.60, then $\angle$ FHO is a right angle, that is, 90°
   Thus a tangent to a circle is at right angles to the radius drawn from the point where the tangent meets the circle.

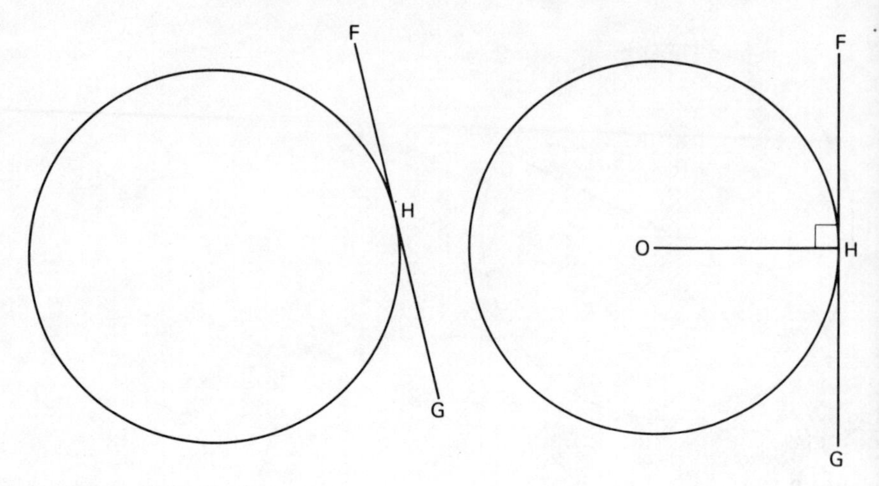

Fig. 9.59 Fig. 9.60

8. A **sector** of a circle is the part of a circle between radii. In Fig. 9.61, the shaded portion OXY is called a sector. If the sector is less than a semicircle, as the shaded portion is in Fig. 9.61, it is called a **minor sector.**

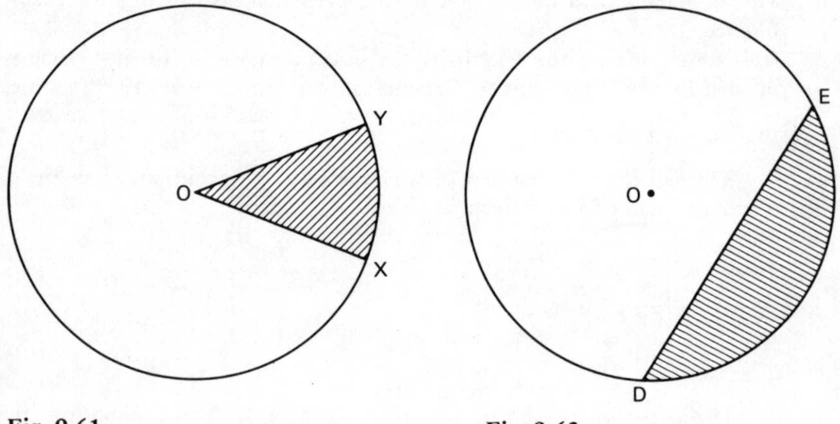

Fig. 9.61 Fig. 9.62

If the sector is greater than a semicircle, as is the unshaded portion of Fig. 9.61, it is called a **major sector.**

9. A **chord** of a circle is any straight line which divides the circle into two parts, and is terminated at each end by the circumference. In Fig. 9.62, the line DE is a chord.
10. A **segment** is the name given to the parts into which a circle is divided by a chord. In Fig. 9.62, DE divides the circle into two parts. Each of the two parts are called segments. If the segment is less than a semicircle it is called a **minor segment.** The shaded area of Fig. 9.62 is therefore a minor

segment. If the segment is greater than a semicircle it is called a **major segment.** The unshaded area of Fig. 9.62 is therefore called a major segment.

11. An **arc** is a portion of the circumference of a circle. In Fig. 9.63, PR is a chord which divides the circle into a minor segment PQR and a major segment PSR.

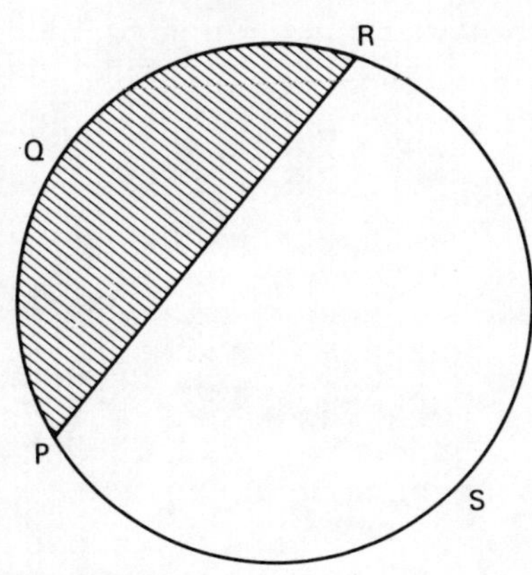

Fig. 9.63

The shorter distance from P to R along the circumference, that is, through Q, is called the **minor arc.** The longer distance from P to R along the circumference, that is, through S, is called the **major arc.**

There is a direct relationship between the length of an arc and the angle subtended at the centre of a circle. In Fig. 9.64 let $\angle$ AOB = 30° and $\angle$ BOC = 30°

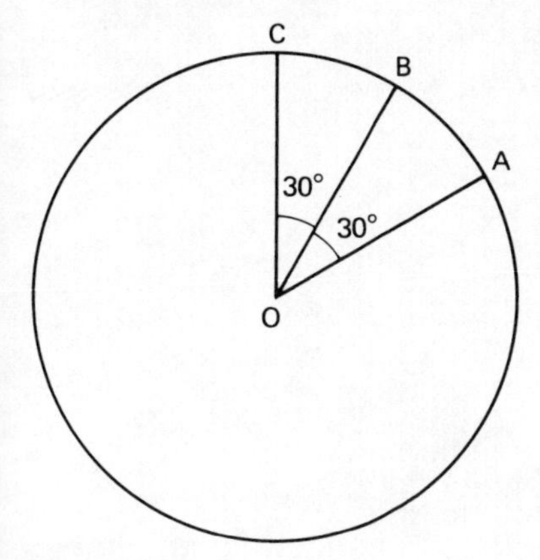

Fig. 9.64

As the angles subtended at the centre of the circle are the same for each of the sectors AOB and BOC, the length of the arc AB is equal to the length of the arc BC, that is, **the arc length is proportional to the angle which it subtends at the centre.** For example the length of arc for a 60° sector is twice that of a 30° sector. If the angle subtended at the centre of a sector is 60°, then, since the circumference subtends an angle of 360°, the length of the arc would, by proportion, be $\frac{60}{360}$, that is $\frac{1}{6}$, of the total circumference.

12. The angle at the centre of a circle, subtended by an arc, is double the angle at the circumference subtended by the same arc, i.e. from fig. 9.65, **∠ AOB = 2 × ∠ ACB.**

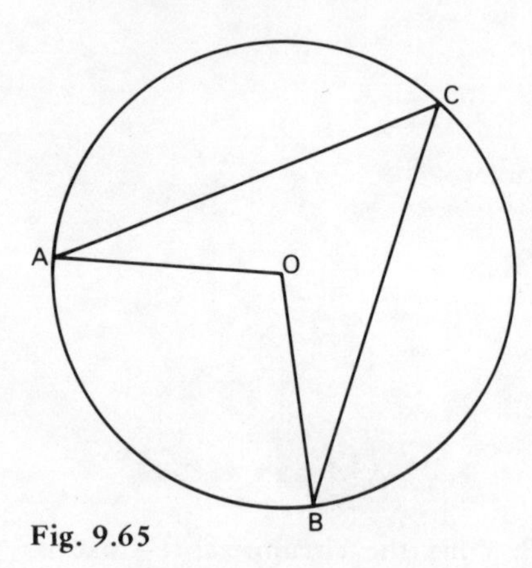

Fig. 9.65

13. An angle may be measured in degrees or **radians.**
A radian is defined as the angle subtended at the centre of a circle by an arc equal in length to the radius.

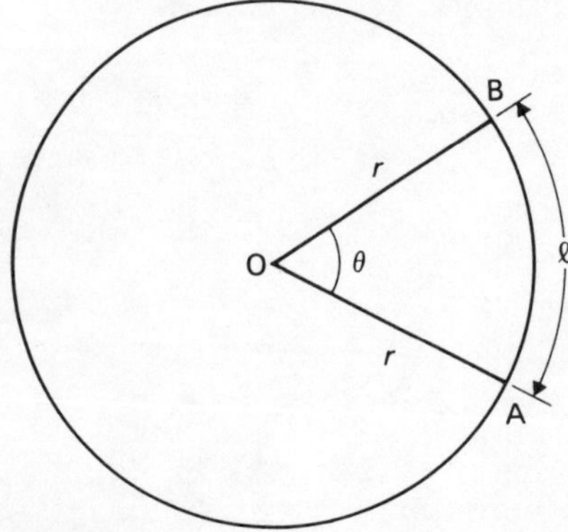

Fig. 9.66

Thus, in Fig. 9.66, when $r = l$ then angle θ is 1 radian. This angle is the unit of measurement in so called circular measure. It is of constant size whatever the length of the radius.

The length of arc, l, for 1 radian is equal to r
The length of arc, l, for 2 radians is equal to $2r$
The length of arc, l, for θ radians is equal to $r\theta$

Therefore generally, $l = r\theta^c$ or $\theta^c = \dfrac{l}{r}$

'θ radians' may be written as 'θ^c', the 'c' standing for circular measure.

The number of radians subtended by the whole circumference of a circle is given by the number of times the radius is contained in the circumference.

Now the circumference $= 2\pi r$. Hence the number of radians in one complete revolution is given by:

$$\frac{2\pi r}{r} = 2\pi = 6.283 \text{ (correct to 3 decimal places)}$$

Hence 2π radians = 360°

$$\textbf{1 radian} = \left(\frac{360}{2\pi}\right)^\circ$$

$$= \mathbf{57.30^\circ} \text{ (correct to 2 decimal places)}$$

There are therefore just over 6 radians contained within a complete revolution.

Whenever conversion from radians to degrees or vice versa is necessary, either of the above relationships may be used, or conversion tables are readily available. Some calculators have the facility to convert from radians to degrees and vice versa at the press of a button.

The following relationships are often used and can be derived from $2\pi^c = 360^\circ$:

$\pi^c = 180^\circ$; $\quad \dfrac{\pi^c}{2} = 90^\circ$; $\quad \dfrac{\pi^c}{3} = 60^\circ$;

$\dfrac{\pi^c}{4} = 45^\circ$; $\quad \dfrac{\pi^c}{6} = 30^\circ$; $\quad \dfrac{\pi^c}{12} = 15^\circ$;

and so on.

Worked problems on the properties of circles

In the following problems it is assumed that $\pi = 3.142$

Problem 1. Find the circumference of a circle of radius 15.0 cm

Circumference, $c = 2\pi$ (radius)

That is $c = 2\pi r$

$= (2)(3.142)(15.0)$

$= 94.26$ cm

Hence the circumference of the circle is **94.26 cm**

 Problem 2. Find the circumference of a circle of diameter 12.0 mm

The circumference $c = \pi d$

$= (3.142)(12.0)$

$= 37.70$ mm

Hence the circumference of the circle is **37.70 mm**

Problem 3. Find the radius of a circle if its perimeter is 54.0 cm

Perimeter = circumference, $c = 54$ cm

$c = 2\pi r$

Therefore $r = \dfrac{c}{2\pi}$

$= \dfrac{54}{(2)(3.142)}$

$r = 8.593$ cm

Hence the radius of the circle is **8.593 cm**

Problem 4. Figure 9.67 shows a tangent XY touching a circle at Y. If the radius of the circle is 50 mm and the distance XY is 152 mm, find the distance OX in centimetres.

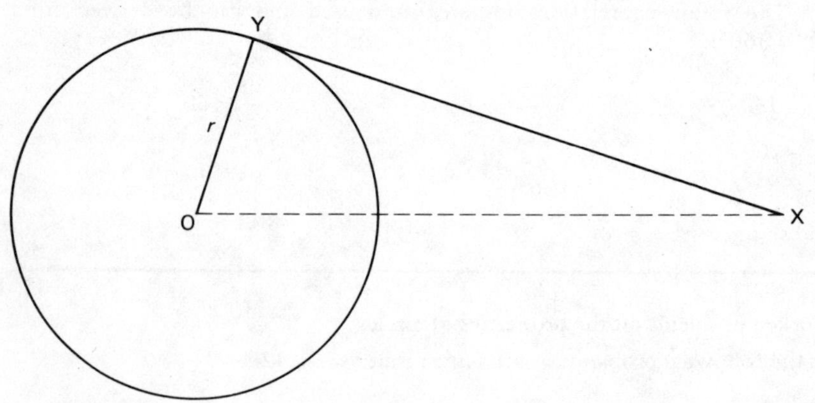

Fig. 9.67

A tangent to a circle is at right angles to a radius drawn from the point of contact, in this case Y. Hence $\angle XYO = 90°$.

Therefore, as XYO is a right-angled triangle, Pythagoras's theorem may be used.

Thus, $(OX)^2 = (XY)^2 + (OY)^2$

$= (152)^2 + (50)^2$

$= 23\,104 + 2\,500$

$= 25\,604$

Therefore $OX = \sqrt{25\ 604}$
$= 160.0$ mm
Hence the distance OX is **16.0 cm**

Problem 5. Without the use of conversion tables convert:
(*a*) 119° to radians (*b*) 73° 33′ to radians
(*c*) 2.681 radians to degrees (*d*) $\frac{3\pi}{7}$ radians to degrees.
Verify the results by using conversion tables.

(*a*) If $180° = \pi$ rad

then $1° = \frac{\pi}{180}$ rad

Therefore $119° = 119\left[\frac{\pi}{180}\right]$ rad

$= 119\left[\frac{3.142}{180}\right]$ rad

$= 2.077$ rad

Therefore 119 degrees = **2.077 radians**

(*b*) 73° 33′ is a mixture of degrees and minutes and it is first necessary to convert to degrees and decimals of a degree.

$33' = \frac{33°}{60} = 0.55°$

Therefore $73°\ 33' = 73.55°$

If $180° = \pi$ rad

then $73.55° = \frac{73.55(\pi)}{180} = \frac{73.55\ (3.142)}{180}$

$= 1.284$ rad

Therefore 73° 33′ = **1.284 radians**

(*c*) π rad $= 180°$

Therefore 1 rad $= \frac{180°}{\pi}$

2.681 rad $= 2.681\left[\frac{180}{\pi}\right]^{\circ}$

$= 153.6°$

$0.6° = (0.6 \times 60)'$

$= 36'$

$153.6° = 153°\ 36'$

Therefore 2.681 radians = **153° 36′**

(*d*) π rad $= 180°$

Therefore 1 rad $= \left[\frac{180}{\pi}\right]^{\circ}$

$\frac{3\pi}{7}$ rad $= \left[\frac{3\pi}{7}\right]\left[\frac{180}{\pi}\right]^{\circ}$

$$= \frac{3}{7}(180)^\circ$$

$$= 77.14^\circ$$

Now $\quad 0.14^\circ = (0.14 \times 60)'$

$$= 8.4'$$

and $\quad 0.4' = (0.4 \times 60)''$

$$= 24''$$

So $\quad 77.14^\circ = 77^\circ\ 8'\ 24''$

Therefore $\frac{3\pi}{7}$ radians = $77^\circ\ 8'\ 24''$

Problem 6. Express the following angles in radians in terms of π:
(*a*) 150° (*b*) 270° (*c*) 37.5° (*d*) 383° 17′ 23″

(*a*) $180^\circ = \pi$ rad

Therefore $150^\circ = \frac{150}{180}\pi$ rad

$$= \frac{5\pi}{6} \text{ rad}$$

Hence $150^\circ = \frac{5\pi}{6}$ radians (or $0.833\,3\,\pi$ radians)

(*b*) $180^\circ = \pi$ rad

Therefore $270^\circ = \frac{270}{180}\pi$ rad

$$= \frac{3\pi}{2} \text{ rad}$$

Hence $270^\circ = \frac{3\pi}{2}$ radians (or $1.5\,\pi$ radians)

(*c*) $180^\circ = \pi$ rad

Therefore $37.5^\circ = \left(\frac{37.5}{180}\right)\pi$ rad

$$= \left(\frac{75}{360}\right)\pi \text{ rad}$$

$$= \left(\frac{5\pi}{24}\right) \text{ rad}$$

Hence $37.5^\circ = \frac{5\pi}{24}$ radians (or $0.208\,3\,\pi$ radians)

(*d*) $23'' = \frac{23'}{60} = 0.38'$

$17.38' = \frac{17.38^\circ}{60} = 0.289\,7^\circ = 0.29^\circ$

Therefore $383^\circ\ 17'\ 23'' = 383.29^\circ = 383.3^\circ$ (to 4 significant figures)
Now $180^\circ = \pi$ rad

Therefore $383.3^\circ = \dfrac{383.3}{180}\pi$ rad

$= 2.129\pi$ rad

Hence $383^\circ\ 17'\ 23'' =$ **$2.129\ \pi$ radians**

Further problems on the properties of circles may be found in the following Section (9.11) (Problems 49–61), page 263.

9.11 Further problems

Angular measurement

1. Add together the following angles:
   (*a*) $15^\circ\ 32'$ and $17^\circ\ 19'$ (*b*) $19^\circ\ 23'\ 42''$ and $68^\circ\ 4'\ 33''$
   (*c*) $26^\circ\ 46'\ 11''$ and $36^\circ\ 38'\ 51''$
   (*d*) $41^\circ\ 10'\ 58''$, $32^\circ\ 11'\ 46''$ and $11^\circ\ 12'\ 13''$
   (*e*) $63^\circ\ 9'\ 25''$, $73^\circ\ 19'\ 37''$ and $22^\circ\ 22'\ 22''$
   (*a*) [$32^\circ\ 51'$] (*b*) [$87^\circ\ 28'\ 15''$] (*c*) [$63^\circ\ 25'\ 2''$]
   (*d*) [$84^\circ\ 34'\ 57''$] (*e*) [$158^\circ\ 51'\ 24''$]
2. Subtract the following angles:
   (*a*) $15^\circ\ 11'\ 18''$ from $24^\circ\ 33'\ 27''$ (*b*) $29^\circ\ 13'\ 29''$ from $47^\circ\ 8'\ 15''$
   (*c*) $4^\circ\ 44'\ 53''$ from $18^\circ\ 39'\ 48''$ (*d*) $34^\circ\ 30'\ 10''$ from $179^\circ\ 23'\ 8''$
   (*e*) $47^\circ\ 53'\ 46''$ from $123^\circ\ 0'\ 0''$
   (*a*) [$9^\circ\ 22'\ 9''$] (*b*) [$17^\circ\ 54'\ 46''$] (*c*) [$13^\circ\ 54'\ 55''$]
   (*d*) [$144^\circ\ 52'\ 58''$] (*e*) [$75^\circ\ 6'\ 14''$]
3. Convert the following angles to degrees and decimals of a degree, to 3 decimal places:
   (*a*) $13^\circ\ 13'$ (*b*) $29^\circ\ 12'\ 35''$ (*c*) $58^\circ\ 58'\ 48''$ (*d*) $146^\circ\ 32'\ 43''$
   (*a*) [13.217] (*b*) [29.210] (*c*) [58.980] (*d*) [146.545]
4. Convert the following angles into degrees, minutes and seconds:
   (*a*) 18.3° (*b*) 27.57° (*c*) 76.817° (*d*) 152.013°
   (*a*) [$18^\circ\ 18'$] (*b*) [$27^\circ\ 34'\ 12''$] (*c*) [$76^\circ\ 49'\ 1''$]
   (*d*) [$152^\circ\ 0'\ 47''$]

Types and properties of angles

5. State the general name given to the following angles:
   (*a*) 79° (*b*) 238° (*c*) 90° (*d*) 151°
   (*a*) [Acute] (*b*) [Reflex] (*c*) [Right angle] (*d*) [Obtuse]
6. Find the angles complementary to the following:
   (*a*) 19° (*b*) $35^\circ\ 42'$ (*c*) $83^\circ\ 41'\ 19''$ (*d*) $14^\circ\ 14'\ 14''$
   (*a*) [71°] (*b*) [$54^\circ\ 18'$] (*c*) [$6^\circ\ 18'\ 41''$] (*d*) [$75^\circ\ 45'\ 46''$]
7. Find the angles supplementary to the following:
   (*a*) 101° (*b*) 32° (*c*) $49^\circ\ 47'\ 9''$ (*d*) $178^\circ\ 53'\ 48''$
   (*a*) [79°] (*b*) [148°] (*c*) [$130^\circ\ 12'\ 51''$] (*d*) [$1^\circ\ 6'\ 12''$]

8.

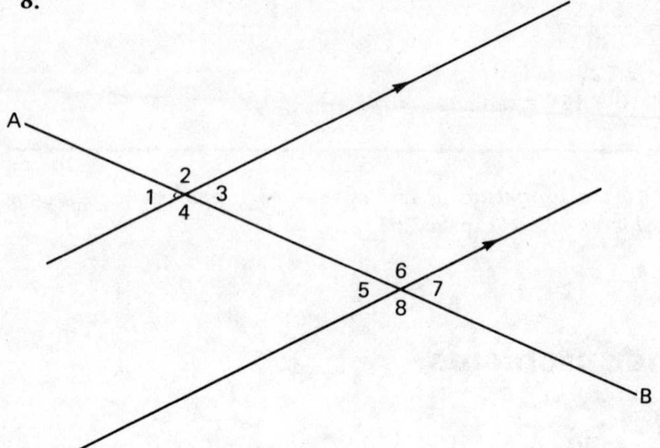

Fig. 9.68

With reference to Fig. 9.68, what is the name given to the line AB [transversal]

Give examples of each of the following:

(*a*) vertically opposite angles (*b*) supplementary angles

(*c*) corresponding angles (*d*) alternate angles.

(*a*) [1 and 3, or 2 and 4, or 5 and 7, or 6 and 8]

(*b*) [1 and 2, or 2 and 3, or 3 and 4, or 4 and 1, or 5 and 6, or 6 and 7, or 7 and 8, or 8 and 5, or 4 and 5, or 3 and 6]

(*c*) [1 and 5, or 4 and 8, or 2 and 6, or 3 and 7]

(*d*) [4 and 6, or 3 and 5]

9. If, in Fig. 9.68, $\angle 1 = 60^\circ$ find all the other numbered angles.

[$\angle 3 = \angle 5 = \angle 7 = 60^\circ$ $\angle 2 = \angle 4 = \angle 6 = \angle 8 = 120^\circ$]

10. In Fig. 9.69, find $\angle x$.

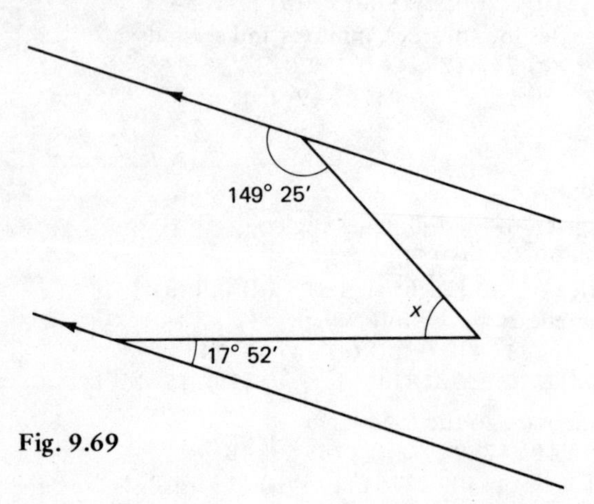

Fig. 9.69 [$48^\circ\ 27'$]

11. XYZ is a straight line and RY is a straight line such that $\angle$ RYZ = 126°. If $\angle$ RYX is bisected by a straight line YS, find $\angle$ XYS. [27°]

12. In Fig. 9.70, find angles x, y and z.

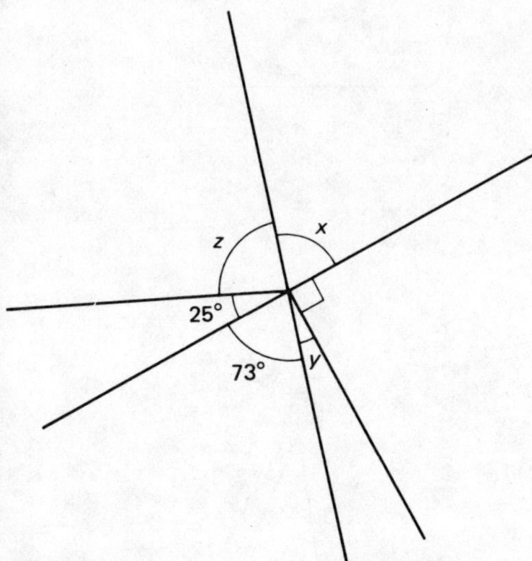

Fig. 9.70

[$x = 73°$ $y = 17°$ $z = 82°$]

13. Which of the following pairs of lines are parallel in Fig. 9.71

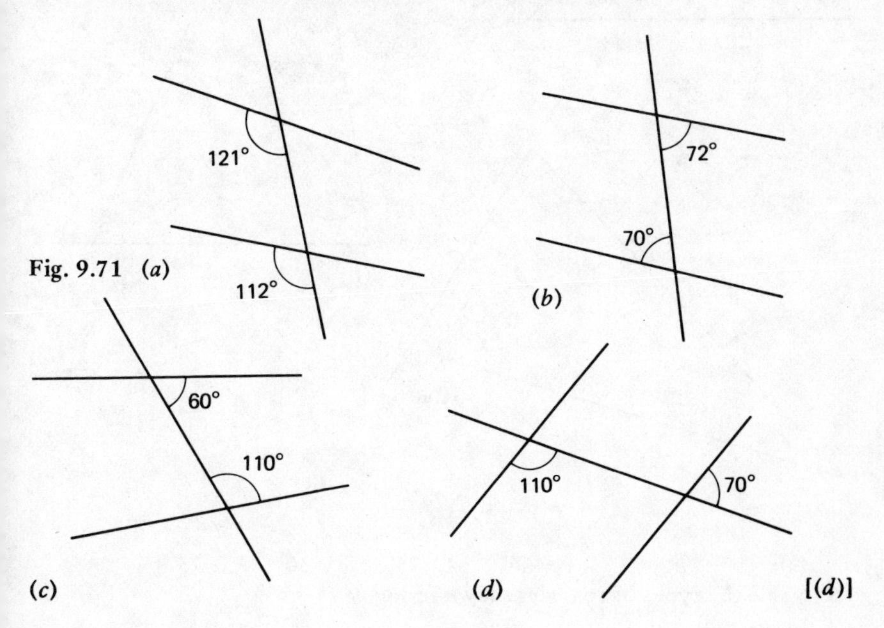

Fig. 9.71 (*a*)

[(*d*)]

 14. Find $\angle\theta$ in Fig. 9.72

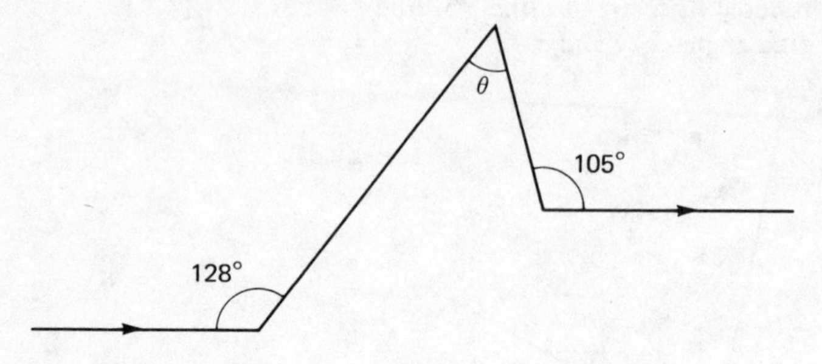

Fig. 9.72

[53°]

Properties of triangles

15. Find angles *a* to *f* in Fig. 9.73

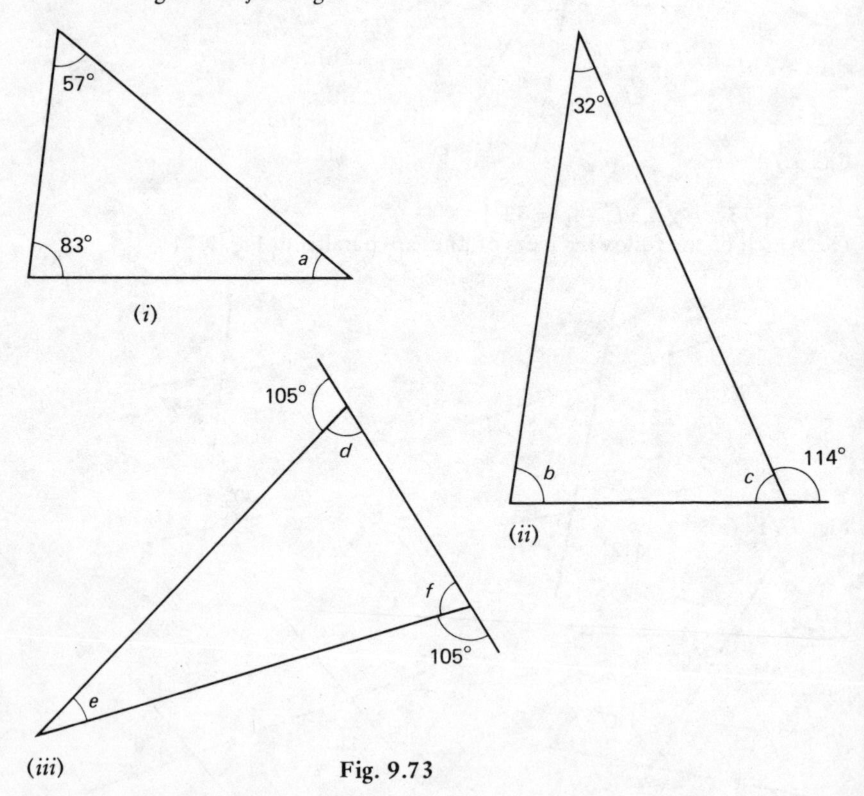

Fig. 9.73

[$a = 40°$ $b = 82°$ $c = 66°$ $d = 75°$ $e = 30°$ $f = 75°$]

16. Name the types of triangles shown in Fig. 9.74

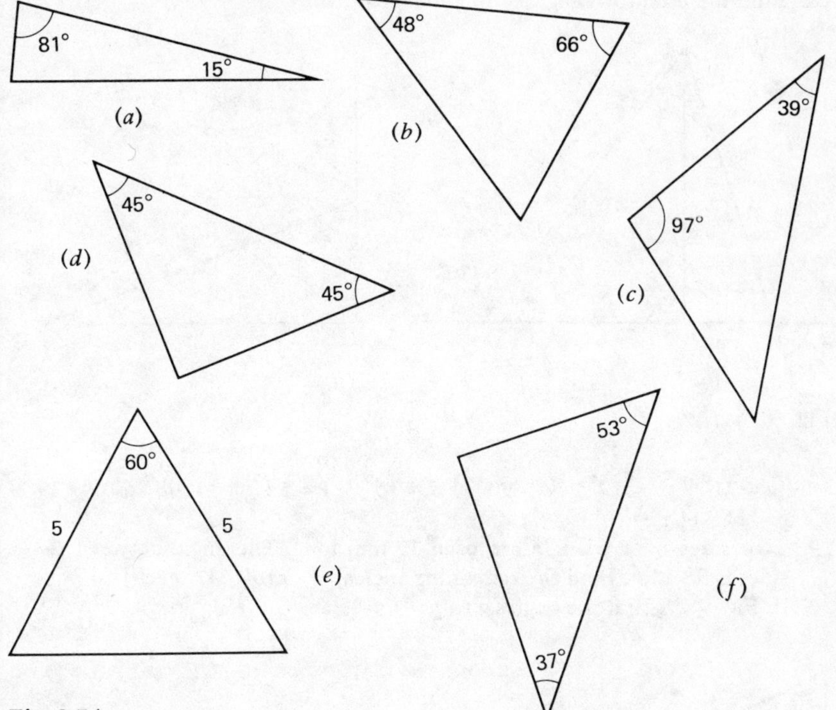

Fig. 9.74

[(*a*) acute-angled scalene triangle (*b*) isosceles triangle
(*c*) obtuse-angled scalene triangle (*d*) right-angled isosceles triangle
(*e*) equilateral triangle (*f*) right-angled triangle]

17. In Fig. 9.75, MNO is an isosceles triangle in which the unequal angle MON is 65°. Find ∠ PMN.

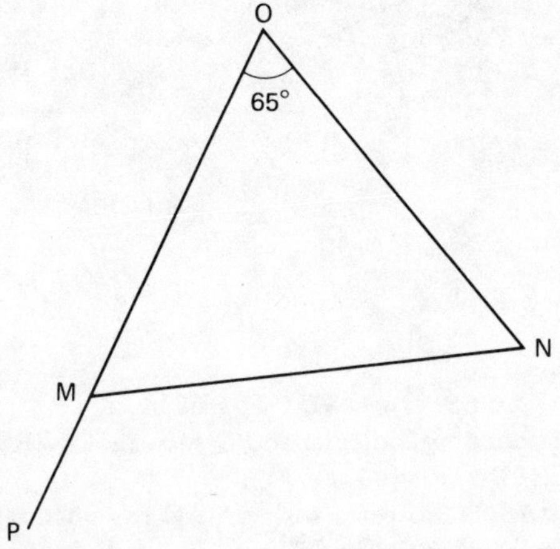

Fig. 9.75 $[122\frac{1}{2}^\circ]$

18. Find the unknown angles a to g in Fig. 9.76

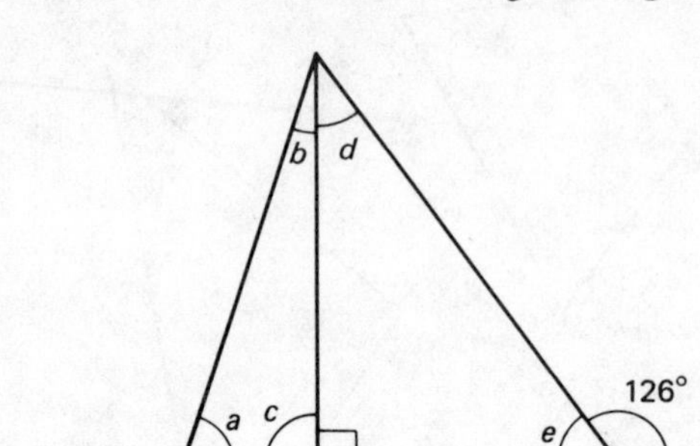

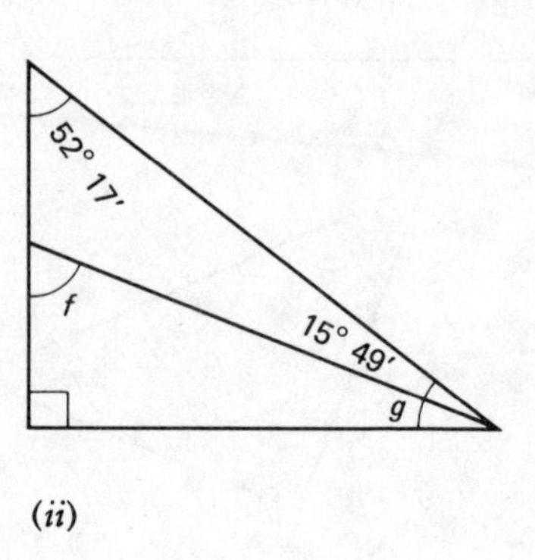

Fig. 9.76 (*i*) (*ii*)

[$a = 71°$ $b = 19°$ $c = 90°$ $d = 36°$ $e = 54°$ $f = 68° \, 6'$ $g = 21° \, 54'$]

19. Two sides of a triangle are each 17 mm long. The angle between these sides is $58° \, 26'$. Find the remaining angles. [$60° \, 47'$ each]

20. In Fig. 9.77, find the angles a to f.

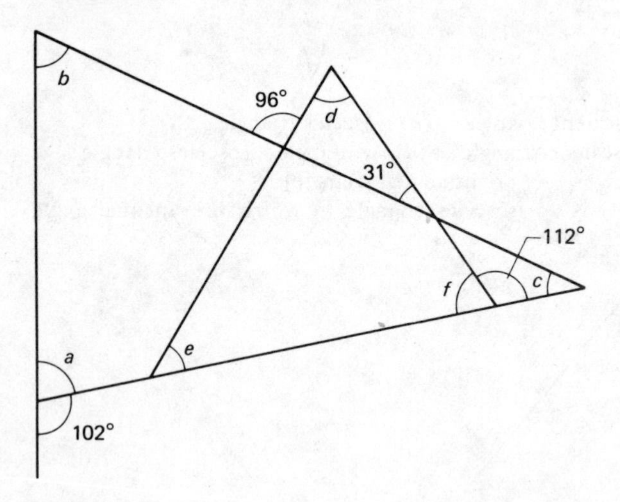

Fig. 9.77

[$a = 78°$ $b = 65°$ $c = 37°$ $d = 65°$ $e = 47°$ $f = 68°$]

21. In a triangle PQR a line SR is drawn perpendicular to PQ such that ∠ PRS is 27°. Calculate the value of ∠ QPR. [63°]

22. Triangle ABC, shown in Fig. 9.78, is isosceles with ∠ BAC the unequal angle. Find the three interior angles of the triangle

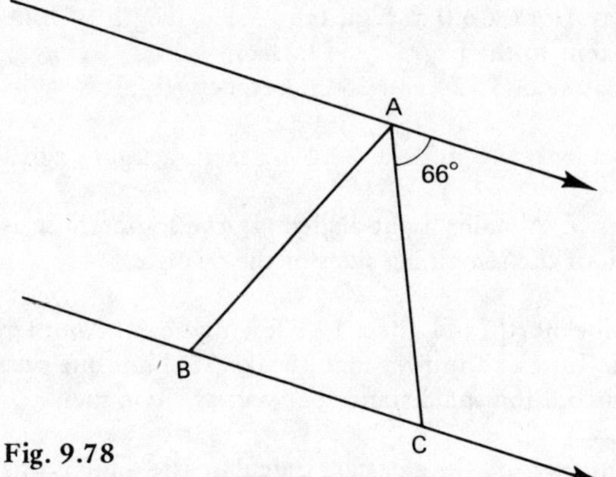

Fig. 9.78

[A = 48°   B = 66°   C = 66°]

23. Triangle JKL has a right angle at K and ∠ KJL is 28°. KL is produced to M. If the bisectors of ∠ JKL and ∠ JLM meet at N find ∠ KNL. [14°]

24. If, in Fig. 9.79, triangle ABC is equilateral, find the interior angles of triangle DCE.

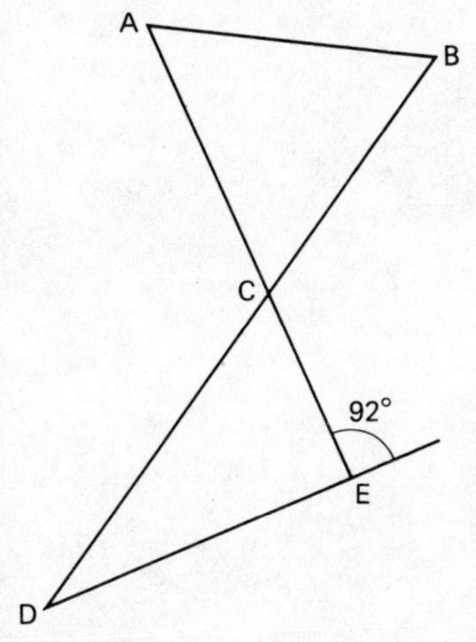

Fig. 9.79

[C = 60°   D = 32°   E = 88°]

Theorem of Pythagoras

25. In a triangle ABC, ∠ A is a right angle, $b = 9.20$ mm and $c = 7.10$ mm. Find side a. [11.62 mm]

26. In a triangle PQR, ∠ Q = 90°, $r = 17.5$ mm and $q = 29.6$ mm. Find side p. [23.87 mm]

27. In a triangle MNQ, ∠ N = 90°, $m = 5.26$ cm and $n = 9.58$ cm. Find side q. [8.007 cm]

28. A tent peg is 4.0 m away from a 6.0 m high tent. What length of rope runs from the top of the tent to the peg? [7.21 m]
29. Show that if a triangle has sides 7, 24 and 25 cm respectively, it is right-angled.
30. A triangle has sides of 9.6 m, 18 m and 20.4 m. Is it a right-angled triangle? [Yes]
31. Triangle ABC is isosceles, ∠ A being right-angled. If the hypotenuse is 11.46 mm find the length of the remaining sides of the triangle. [8.103 mm each]
32. A man walks 15.0 km due north and then 19.0 km due east. Another man, starting at the same time as the first man, walks 25.0 km due east and then 9.00 km due north. Find the distance between the two men. [8.485 km]
33. Figure 9.80 shows a plan view of a kite design. Calculate the dimensions d and l

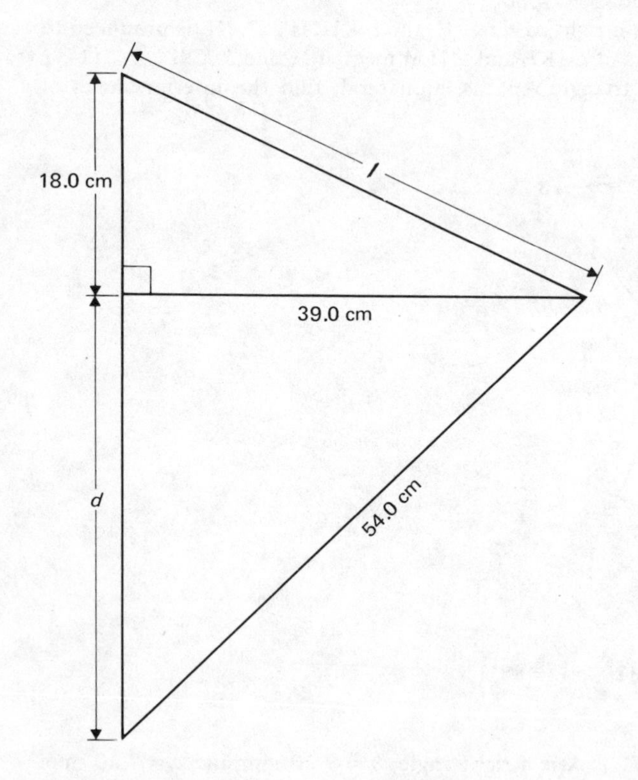

Fig. 9.80

[d = 37.35 cm l = 42.95 cm]

34. A ladder 4.0 m long is placed against a wall with its foot 1.5 m from the wall. How far up the wall does the ladder reach? If the foot of the ladder

is now pulled 1.0 m further away from the wall, calculate how far the top of the ladder falls. [3.71 m, 59 cm]

35. Two aircraft leave an airfield at the same time. One travels due south at an average speed of 350 km/h and the other travels due west at an average speed of 250 km/h. Find their distance apart after 2 hours 30 minutes. [1 075 km]

Congruent triangles

36. State which of the following pairs of triangles in Fig. 9.81 are congruent and name them in correct sequence.

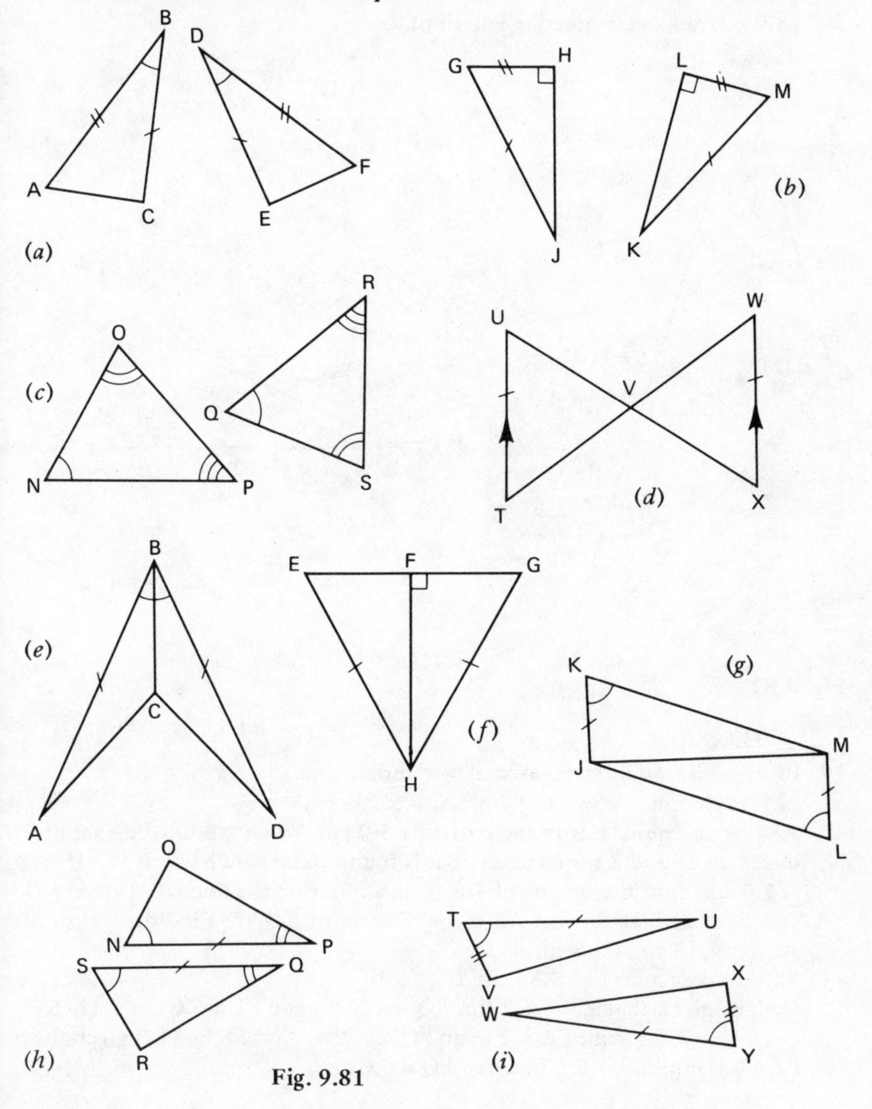

Fig. 9.81

[(*a*) Congruent ABC, FDE (*S.A.S.*) (*b*) Congruent JGH, KML (*R.H.S.*)
(*c*) Not congruent (*d*) Congruent TUV, WXV (*A.S.A.*)
(*e*) Congruent ABC, DBC (*S.A.S.*) (*f*) Congruent EFH, GFH (*R.H.S.*)
(*g*) Congruent LMJ, KJM (*A.S.A.*) (*h*) Congruent NOP, SRQ (*A.S.A.*)
(*i*) Congruent TUV, YWX (*S.A.S.*)]

37. In triangle MNO, MN = NO and P and Q are points on MN and NO respectively such that MP = OQ. Show that triangles MQN and OPN are congruent.

Similar triangles

38. In Fig. 9.82, determine the length of a.

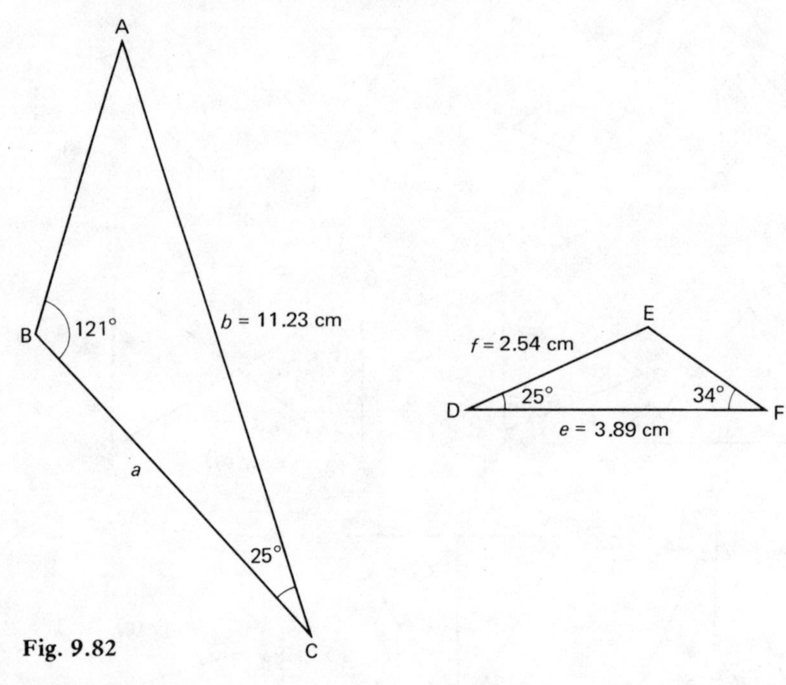

Fig. 9.82

[7.333 cm]

39. In Fig. 9.83, find the lengths of a, c and d.
[a = 15.38 cm c = 9.468 cm d = 5.581 cm]

40. ABC is an equilateral triangle of side 3.0 cm. When AB and AC are produced to D and E respectively, DE is found to be parallel with BC. If AD is 7.0 cm, find the length of DE. F is a point on the line DE, between D and E, such that the line AF is the bisector of ∠DAE. Find the length of AF. [7 cm, 6.06 cm]

41. With reference to Fig. 9.84, find
(*a*) TR, given that PS = 4.0 cm, SQ = 2.5 cm and PT = 5.0 cm (*b*) ST, given that PT = 4 m, TR = 2 m and QR = 9 m (*c*) PS and SQ given that PT = 4.0 mm, TR = 7.0 mm and PQ = 9.0 mm

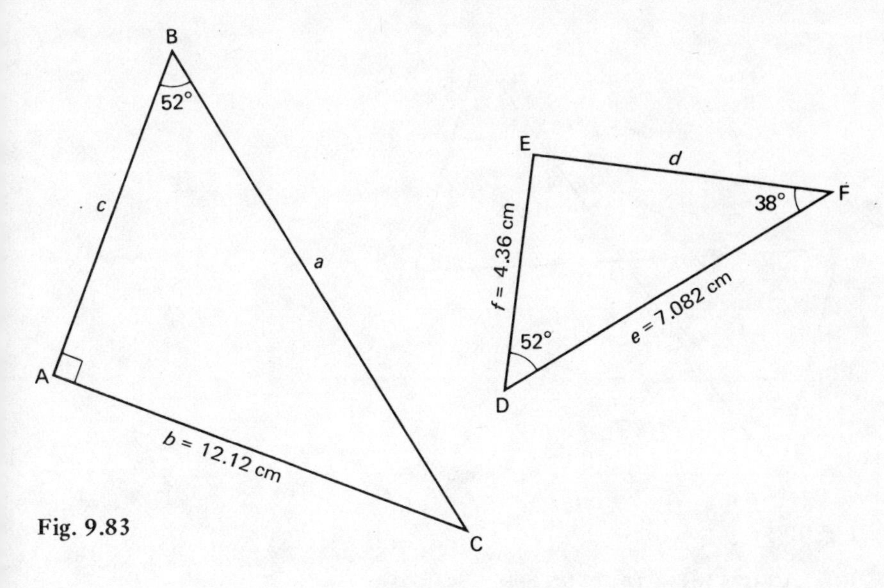

Fig. 9.83

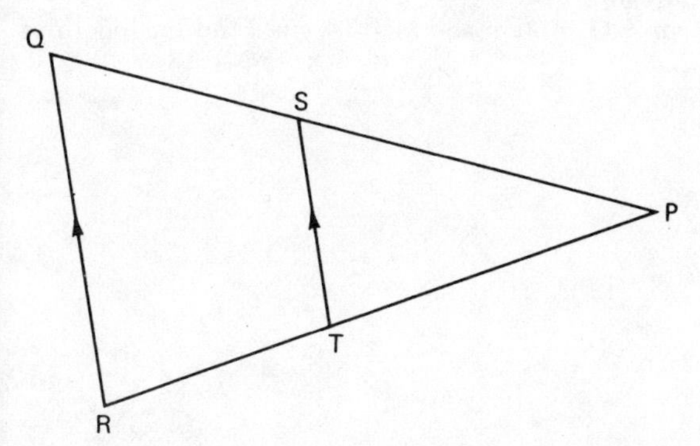

Fig. 9.84

(*a*) [3.125 cm] (*b*) [6 m] (*c*) [SP = 3.27 m, SQ = 5.73 m]

42. In a triangle ABC, AB = 3.5 cm, BC = 2.4 cm and AC = 3.2 cm. DE is drawn parallel to BC so that AD = 2.8 cm. (D lies on AB and E lies on AC.) Find the lengths of the remaining sides of triangle ADE.
[DE = 1.92 cm, AE = 256 cm]

43. The triangle MNO is right-angled at N. If MN = 3.40 cm and NO = 1.80 cm find NP, the length of the perpendicular to the hypotenuse and also the lengths of MP and PO.
[NP = 1.591 cm PO = 0.842 cm MP = 3.005 cm]

44. In Fig. 9.85, find:
(*a*) the length of OQ when MN = 8.0 m, PQ = 10.0 m and MO = 3.0 m
(*b*) the length of PQ when NO = 1.8 cm, PO = 5.4 cm, and MN = 10.5 cm.

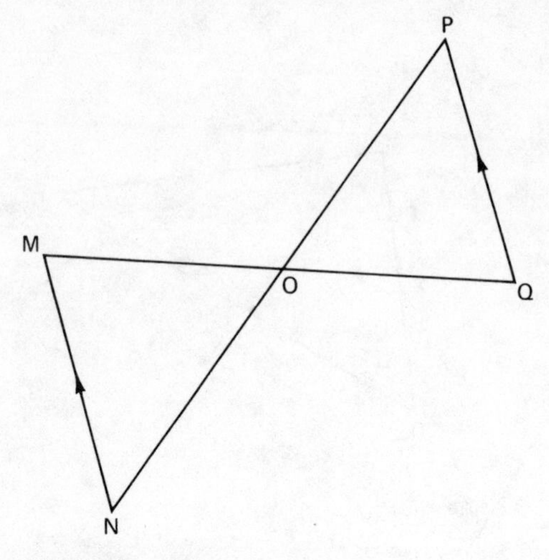

Fig. 9.85

(*a*) [3.75 m] (*b*) [31.5 cm]

45. In Fig. 9.86, OQ = 9 cm, NO = 6 cm and MN = 4 cm. Find the length of NR.

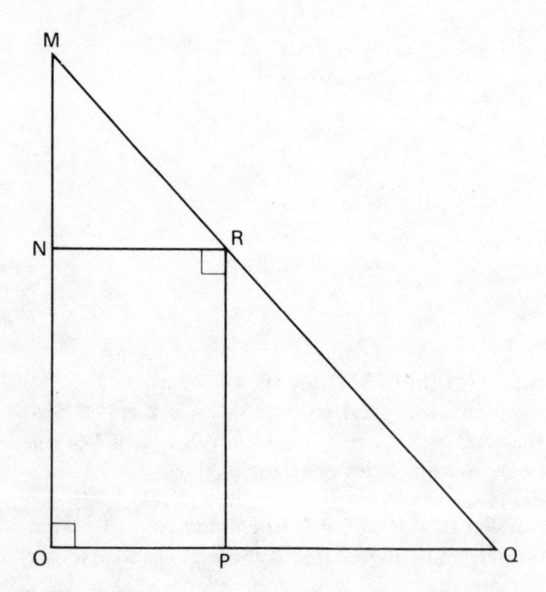

Fig. 9.86

[3.6 cm]

46. If, in Fig. 9.87, WR = 6.0 mm, RX = 4.0 mm and SR = 5.0 mm, find the length of RZ, SZ and WX.

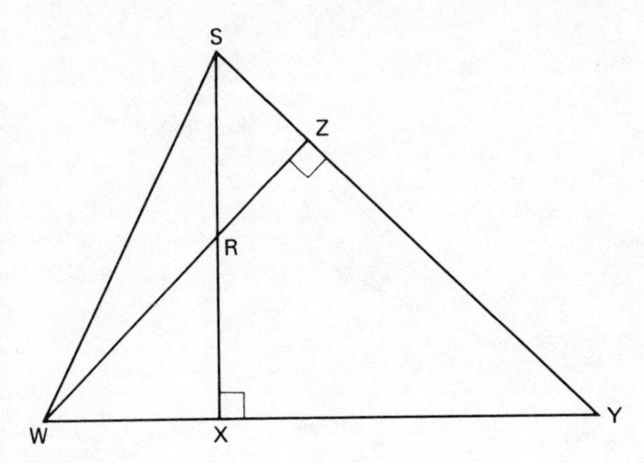

Fig. 9.87

[RZ = 3.33 mm SZ = 3.73 mm WX = 4.47 mm]

47. If the sides of a triangle are in the ratio of $1:\sqrt{3}:2$ find the length of the perimeter of the triangle if its shortest side is 7.50 cm. [35.49 cm]

Construction of triangles

48. Construct the following triangles:
(*a*) $a = 9$ cm, $b = 7$ cm and $c = 5$ cm
(*b*) $a = 4.5$ cm, $b = 6.25$ cm and $\angle C = 62^\circ$
(*c*) $a = 53$ mm, $\angle C = 49^\circ$ and $\angle B = 81^\circ$
(*d*) $c = 3.8$ cm, $\angle A = 125^\circ$ and $\angle C = 20^\circ$
(*e*) $a = 48$ mm, $\angle B = 90^\circ$, hypotenuse = 75 mm

Circles

In the following questions assume $\pi = 3.142$ where necessary.

49. Calculate the length of the circumference of a circle of radius 6.50 cm [40.85 cm]
50. If the diameter of a circle is 65.86 mm calculate the length of the circumference. [206.9 mm]
51. Find the length of the radius of a circle whose circumference is 11.48 cm. [1.827 cm]
52. Calculate the length of the diameter of a circle whose perimeter is 72.40 cm [23.04 cm]
53. How many degrees are there in:
(*a*) 1 revolution (*b*) $3\frac{1}{2}$ revolutions (*c*) a semicircle (*d*) a quadrant.
(*a*) [360°] (*b*) [1 260°] (*c*) [180°] (*d*) [90°]
54. From a point A, a tangent to a circle is drawn, touching the circle at the point B (see Fig. 9.88). If the diameter of the circle is 6.0 cm and the length of AB is 9.0 cm find the distance from A to the centre of the circle, O.

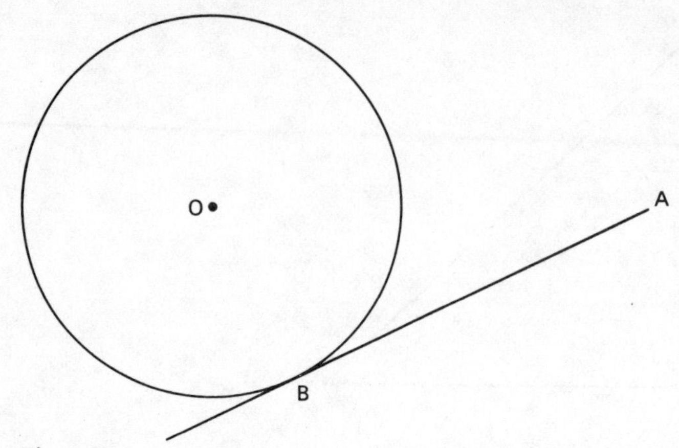

Fig. 9.88

[9.49 cm]

55. A tangent XY meets a circle at Y as shown in Fig. 9.89. The line YZ divides the angle XYO such that OZ = XZ. If the diameter of the circle is 15.0 cm find the length of the tangent XY.

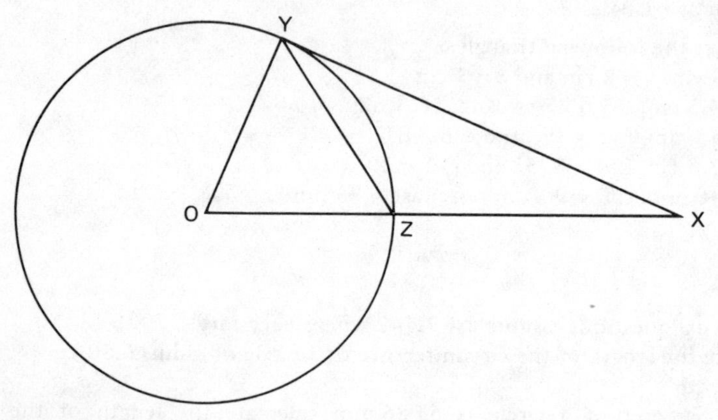

Fig. 9.89

[12.99 cm]

56. In Fig. 9.90, find the length of x.
[20 cm]

57. Without the use of conversion tables convert the following angles from degrees to radians in terms of π:
(*a*) 90° (*b*) 225° (*c*) 75° (*d*) 60° (*e*) 420° (*f*) $1\ 110^\circ$
(*g*) $263^\circ\ 15'\ 25''$
Verify the results obtained by using conversion tables.
(*a*) $\left[\frac{\pi}{2} \text{ or } 0.5\pi\right]$ (*b*) $\left[\frac{5\pi}{4} \text{ or } 1.25\pi\right]$ (*c*) $\left[\frac{5\pi}{12} \text{ or } 0.416\ 7\pi\right]$

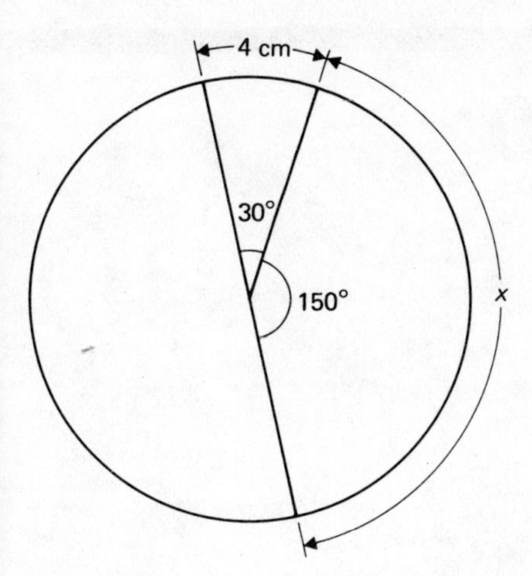

Fig. 9.90

(*d*) $\left[\frac{\pi}{3} \text{ or } 0.333\,3\pi\right]$ (*e*) $\left[\frac{7\pi}{3} \text{ or } 2.333\pi\right]$ (*f*) $\left[\frac{37\pi}{6} \text{ or } 6.167\pi\right]$
(*g*) $[1.463\pi]$

58. Without the use of conversion tables convert the following angles from degrees to radians:
(*a*) 65° (*b*) 34° 17′ (*c*) 111° 5′ (*d*) 251° 52′ (*e*) 192° 15′
(*f*) 323° 47′ 15″
Verify the results obtained by using conversion tables.
(*a*) [1.135^c] (*b*) [0.598 4^c] (*c*) [1.939^c] (*d*) [4.396^c]
(*e*) [3.356^c] (*f*) [5.652^c]

59. Without the use of conversion tables convert the following angles from radians to degrees:
(*a*) 2π (*b*) $\frac{\pi}{2}$ (*c*) $\frac{5\pi}{6}$ (*d*) $\frac{5\pi}{3}$ (*e*) $\frac{9\pi}{8}$ (*f*) $\frac{5\pi}{9}$
(*a*) [360°] (*b*) [90°] (*c*) [150°] (*d*) [300°] (*e*) [$202\frac{1}{2}$°]
(*f*) [100°]

60. Without the use of conversion tables convert the following angles from radians to degrees, minutes and seconds:
(*a*) 2.48 (*b*) 3.912 (*c*) 0.013 2 (*d*) 17.82 (*e*) 5.75 (*f*) 7.325
Verify the results obtained by using conversion tables.
(*a*) [142° 4′ 30″] (*b*) [224° 6′ 43″] (*c*) [0° 45′ 22″]
(*d*) [1 020° 52′ 42″] (*e*) [329° 24′ 29″] (*f*) [419° 38′ 14″]

61. The lines AOC and BOD are diameters of the circle ABCD, centre 0. Given that the angle ABD = 36°, calculate angles OBC, OAB, AOD and DOC. [54°; 36°; 72°; 108°]

Chapter 10

Mensuration

10.1 Definition

Mensuration is a branch of mathematics concerned with the determination of lengths, areas and volumes.

10.2 Properties of quadrilaterals

A **polygon** is a closed plane figure bounded by straight lines. A polygon which has four sides is called a **quadrilateral**. Figure 10.1 shows some typical quadrilaterals.

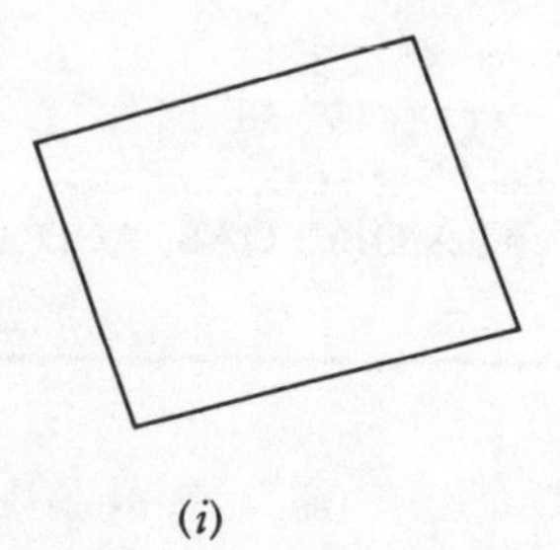

(*i*)

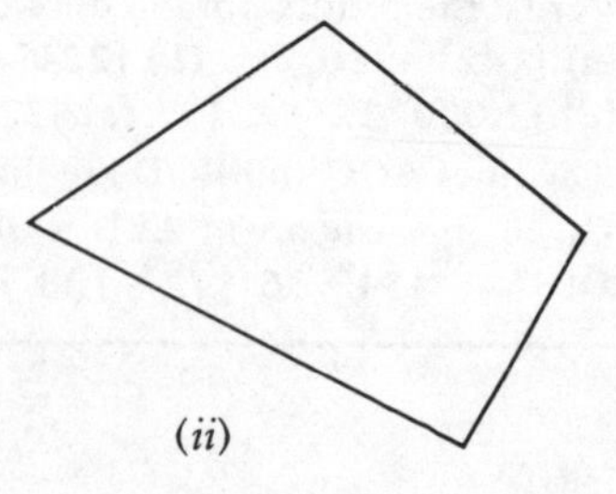

(*ii*)

Fig. 10.1

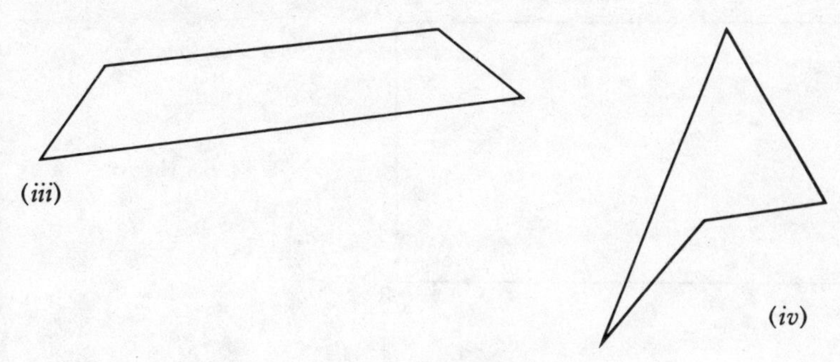

Fig. 10.1

If the opposite corners of a quadrilateral are joined by a straight line, two triangles are produced. Figure 10.2 shows a quadrilateral ABCD. If A is joined to C by a straight line two triangles ABC and ADC, are produced.

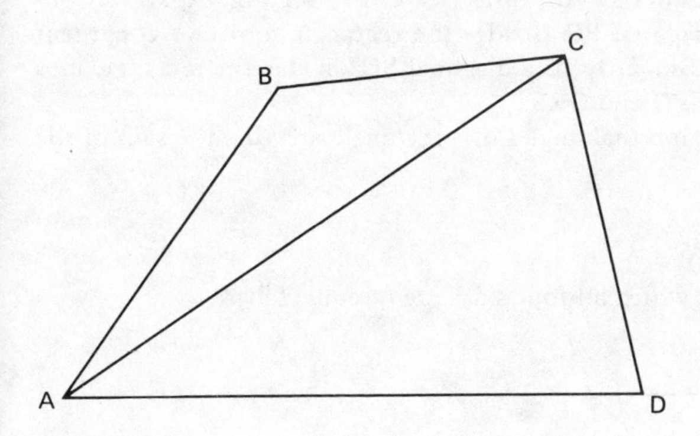

Fig. 10.2

There are five types of quadrilaterals, which are given special names:

1. Rectangle
2. Square
3. Parallelogram
4. Rhombus
5. Trapezium

1. A rectangle

A **rectangle** is a plane rectilinear four-sided figure with all four of its internal angles equal to 90°. (The word **rectilinear** means it is composed entirely of straight lines.)

In Fig. 10.3 a rectangle PQRS is shown. Side PS is parallel to QR and side PQ is parallel to SR. Also the length of PS is equal to the length of QR and the length of PQ is equal to the length of SR.

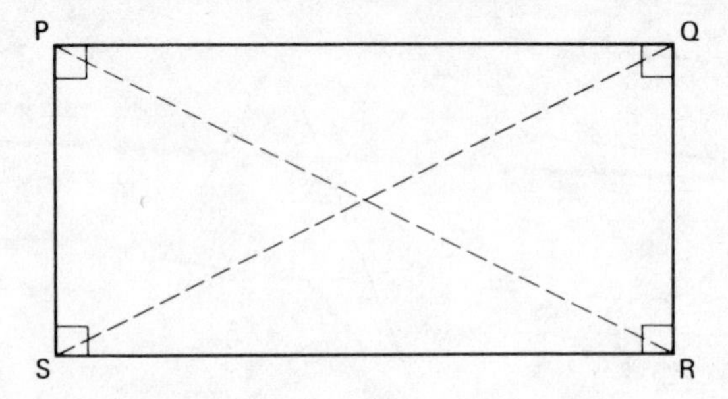

Fig. 10.3

A **diagonal** is the name given to a straight line joining the opposite corners of a quadrilateral. Thus the broken lines PR and SQ in Fig. 10.3 are called diagonals of the rectangle. In a rectangle the two diagonals are equal in length. Thus, PR = SQ in Fig. 10.3.

If the sides of a rectangle are known, the length of the diagonal may be calculated using the theorem of Pythagoras (see Ch. 9, Section 9.6).

In Fig. 10.3, the diagonal PR divides the rectangle into two congruent triangles PSR and PQR. Similarly, the diagonal SQ divides the rectangle into two congruent triangles SPQ and QRS.

As each of the four internal angles of a rectangle are 90°, the sum of the interior angles is 360°.

2. A square

A **square** is a rectangle in which all four sides are of equal length.

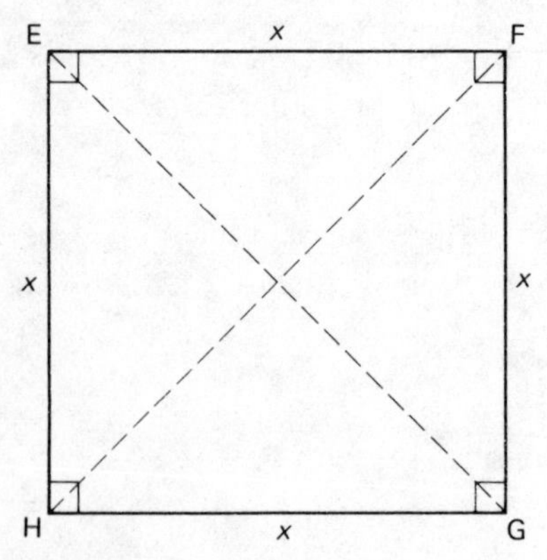

Fig. 10.4

Figure 10.4 shows a square EFGH where the sides are all equal to length x, and each of the four angles is equal to 90°. The lengths of the two diagonals EG and FH are equal. In a square the diagonals bisect each of the right angles. The triangle EGH is congruent to triangle GEF and the triangle EFH is congruent to triangle GHF. Using Pythagoras's theorem on triangle EGH gives:

$$(EG)^2 = (EH)^2 + (HG)^2$$

Therefore $(EG)^2 = x^2 + x^2$

$$= 2x^2$$

$$EG = \sqrt{(2x^2)}$$

$$EG = (\sqrt{2})x$$

Similarly the diagonal **FH** = $(\sqrt{2})x$

Thus the length of a diagonal of a square is always equal to $\sqrt{2}$ (i.e. 1.414 to 3 decimal places) times the length of a side.

The sum of the interior angles of a square is again 360°.

3. A parallelogram

A **parallelogram** is a plane rectilinear four-sided figure with its opposite sides parallel. From this definition it is evident that the rectangle and the square are both particular kinds of parallelogram with additional properties.

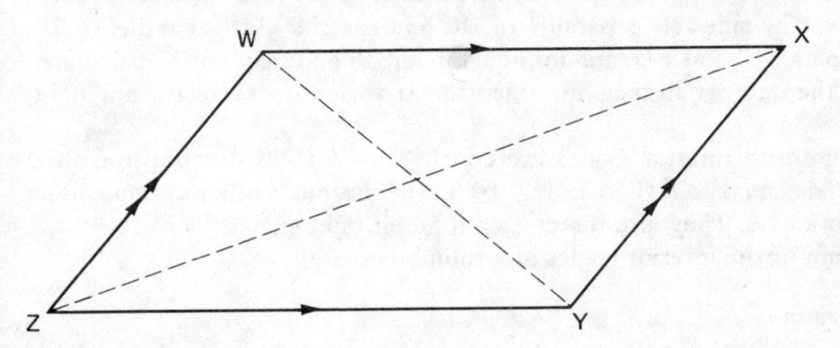

Fig. 10.5

Figure 10.5 shows a typical parallelogram WXYZ where the side WX is parallel to ZY and the side WZ is parallel to XY. The length of the side WX is equal to side ZY and also the length of side WZ is equal to XY.

The lengths of the diagonals WY and XZ are not equal to each other except in the special cases of a rectangle and a square, where the four interior angles are each equal to 90°.

The theorem of Pythagoras cannot be used to calculate the lengths of the diagonals since the triangles involved are not right-angled.

Since the opposite sides of a parallelogram are parallel the opposite interior angles are equal. (This may be deduced from Ch. 9, Section 9.4, item 5, regarding transversals of parallel lines.) Thus $\angle$ ZWX is equal to $\angle$ XYZ and $\angle$ WXY is equal to $\angle$ YZW in Fig. 10.5. The sum of the interior angles of a parallelogram is 360°.

4. A rhombus

A **rhombus** is a plane rectilinear four-sided figure in which all four sides are of equal length and the opposite sides are parallel. A rhombus is thus a parallelogram having additional properties.

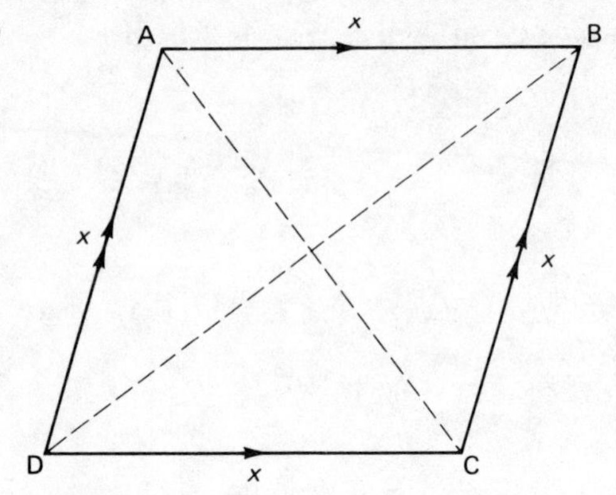

Fig. 10.6

Figure 10.6 shows a typical rhombus where each of the four sides are equal in length to x. The side AB is parallel to DC and the side AD is parallel to BC. The diagonals, AC and BD, are not equal in length and may not be calculated using the theorem of Pythagoras since the triangles involved are not right-angled.

The opposite interior angles are equal. Thus $\angle$ DAB is equal to $\angle$ BCD and $\angle$ ABC is equal to $\angle$ CDA in Fig. 10.6. The diagonals of a rhombus bisect the interior angles. They also bisect each other at right angles.

The sum of the interior angles of a rhombus is 360°.

5. A trapezium

A **trapezium** is a plane rectilinear four-sided figure which has a pair of parallel lines which are not of equal length.

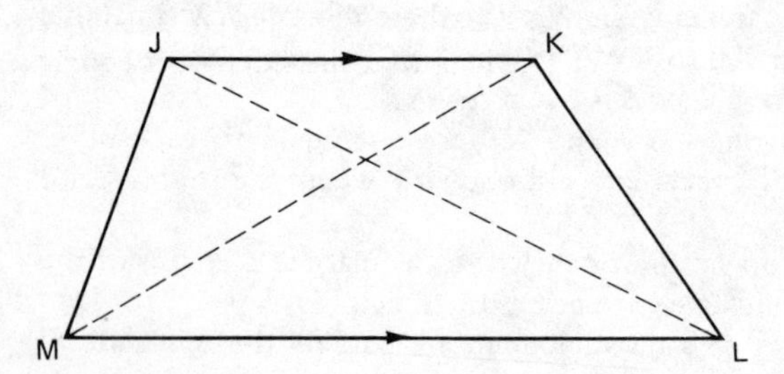

Fig. 10.7

Figure 10.7 shows a typical trapezium JKLM, where JK is parallel to ML. The lengths of the diagonals JL and KM are not equal, except in the special case when $\angle$ LMJ is equal to $\angle$ MLK and $\angle$ MJK is equal to $\angle$ LKJ.

The lengths of the diagonals may not be calculated by the theorem of Pythagoras since the triangles involved are not right-angled.

The sum of the interior angles of a trapezium is 360°.

10.3 Areas of plane figures

Area is a measure of the size or extent of a plane surface.

1. A square

Figure 10.8 shows a square ABCD of side 4 units.

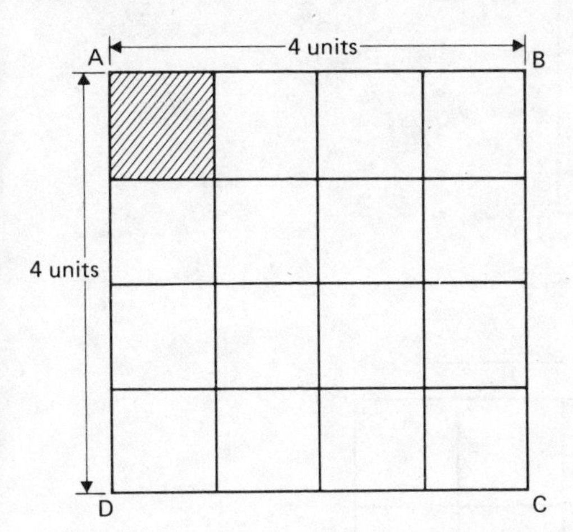

Fig. 10.8

[The SI unit of length is the metre (m). However, as we have already seen, decimal multiples such as millimetres (mm), centimetres (cm), or kilometres (km), are acceptable and much used.]

The area of the square is given by the number of unit squares it contains. In Fig. 10.8 a one unit square, where each side is of one unit length, is shown shaded. Thus the number of such unit squares is 16 and the area of the square is said to be 16 square units.

If the unit of length is centimetres then the unit of area will be square centimetres, written as cm^2.

Similarly, if the unit of length is metres, then the unit of area will be square metres, written as m^2.

In the above example the answer of 16 square units could have been obtained by multiplying the sides of the square together, i.e. 4 units × 4 units = 16 square units. **The area of a square is the product of the lengths of two of its sides.** This method of obtaining the area is true for any size square.

If any square EFGH has each of its sides x units in length as shown in Fig. 10.9, then the area $A = x^2$ **square units.**

2. A rectangle

Figure 10.10 shows a rectangle JKLM, whose sides are 5 units and 3 units long.

The area of the rectangle is the number of unit squares it contains. Then the area of JKLM is 15 square units. The area of a rectangle may be calculated

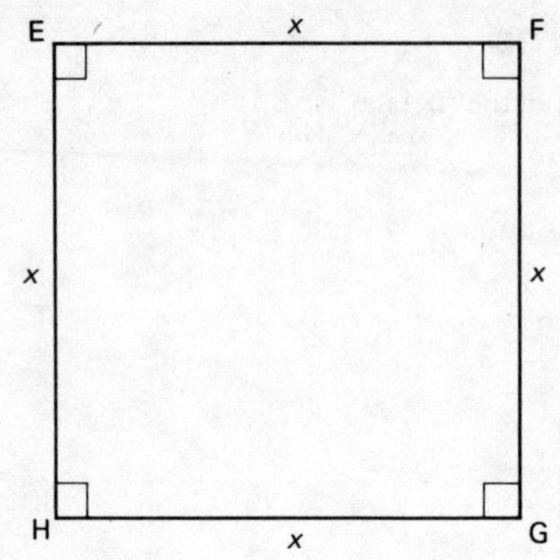

Fig. 10.9

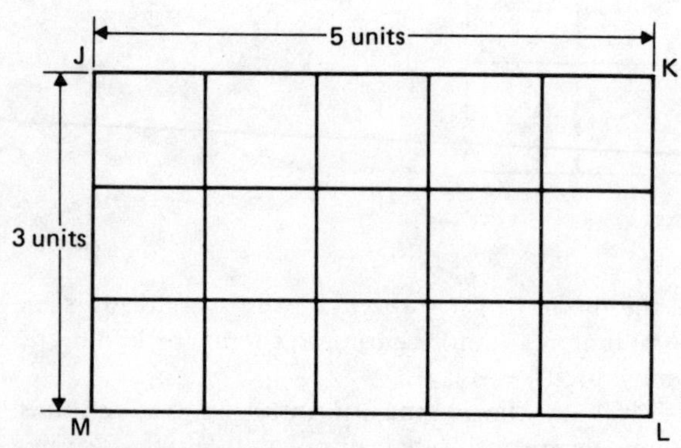

Fig. 10.10

by multiplying together the lengths of two perpendicular sides, that is, $5 \times 3 = 15$ square units. If any rectangle NOPQ has sides b and l units in length as shown in Fig. 10.11 then area A is given by:

$A = b\,l$ **square units.**

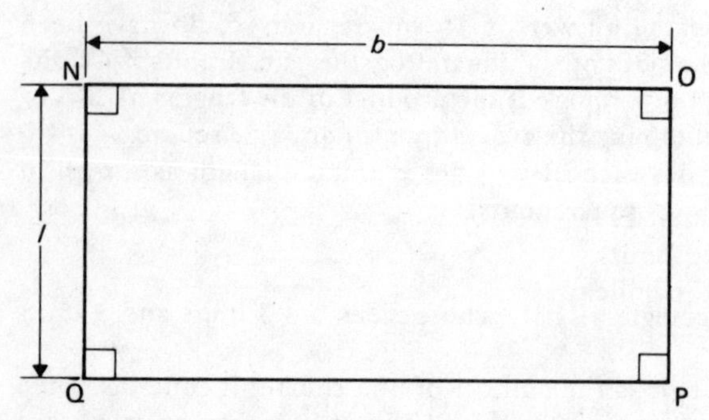

Fig. 10.11

3. A parallelogram

Figure 10.12 shows a parallelogram ABCD whose sides are a and b.

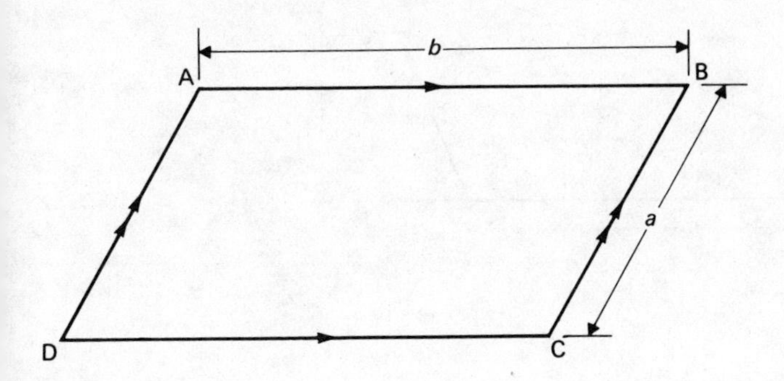

Fig. 10.12

DF and CE are constructed perpendicular to AB and AB (produced), as shown in Fig. 10.13.

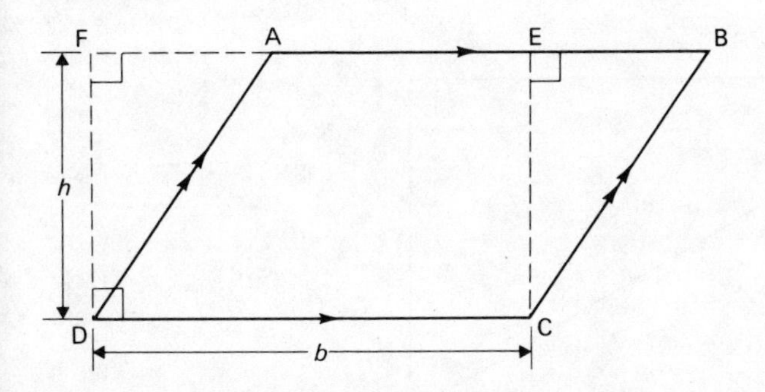

Fig. 10.13

Now AD = BC and FD = EC and ∠ AFD = ∠ BEC = 90°. Hence triangles AFD and BEC are congruent (right angle, hypotenuse and one other side, that is, *R.H.S.*, see Ch. 9, Section 9.7). Thus the area of triangle FAD is equal to the area of triangle EBC. Therefore, in Fig. 10.13, the area of parallelogram ABCD is exactly the same as the rectangle FECD.

Let FD = EC = h units. Then, area of rectangle FECD = **area of parallelogram ABCD = $b\,h$ square units.**

Hence, **the area of a parallelogram is given by the length of the base multiplied by the perpendicular height.**

4. A triangle

Figure 10.14 shows a parallelogram EFGH with a diagonal HF.

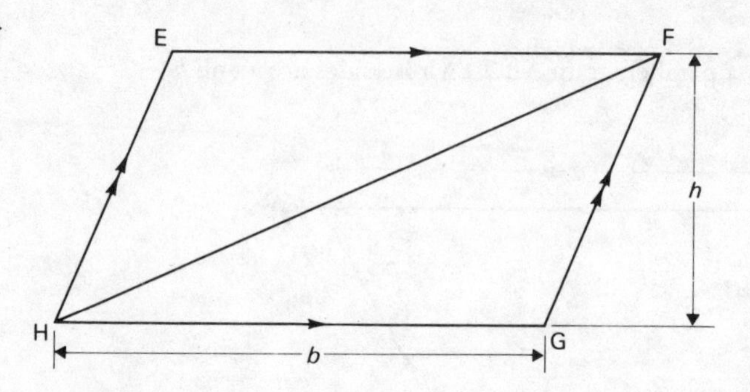

Fig. 10.14

Since triangles EFH and GHF are congruent, their areas are the same.

Therefore, area of triangle EFH = area of triangle GHF
$= \frac{1}{2}$ area of parallelogram EFGH
$= \frac{1}{2}\, b\, h$ square units

Similarly, Fig. 10.15 shows a rectangle ABCD with a diagonal AC.

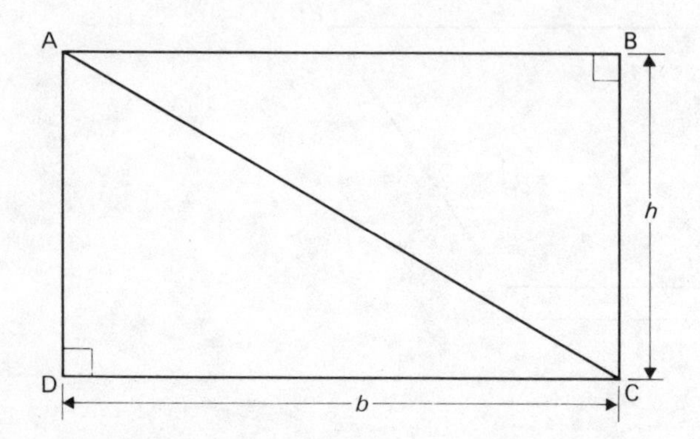

Fig. 10.15

Since triangles ADC and CBA are congruent, their areas are the same.

Therefore, area of triangle ADC = area of triangle CBA
$= \frac{1}{2}$ area of rectangle
$= \frac{1}{2}\, b\, h$ square units

A general expression for the area (A) of any triangle is given by: $A = \frac{1}{2} \times$ **base** $\times$ **perpendicular height**.

5. A trapezium

Figure 10.16 shows a trapezium MNOP with a diagonal NP.

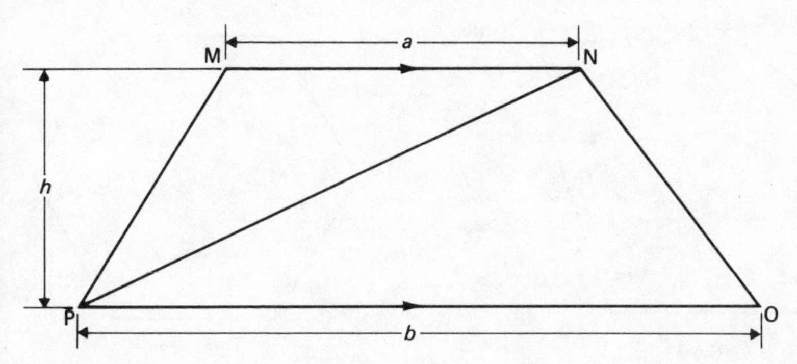

Fig. 10.16

Area of trapezium MNOP = area of triangle MNP + area of triangle NOP. The area of any triangle = $\frac{1}{2}$ base × perpendicular height.

$$\text{Hence the area of triangle MNP} = \tfrac{1}{2}\,a\,h$$
$$\text{and the area of triangle NOP} = \tfrac{1}{2}\,b\,h$$
$$\text{Therefore the area of trapezium MNOP} = \tfrac{1}{2}\,a\,h + \tfrac{1}{2}\,b\,h$$
$$= \tfrac{1}{2}\,(a + b)h \text{ square units}$$

Hence **the area of any trapezium = $\frac{1}{2}$ (sum of the parallel sides) × (perpendicular distance between the parallel sides).**

6. Pentagon, hexagon, heptagon and octagon

There are many plane rectilinear figures with more than four sides. The first four of these **polygons** are defined as follows:

(*i*) A **pentagon** is a five-sided polygon
(*ii*) A **hexagon** is a six-sided polygon
(*iii*) A **heptagon** is a seven-sided polygon
(*iv*) An **octagon** is an eight-sided polygon

If the term 'regular' is applied to any of the above figures it means that all the sides in the polygon are equal. For example, a 'regular hexagon' means that all six sides are of equal length. Figure 10.17 shows typical examples of regular polygons.

To find the area of any of the regular polygons shown in Fig. 10.17 the procedure adopted is always the same. The bisectors of any two interior angles are drawn to meet at the centre. (The bisectors of each of the interior angles of a regular polygon will meet at the centre.) A radius is drawn from the centre of the figure to each of the vertices. This produces triangles the areas of which are equal and may be calculated. For example, in the hexagon shown in Fig. 10.18 six equal triangles have been produced. Thus the area of the hexagon is equal to six times the area of one of the triangles. (The formula, area = $\frac{1}{2}$ base × perpendicular height, is used to calculate the area of a triangle.)

7. A circle

Figure 10.19 shows a circle of radius r divided into eight equal sectors.

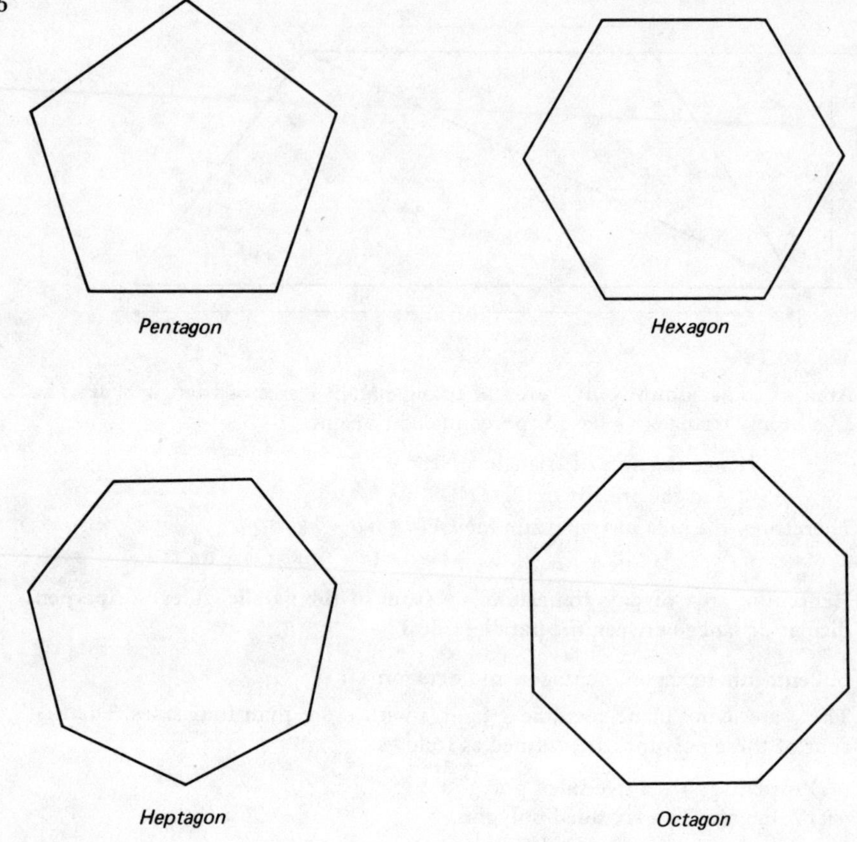

Fig. 10.17

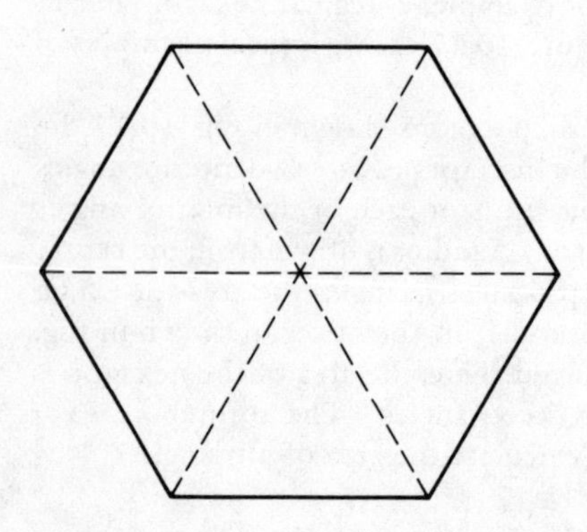

Fig. 10.18

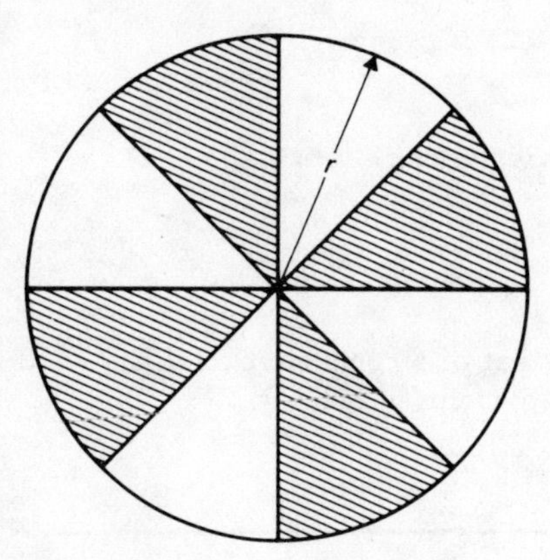

Fig. 10.19

Let the sectors be cut out and arranged as in Fig. 10.20.

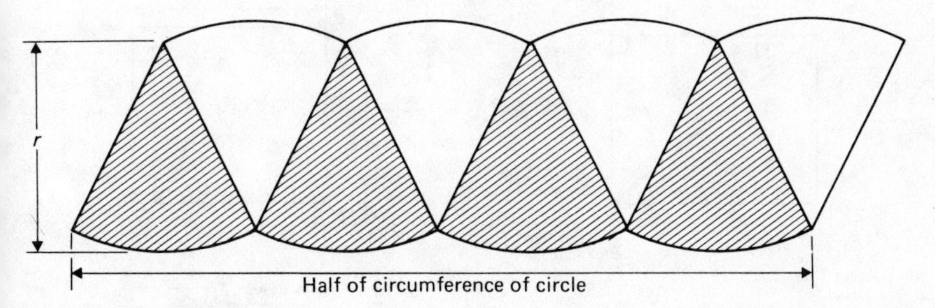

Fig. 10.20

The area approximates to a parallelogram of length πr (since the circumference of a circle is $2\pi r$) and perpendicular height r. If instead of eight sectors a very large number of sectors were taken the arc curvature of each sector would become less prominent, until in the limit, the sides of the shape shown in Fig. 10.20 would appear to be straight lines as shown in Fig. 10.21.

Fig. 10.21

 The area A of the parallelogram of Fig. 10.21 is given by:

$A = (\pi r)r = \pi r^2$

Hence the area A of any circle of radius r is given by:

$$A = \pi r^2 = \pi\left(\frac{d}{2}\right)^2 = \frac{\pi d^2}{4} \text{ square units}$$

where d is the diameter of the circle.

The area of a **semicircle** is one half that of a complete circle. The word 'semi' means 'half'. Hence the area of a semicircle, of radius r, $= \frac{1}{2}\pi r^2$.

Worked problems on areas of plane figures

Problem 1. State the types of quadrilateral shown in Fig. 10.22 and find the angles a to l

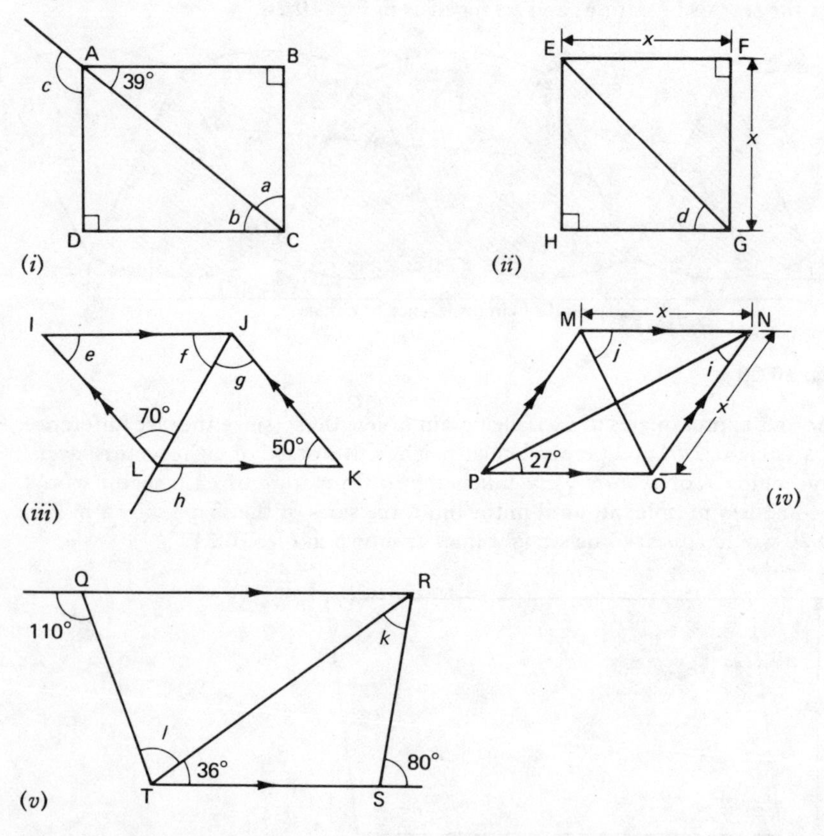

Fig. 10.22

(*i*) ABCD is a rectangle.

In triangle ABC, $39^\circ + 90^\circ + \angle a = 180^\circ$ (angles in a triangle add up to 180°)
Therefore $\angle a = 180^\circ - 90^\circ - 39^\circ$
$\angle a = \mathbf{51^\circ}$
$\angle b = \angle$ BAC (alternate angles between parallels AB and DC)
$\angle b = \mathbf{39^\circ}$
$\angle c = \angle$ ADC + $\angle$ DCA (external angle of a triangle equals the sum of the interior opposite angles)
$\angle c = 90^\circ + 39^\circ$
$\angle c = \mathbf{129^\circ}$

(*ii*) EFGH is a square.
The diagonals of a square bisect each of the right angles. Hence the diagonal EG bisects $\angle$ HEF and $\angle$ FGH.

Therefore $\angle d = \frac{90^\circ}{2}$
$\angle d = \mathbf{45^\circ}$

(*iii*) IJKL is a parallelogram.
$\angle e = \mathbf{50^\circ}$ (since opposite interior angles of a parallelogram are equal)
In triangle IJL $\angle e + \angle f + 70^\circ = 180^\circ$ (angles in a triangle add up to 180°)
Therefore $\angle f = 180^\circ - 70^\circ - 50^\circ$
$\angle f = \mathbf{60^\circ}$
$\angle g = \mathbf{70^\circ}$ (alternate angles between the parallel lines IL and JK)
$\angle$ JLK = $\angle f = 60^\circ$ (alternate angles between the parallel lines IJ and LK)
$\angle h + 60^\circ = 180^\circ$ (angles on a straight line equal 180°)
Therefore $\angle h = \mathbf{120^\circ}$

(*iv*) MNOP is a rhombus.
The diagonals of a rhombus bisect the interior angles, and opposite interior angles are equal.
Thus $\angle$ OPN = $\angle$ MPN = $\angle$ ONP = $\angle$ MNP = 27°
Therefore $\angle i = \mathbf{27^\circ}$
If $\angle$ MPN = $\angle$ MNP = 27° then $\angle$ PMN is given by
$27^\circ + 27^\circ + \angle$ PMN = 180° (angles in a triangle add up to 180°)
Therefore $\angle$ PMN = $180^\circ - 27^\circ - 27^\circ$
= 126°
As the diagonal MO bisects $\angle$ PMN, then

$\angle j = \frac{126^\circ}{2} = \mathbf{63^\circ}$

(*v*) QRST is a trapezium.
$\angle k + 36^\circ = 80^\circ$ (external angle of a triangle is equal to the sum of the two interior opposite angles)
Therefore $\angle k = 80^\circ - 36^\circ$
$\angle k = \mathbf{44^\circ}$
$\angle$QRT = 36° (alternate angles between parallel lines QR and TS)
Therefore $\angle l + 36^\circ = 110^\circ$ (external angle of a triangle is equal to the sum of the two interior opposite angles)

$\angle l = 110^\circ - 36^\circ$

$\angle l = 74^\circ$

Problem 2. Find the areas of the figures shown in Fig. 10.23. Assume π = 3.142 where necessary.

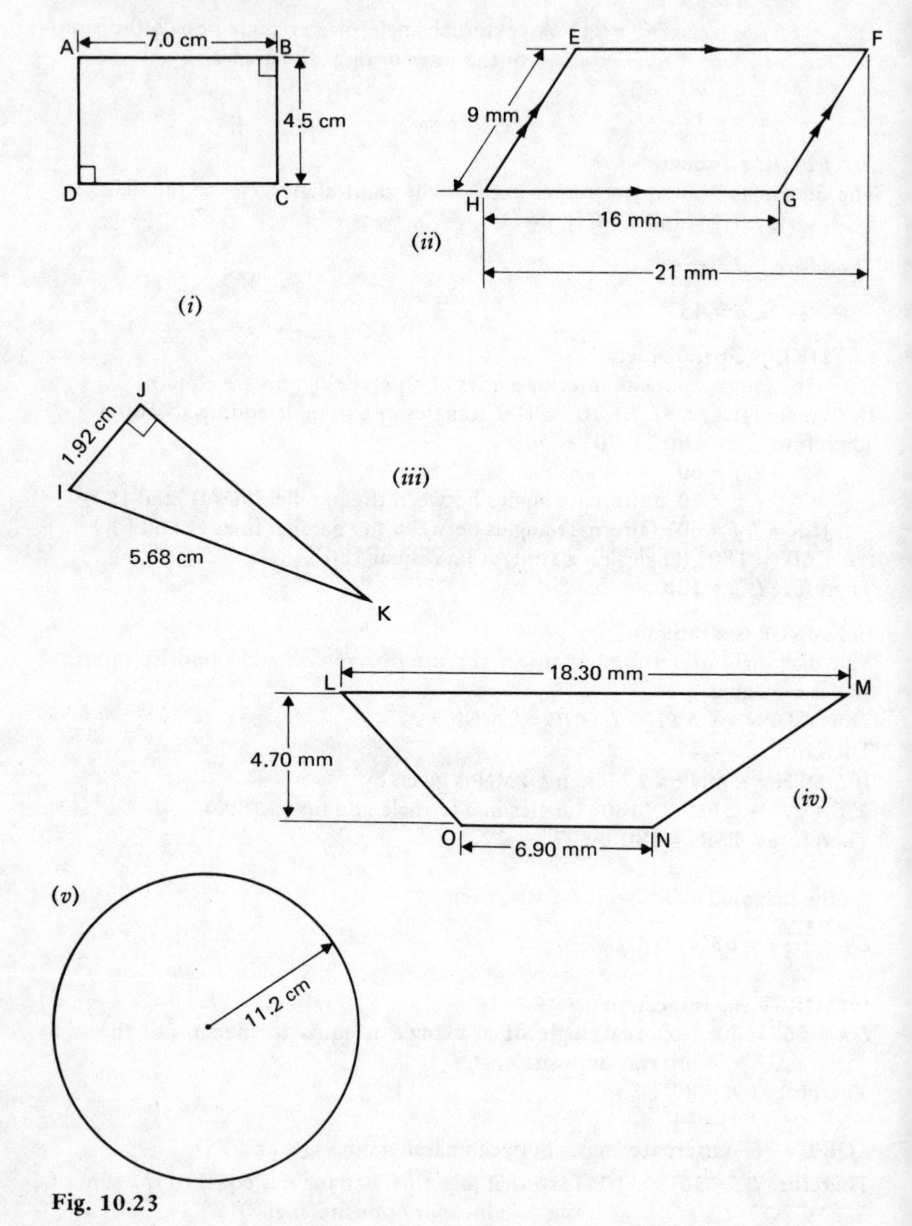

Fig. 10.23

(*i*) Area of a rectangle = length × breadth
Therefore **area of ABCD** = 7.0 cm × 4.5 cm
= **31.5 cm²**

(*ii*) Area of a parallelogram = base × perpendicular height.
The perpendicular height, h, may be found using Pythagoras's theorem.

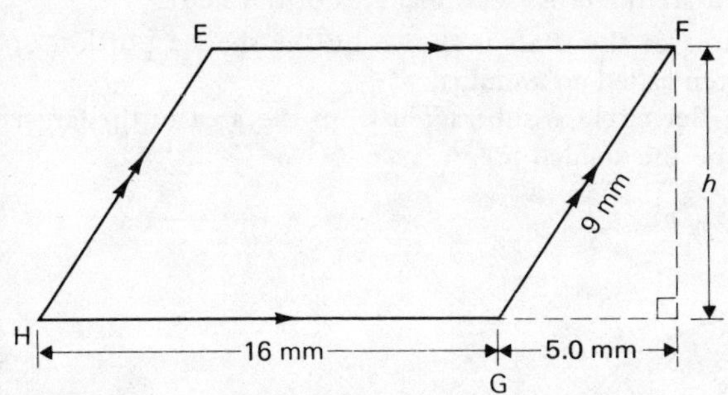

Fig. 10.24

From Fig. 10.25, $9^2 = 5^2 + h^2$
Therefore $h^2 = 9^2 - 5^2$
$= 81 - 25$
$= 56$
$h = \sqrt{56} = \pm 7.48$ mm
$h = 7.48$ mm (-7.48 can be neglected)
Therefore **area of EFGH** = 16 mm × 7.48 mm
= **120 mm²**

(*iii*) Area of a triangle = $\frac{1}{2}$ base × perpendicular height.
If IJ = 1.92 cm is assumed to be the base of the triangle, then the perpendicular height needs to be found. In triangle IJK, JK is the perpendicular height since ∠ IJK is right-angled.
By Pythagoras's theorem $(IK)^2 = (IJ)^2 + (JK)^2$
Therefore $(5.68)^2 = (1.92)^2 + (JK)^2$
$(JK)^2 = (5.68)^2 - (1.92)^2$
$= 32.26 - 3.686$
$= 28.57$
$JK = \sqrt{28.57}$
$= \pm 5.345$ cm
JK = 5.345 cm (-5.345 can be neglected)
Therefore **area of IJK** = $\frac{1}{2}$ (1.92) (5.345)
= **5.131 cm²**

(*iv*) Area of a trapezium = $\frac{1}{2}$ × (sum of the parallel sides) × (perpendicular distance between parallel sides).

Therefore **area of LMNO** = $\frac{1}{2}$ (18.30 + 6.90) (4.70)
= $\frac{1}{2}$ (25.20) (4.70)
= **59.22 mm²**

(*v*) Area of a circle = πr^2

Therefore **area of circle with radius 11.2 cm** = $(3.142)\ (11.2)^2$

$= \mathbf{394.1\ cm^2}$

Problem 3. A hollow shaft has an outside diameter of 6.42 cm and an inside diameter of 2.15 cm. Calculate the cross-sectional area of the shaft.

The cross-sectional area of the shaft is shown by the shaded portion of Fig. 10.25. This area is often called an **annulus.**

If the area of the smaller circle is subtracted from the area of the larger circle the remainder will be the shaded area.

Now the area of a circle = $\pi r^2 = \dfrac{\pi d^2}{4}$

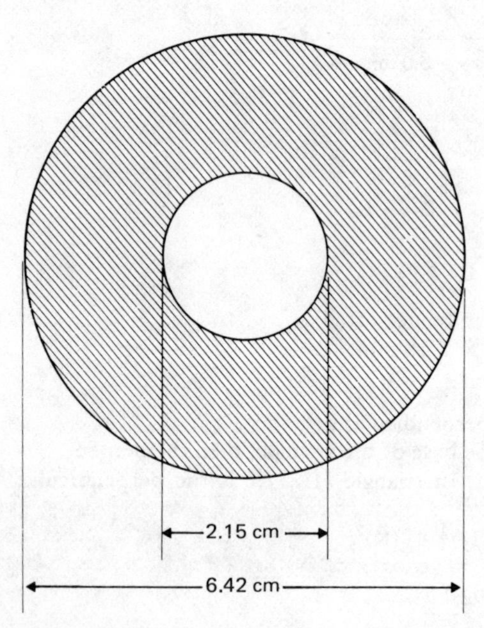

Fig. 10.25

Therefore **cross-sectional area of shaft** $= \dfrac{\pi}{4}(6.42)^2 - \dfrac{\pi}{4}(2.15)^2$

$$= \frac{\pi}{4}[(6.42)^2 - (2.15)^2]$$

$$= \frac{3.142}{4}[41.22 - 4.62]$$

$$= \frac{3.142}{4}(36.60)$$

$$= \mathbf{28.75\ cm^2}$$

Problem 4. An open water tank has a rectangular base and sides and is made from steel. The sides are 5 m long by 78.0 cm wide and 37.5 cm high. Find the area of the steel plate required to make the tank in (*a*) cm^2 and (*b*) m^2.

Before the areas can be calculated all measurements should be in the same units. Thus for (*a*) the measurements of the tank are 500 cm by 78.0 cm by 37.5 cm as shown in Fig. 10.26.

Area of base of tank = $78.0 \times 500 = 39\,000\ cm^2$

Area of 2 ends of tank = $2(78.0 \times 37.5) = 2(2\,925)\ cm^2$

Area of 2 sides of tank = $2(37.5 \times 500) = 2(18\,750)\ cm^2$

Therefore **area of metal required** = $39\,000 + 2(2\,925) + 2(18\,750)\ cm^2$

$= \mathbf{82\,350\ cm^2}$

Now 100 cm = 1 m

and $100\ cm \times 100\ cm = 1\ m^2$

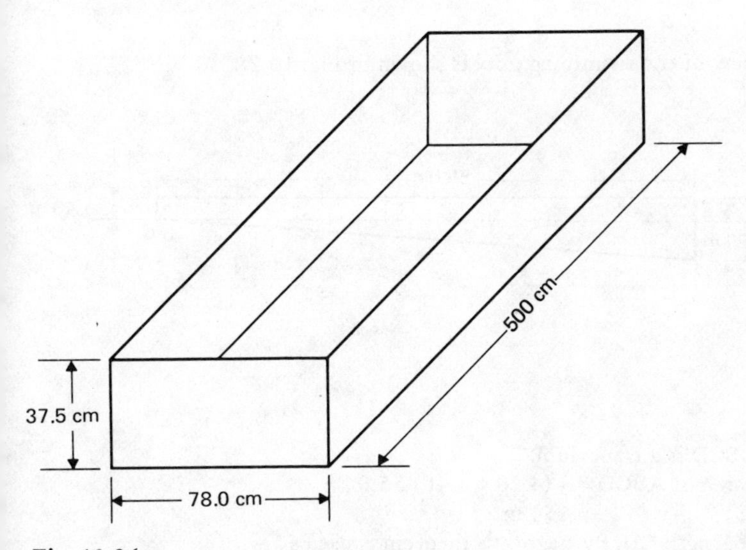

Fig. 10.26

Therefore $10\,000\ cm^2 = 1\ m^2$

$$82\,350\ cm^2 = \frac{82\,350}{10\,000}\ m^2 = 8.235\ m^2$$

Thus the area of steel required is **82 350 cm²** or **8.235 m²**

Problem 5. A swimming pool is 55.0 m long and 10.0 m wide. The perpendicular depth at the deep end is 4.20 m and at the shallow end 140.0 cm, the slope being uniform. The pool needs two coats of a protective paint inside. Find how many litres of paint will be required if 1 litre covers 12 m^2.

The dimensions of the swimming pool are shown in Fig. 10.27 in metre units.

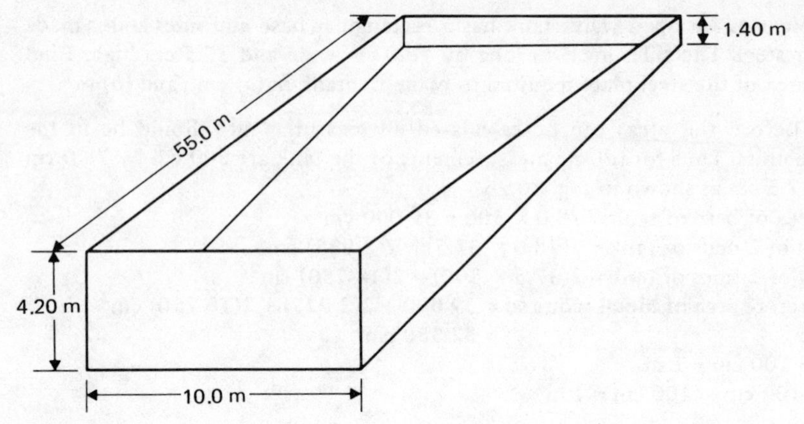

Fig. 10.27

The side view of the swimming pool is shown in Fig. 10.28

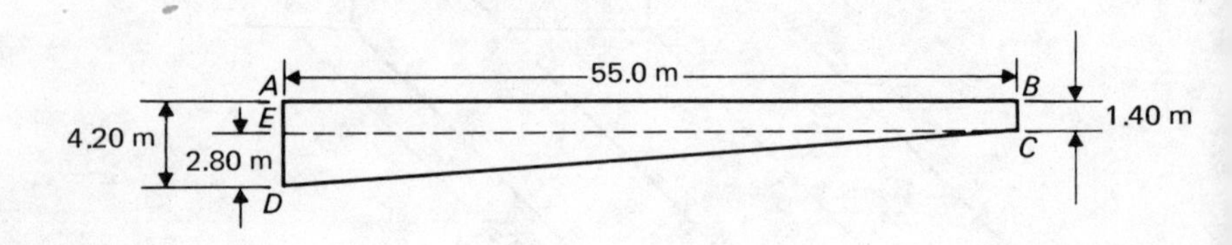

Fig. 10.28

The area ABCD is a trapezium.

Therefore area of ABCD = $\frac{1}{2}$ (4.20 + 1.40) 55.0

$= 154 \text{ m}^2$

To find the length CD, Pythagoras's theorem is used:

$$(CD)^2 = (ED)^2 + (CE)^2$$

Therefore $(CD)^2 = (2.80)^2 + (55.0)^2$

$= 7.84 + 3\,025$

$= 3\,032.84 = 3\,033$ (correct to 4 significant figures)

$CD = \sqrt{3\,033} = \pm 55.07$

$= 55.07$ m (-55.07 can be neglected)

The area of the bottom of the pool = (55.07) (10.0) = 550.7 m^2

The area of the two sides of the pool = 2(154) = 308 m^2

The area of the deep end face of the pool = 4.20(10.0) = 42 m^2

The area of the shallow end face of the pool = 1.40(10.0) = 14 m^2

The total area of the pool = (550.7 + 308 + 42 + 14) m^2

= 914.7 m^2

Two coats of paint would therefore cover 2 × 914.7, that is, 1 829.4 m^2. If 1 litre of paint covers 12 m^2 then the amount of paint required is

$\frac{1\,829.4}{12}$ litres = 152.45 litres

= 152.5 litres (to 4 significant figures)

Thus the amount of paint required is **152.5 litres.**

Problem 6. Find the area of a regular octagon if each side is 4.14 cm and the width across the flats is 10.0 cm.

An octagon is an eight-sided polygon. If radii are drawn from the centre of the polygon to the vertices then eight equal triangles are produced. This is shown in Fig. 10.29.

Each of the eight triangles have a base of 4.14 cm and a perpendicular height of 5.0 cm.

Hence the area of one triangle = $\frac{1}{2}$ (4.14) (5.0)

= 10.35 cm^2

Therefore **area of the octagon** = 8(10.35)

= **82.80 cm^2**

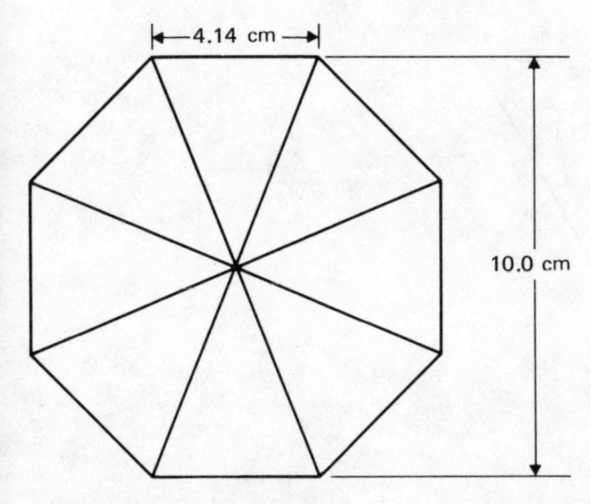

Fig. 10.29

Problem 7. Find the area of a regular hexagon which has sides 6.0 cm long.

A hexagon is a six-sided polygon. If radii are drawn from the centre of the polygon to the vertices then six equal triangles are produced. This is shown in Fig. 10.30.

The angle subtended at the centre by each triangle is $\frac{360^\circ}{6} = 60^\circ$. The two other angles in the triangle add up to 120° and are equal to each other. Therefore each triangle has three 60° angles, that is, each of the triangles is equilateral.

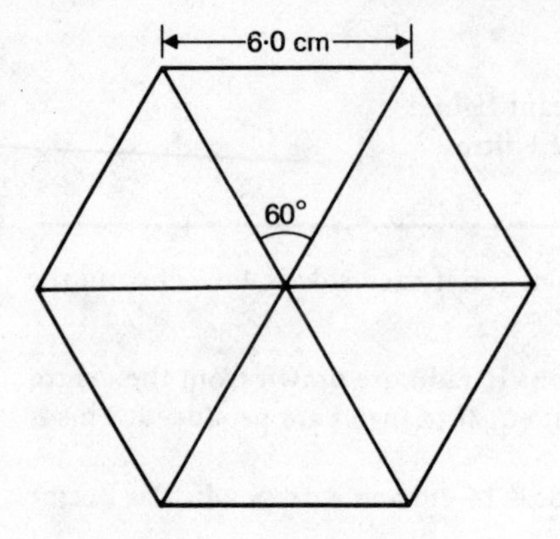

Fig. 10.30

To find the area of one of the equilateral triangles, a perpendicular is drawn from one vertex to bisect the opposite side as shown in Fig. 10.31.

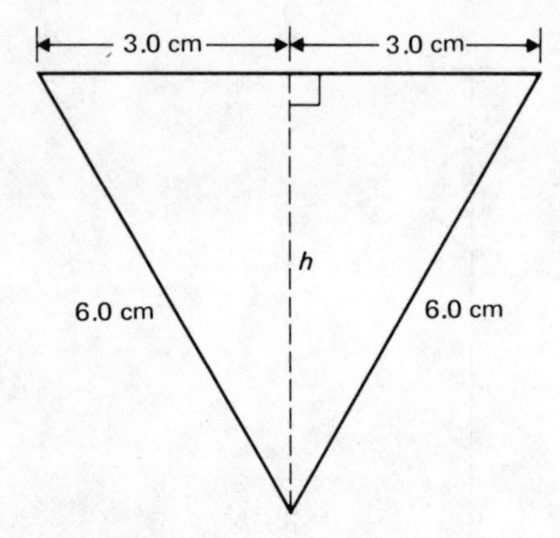

Fig. 10.31

Using Pythagoras's theorem:

$$(6.0)^2 = (3.0)^2 + h^2$$

Therefore $h^2 = (6.0)^2 - (3.0)^2$

$$= 36 - 9$$

$$= 27$$

$$h = \sqrt{27}$$

$$= \pm 5.20 \text{ cm}$$

$h = 5.20$ cm (-5.20 can be neglected)

Hence the area of one triangle $= \frac{1}{2}(6.0)(5.20)$

$$= 15.60 \text{ cm}^2$$

Hence **the area of the hexagon** $= 6(15.60)$

$$= \mathbf{93.6 \text{ cm}^2}$$

Further problems on areas of plane figures may be found in Section 10.6 (Problems 1–23), page 302.

10.4 Volumes of regular solids

The **volume** of any solid is a measure of the space occupied by the solid. If a cube measures 4 units by 4 units by 4 units, as shown in Fig. 10.32, then the volume is found by counting the number of unit cubes contained within the space occupied by the whole cube.

Thus the volume of the cube shown is 64 cubic units. This volume could be obtained by multiplying the three perpendicular sides of the cube together, that is, 4 units × 4 units × 4 units = 64 cubic units. In general, if each side of a cube is of length x units then the volume of the cube is x^3 cubic units.

If the unit of length is the centimetre, then the unit of volume will be cm × cm × cm, that is, cubic centimetres, written as cm^3.

Similarly, if the unit of length is the metre then the unit of volume will be m × m × m, that is, cubic metres, written as m^3.

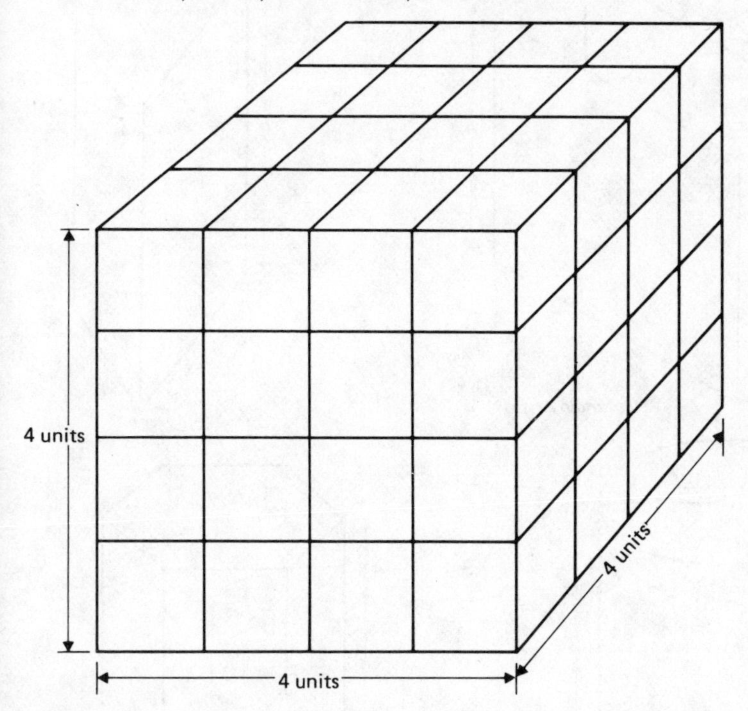

Fig. 10.32

1. Prisms

A prism is a solid with a constant cross-section, and with the two ends parallel.

The shape of the end is used to describe the prism. Figure 10.33 shows four typical prisms.

The volume V of a prism of cross-sectional area A and perpendicular height h is given by

$V = A\,h$ **cubic units**

This formula for volume is true for any prism irrespective of the shape of the base. So, referring to Fig. 10.33:

(*a*) **Volume of rectangular prism** $= (a\,b)\,h$
$= a\,b\,h$ **cubic units**

(*b*) **Volume of triangular prism** $= (\frac{1}{2}\,xy)\,h$
$= \frac{1}{2}\,xyh$ **cubic units**

(*c*) **Volume of cylinder** $= (\pi r^2)\,h$
$= \pi r^2\,h$ **cubic units**

(*d*) **Volume of hexagonal prism** = **(Area of hexagon)** h **cubic units.**

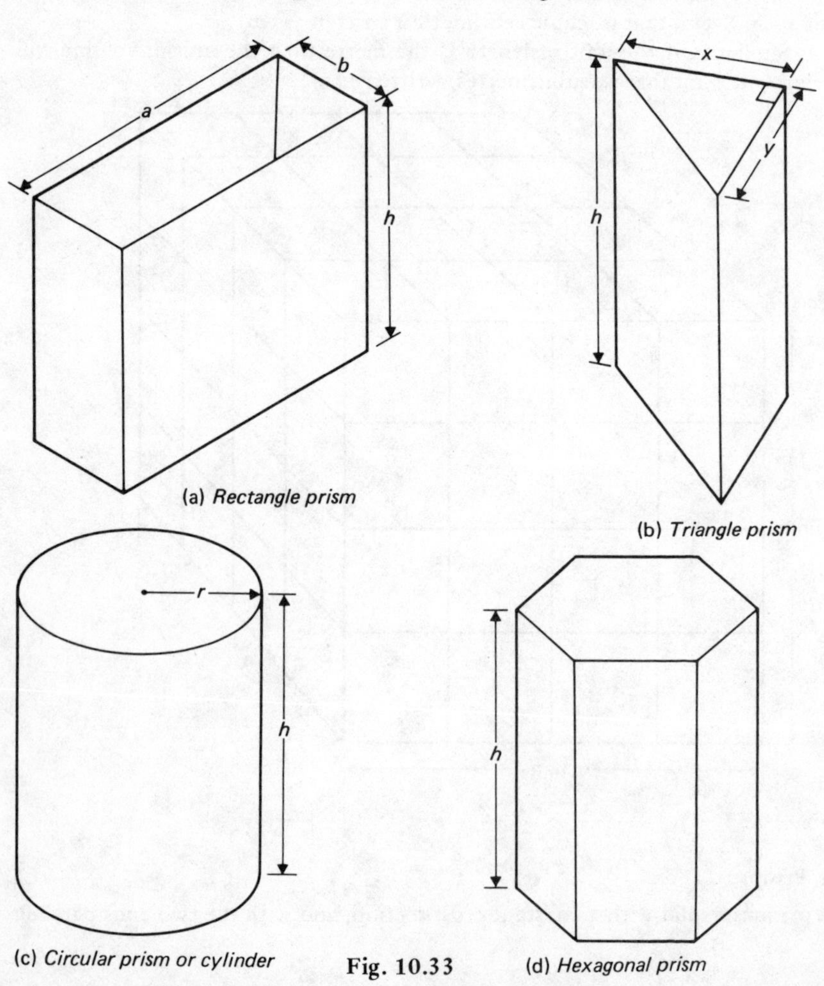

(a) *Rectangle prism*

(b) *Triangle prism*

(c) *Circular prism or cylinder*

Fig. 10.33

(d) *Hexagonal prism*

2. Pyramids

A solid with a plane end and straight sides meeting in a point is called a pyramid. As in the case of prisms, the shape of the plane end is used to describe the pyramid. Figure 10.34 shows four typical pyramids.

The volume V of any pyramid of base area A and perpendicular height h is given by

$$V = \frac{1}{3} A\,h \text{ cubic units}$$

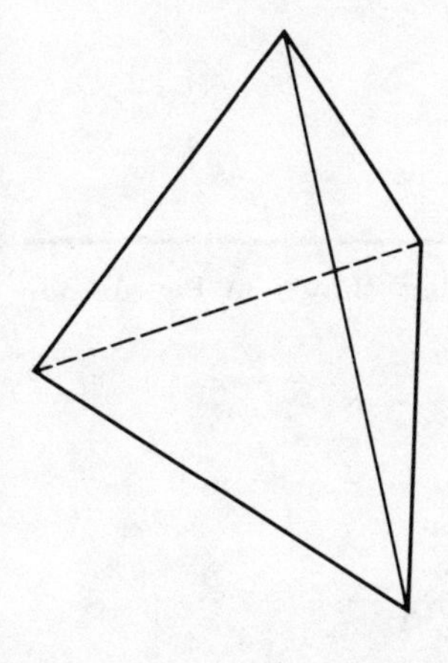

(a) *Triangular pyramid*

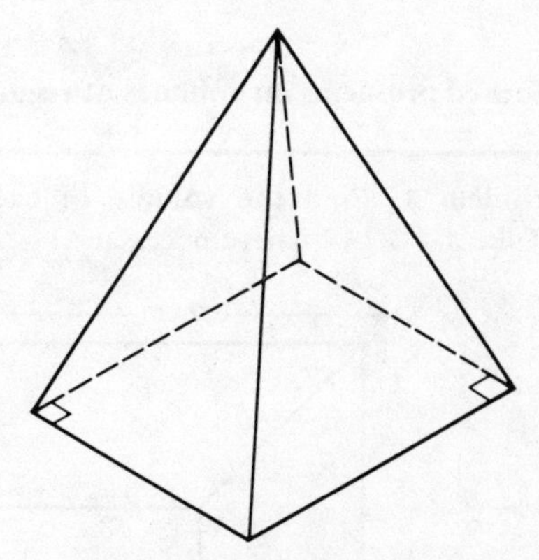

(b) *Rectangular pyramid*

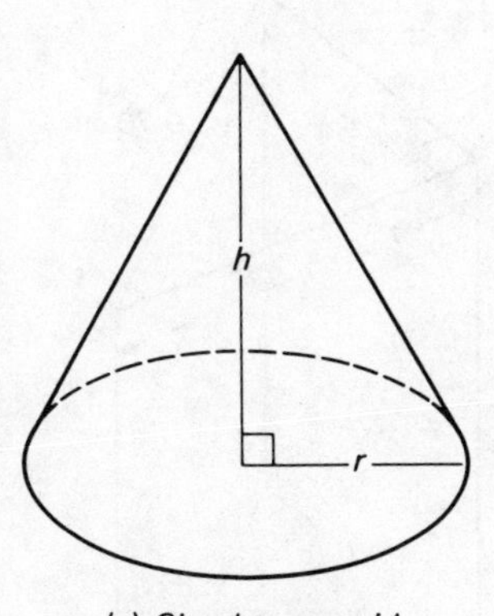

(c) *Circular pyramid or cone*

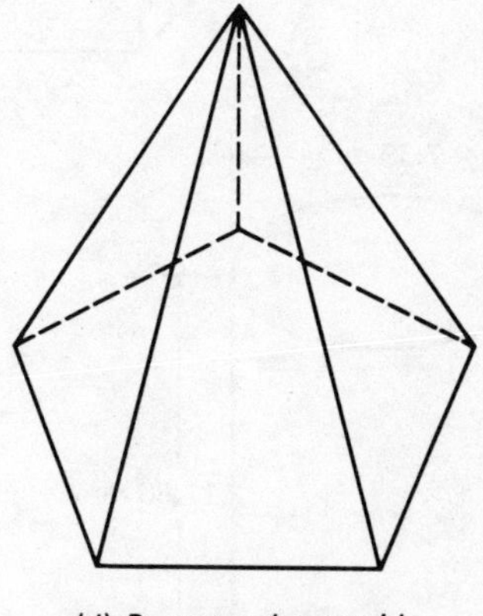

(d) *Pentagonal pyramid*

Fig. 10.34

A circular pyramid is given the special name of a **cone.** The volume V of a cone is given by

$$V = \frac{1}{3}(\pi r^2)h \text{ cubic units}$$

since the base is a circle of area πr^2.

3. Spheres

The volume V of a sphere of radius r is given by

$$V = \frac{4}{3}\pi r^3 \text{ cubic units.}$$

Worked problems on volumes of regular solids

Problem 1. Find the volume of each of the solids shown in Fig. 10.35. (Take π = 3.142 where necessary.)

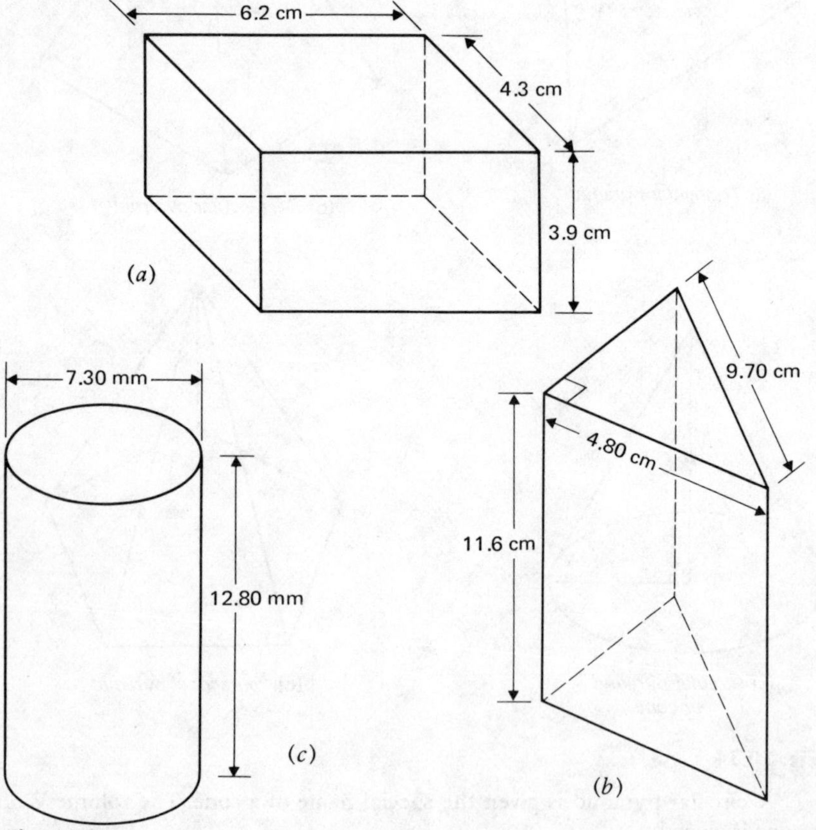

Fig. 10.35

(a) The solid is a **rectangular prism.**
Volume = area of base × height (any of the faces can be termed the base).
Therefore **volume** = (6.2) (4.3) (3.9)
$= \mathbf{104\ cm^3}$

(b) The solid is a **triangular prism.**
Volume = area of base × perpendicular height
Area of a triangle = $\frac{1}{2}$ base × perpendicular height
If the base of the triangle is taken as 4.80 cm then the perpendicular height x is required (see Fig. 10.36). Using Pythagoras's theorem

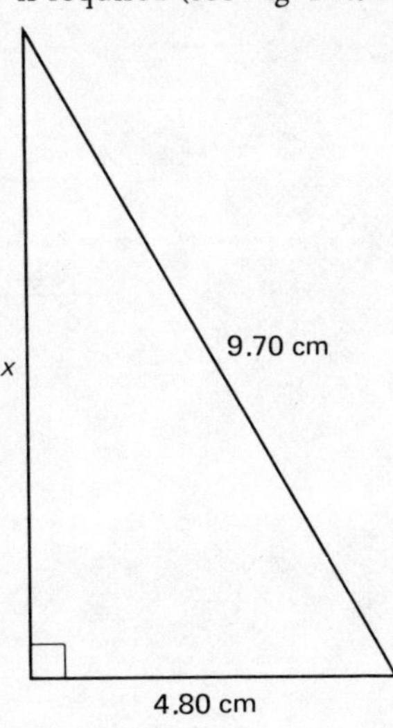

Fig. 10.36

$$(9.70)^2 = (4.80)^2 + x^2$$

Therefore
$$\begin{aligned} x^2 &= (9.70)^2 - (4.80)^2 \\ &= 94.09 - 23.04 \\ &= 71.05 \\ x &= \sqrt{71.05} = \pm 8.43 \\ &= 8.43 \text{ cm } (-8.43 \text{ can be neglected}) \end{aligned}$$

Hence the area of the triangular base $= \frac{1}{2}(4.8)\ (8.43)$
$= 20.23\ \text{cm}^2$

Therefore **volume of triangular prism** = (20.23) (11.6)
$= \mathbf{234.7\ cm^3}$

(c) The solid is a **circular prism** or **cylinder.**
Volume of cylinder $= \pi r^2 h$

$$= \frac{\pi d^2 h}{4}$$

$$= \frac{(3.142)\ (7.30)^2\ (12.80)}{4}$$

$$= \mathbf{535.8\ mm^3}$$

 Problem 2. Calculate the volume of the solid prism shown in Fig. 10.37.

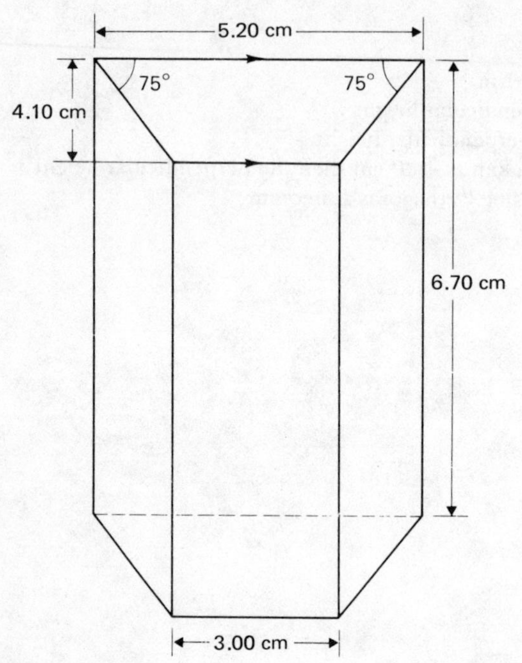

Fig. 10.37

The prism has a trapezoidal base.

Volume of a prism = area of base × perpendicular height.

Area of trapezium = $\frac{1}{2}$ (sum of parallel sides) × (perpendicular distance between them)

$= \frac{1}{2}(5.20 + 3.00)(4.10)$

$= \frac{1}{2}(8.20)(4.10)$

$= 16.81 \text{ cm}^2$

Therefore **volume of prism** $= (16.81)(6.70)$

$= \mathbf{112.6 \text{ cm}^3}$

Problem 3. Find the volume of the square pyramid shown in Fig. 10.38.

Volume of a pyramid = $\frac{1}{3}$ (area of base) × (perpendicular height)

Therefore **volume of pyramid** $= \frac{1}{3}(4.60)(4.60)(11.2)$

$= \mathbf{79.00 \text{ cm}^3}$

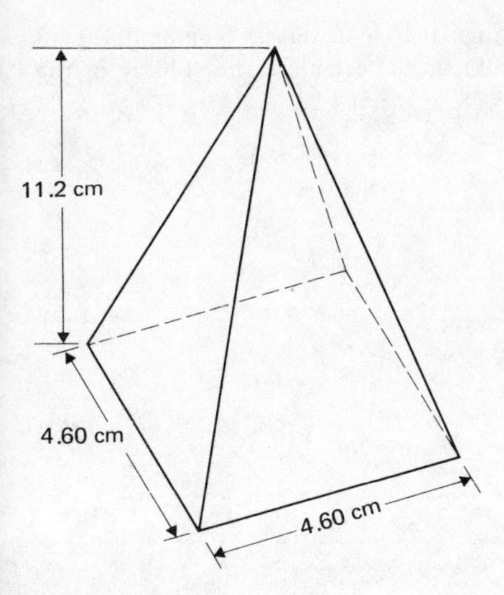

Fig. 10.38

Problem 4. Find the volume of a cone of radius 6.0 cm and perpendicular height 8.0 cm.

The cone is shown in Fig. 10.39.

Volume of cone $= \frac{1}{3}$ (area of base) × (perpendicular height)
$= \frac{1}{3}\pi r^2 h$
$= \frac{1}{3}(3.142)(6.0)^2(8.0)$
$= \mathbf{301.6\ cm^3}$

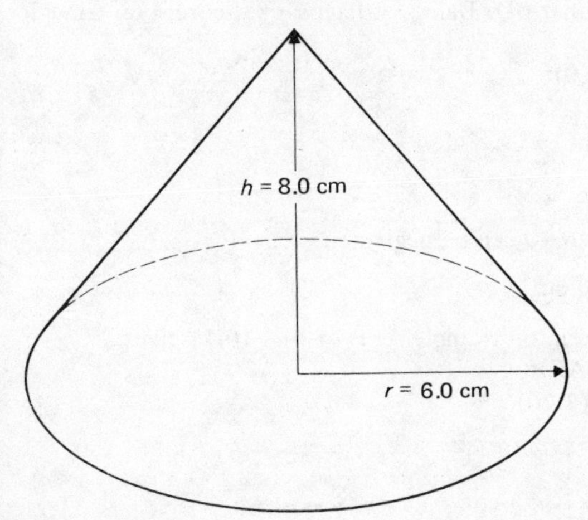

Fig. 10.39

 Problem 5. A pyramid stands on a rectangular base of sides 5.0 cm and 4.0 cm. If the length of each sloping edge is 11.0 cm, calculate the volume of the pyramid.

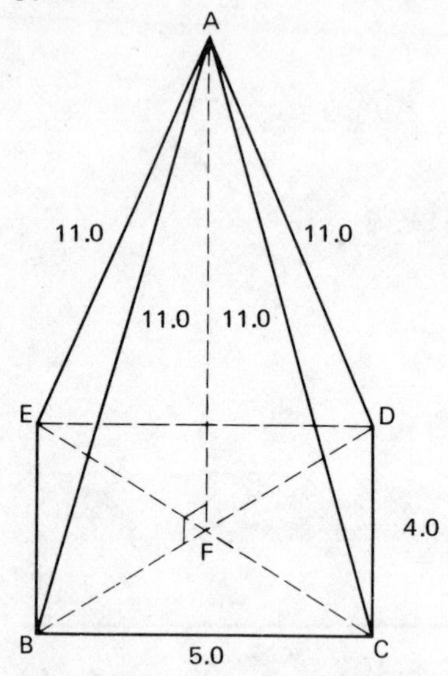

Fig. 10.40

Figure 10.40 shows the rectangular pyramid with equal sloping edges, the dimensions being in centimetres.
Volume of a pyramid = $\frac{1}{3}$ (area of base) × (perpendicular height)

To find the volume of a pyramid it is therefore necessary to find the perpendicular height. This is shown as AF in Fig. 10.40. AF is one side of triangle AFB shown in Fig. 10.41.

FB is half the length of diagonal BD. Using Pythagoras's theorem on triangle BCD of Fig. 10.40 gives

$$BD^2 = (5.0)^2 + (4.0)^2$$
$$= 25 + 16$$
$$= 41$$

Therefore $BD = \sqrt{41}$

$$= \pm 6.40 \text{ cm}$$
$$= 6.40 \text{ cm } (-6.40 \text{ can be neglected})$$

Therefore $FB = \dfrac{6.40}{2} = 3.20$ cm

Now, using Pythagoras's theorem on triangle AFB of Fig. 10.41 gives

$$(11.0)^2 = (3.20)^2 + (AF)^2$$

Therefore $AF^2 = (11.0)^2 - (3.20)^2$

$$= 121 - 10.24$$
$$= 110.8$$

Therefore $AF = \sqrt{110.8} = \pm 10.53$ cm

$$= 10.53 \text{ cm } (-10.53 \text{ can be neglected})$$

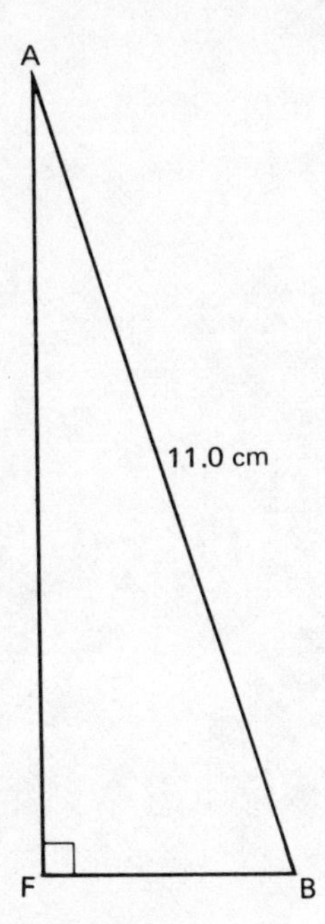

Fig. 10.41

Hence the perpendicular height of the pyramid is 10.53 cm.
Therefore **volume of pyramid** $= \frac{1}{3}(5.0)(4.0)(10.53)$
$= \mathbf{70.20\ cm^3}$

Problem 6. If a sphere has a diameter of 54.0 mm, find its volume.

If the diameter is 54.0 mm then the radius $= \dfrac{54.0}{2}$
$= 27.0$ mm

Volume of a sphere $= \frac{4}{3}\pi r^3$
$= \frac{4}{3}(3.142)(27.0)^3 = \mathbf{82\ 460\ mm^3}$

As 1 cm = 10 mm
$1\ cm^3 = 1\ 000\ mm^3$

Therefore **volume of sphere** $= \dfrac{82\ 460}{1\ 000} = \mathbf{82.46\ cm^3}$

Problem 7. Figure 10.42 shows a 2 m section of metal rod. Find its volume in cm^3 and m^3 each correct to 4 significant figures.

The section of metal is a prism where the end comprises a rectangle and a semicircle. Since the radius of the semicircle is 6 cm, the dimensions of the rectangle are 8 cm by 12 cm.

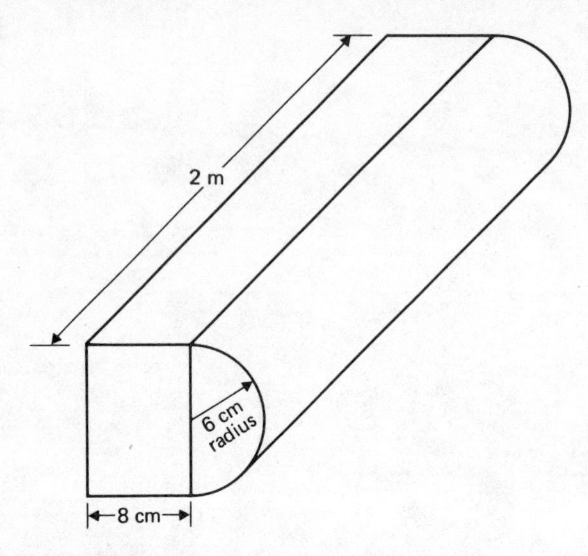

Fig. 10.42

Thus, area of rectangle $= 8 \times 12 = 96\ \text{cm}^2$

area of semicircle $= \frac{1}{2}\pi r^2 = \frac{1}{2}(3.142)(6)^2$

$= 56.56\ \text{cm}^2$

Hence area of the end of the prism $= 96 + 56.56$

$= 152.56\ \text{cm}^2$

$= 152.6\ \text{cm}^2$ (to 4 significant figures)

Therefore volume of metal section $= (152.6)(200)$

$= 30\,520\ \text{cm}^3$

Now 100 cm = 1 m

Therefore $100\ \text{cm} \times 100\ \text{cm} = 1\ \text{m} \times 1\ \text{m} = 1\ \text{m}^2$

and $100\ \text{cm} \times 100\ \text{cm} \times 100\ \text{cm} = 1\ \text{m} \times 1\ \text{m} \times 1\ \text{m} = 1\ \text{m}^3$

That is, $1\,000\,000\ \text{cm}^3 = 1\ \text{m}^3$

Therefore $30\,520\ \text{cm}^3 = \dfrac{30\,520}{1\,000\,000}\ \text{m}^3$

$= 0.030\,52\ \text{m}^3$

Thus the volume of the metal section is **30 520 cm³** or **0.030 52 m³**.

Problem 8. A rectangular piece of metal with dimensions 5.20 cm by 8.60 cm by 11.30 cm is melted down and recast into a square pyramid of perpendicular height 1.20 m. Find the length of the sides of the base of the pyramid.

Volume of rectangular prism of metal $= 5.20 \times 8.60 \times 11.30$

$= 505.3\ \text{cm}^3$

Volume of a pyramid $= \frac{1}{3}$ (area of base) $\times$ (perpendicular height)

Assuming that there is no waste of metal, the volume of the pyramid is $505.3\ \text{cm}^3$ and its perpendicular height is 120 cm.

Therefore $505.3 = \frac{1}{3}$ (area of base) 120

Therefore area of base $= \frac{3(505.3)}{120}$

$= 12.63 \text{ cm}^2$

As the base of the pyramid is square the length of a side will be $\sqrt{12.63} = 3.554$ cm.

Hence **the length of side of the base of the pyramid is 3.554 cm.**

Problem 9. A block of copper weighing 30 kg is drawn out to make 400 m of wire of uniform circular cross-section. Given that 1 cm^3 of copper weighs 8.91 g calculate:

(*a*) the volume of the copper in cm^3;

(*b*) the area of the cross-section of the wire in cm^2;

(*c*) the radius of the cross-section of the wire in mm.

(*a*) If 8.91 g of copper has a volume of 1 cm^3

then 1 g of copper has a volume of $\frac{1}{8.91} \text{ cm}^3$

hence **30 000 g of copper has a volume of** $\frac{30\,000}{8.91} \text{ cm}^3$

$= \mathbf{3\,367 \text{ cm}^3}$

(*b*) The wire may be regarded as a very long circular prism or cylinder.

Hence the volume of wire = (area of cross-section) × (length of wire)

Therefore 3 367 = (area of cross-section) (40 000)

Therefore **area of cross section** $= \frac{3\,367}{40\,000} \text{ cm}^2$

$= \mathbf{0.084 \text{ cm}^2}$

(*c*) Area of circle $= \pi r^2$.

Therefore $0.084 \text{ cm}^2 = 8.4 \text{ mm}^2 = \pi r^2$

Therefore $r^2 = \frac{8.4}{3.142} = 2.673$

$r = \sqrt{2.673} = \pm 1.635$ mm

$= 1.635$ mm (-1.635 can be neglected)

Hence **the radius of cross section is 1.635 mm.**

Problem 10. A solid metal cylinder of diameter 10 cm and height 25 cm is melted down and recast into the shape shown in Fig. 10.43 with 15% of the metal wasted in the process. The new shape is a hemisphere attached to the circular base of a cone. If the radius of the cone is to be equal to the height of the cone find the overall length l of the new shape.

 The volume of a cylinder = $\pi r^2 h$. If the diameter of the cylinder is 10 cm the radius is 5 cm.

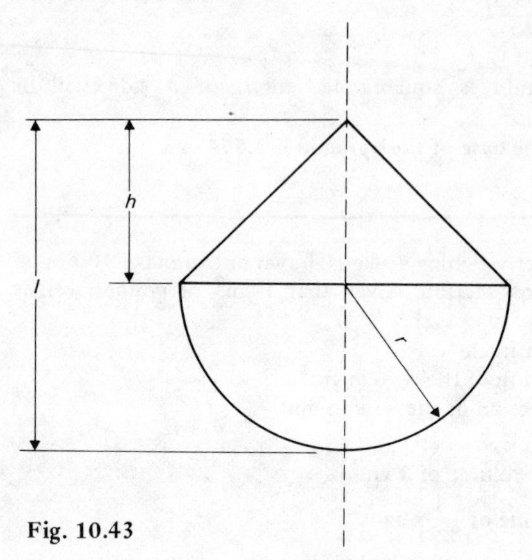

Fig. 10.43

Volume of cylinder = $\pi(5)^2(25)$
$= 625\pi \text{ cm}^3$

If 15% of the metal is lost then 85% of 625π forms the volume of the new shape. That is, volume of new shape = $(0.85)(625\pi) \text{ cm}^3$.

The volume of Fig. 10.43 consists of:

(*a*) a hemisphere (i.e. half a sphere).
Therefore volume of hemisphere = $\frac{1}{2}(\frac{4}{3}\pi r^3)$
$= \frac{2}{3}\pi r^3$

(*b*) a cone.
Volume of cone = $\frac{1}{3}\pi r^2 h$. But $h = r$,
Therefore volume of cone = $\frac{1}{3}\pi r^2 r$
$= \frac{1}{3}\pi r^3$

Therefore the total volume of the new shape = $\frac{2}{3}\pi r^3 + \frac{1}{3}\pi r^3$
$= \pi r^3 \text{ cm}^3$

Hence $(0.85)(625\pi) = \pi r^3$
$r^3 = (0.85)(625)$
$= 531.25$
$= 531.3$ (to 4 significant figures)
$r = \sqrt[3]{531.3}$
$= 8.099$ cm

The overall length l of the new shape is $h + r$
that is, $2r = 2(8.099)$
$= 16.198$ cm
= **16.20 cm** (to 4 significant figures)

Further problems on volumes of regular solids may be found in Section 10.6 (Problems 24–46), page 304.

10.5 Areas and volumes of similar shapes

Areas

Figure 10.44 shows two squares, one of which has sides three times as long as the other. Area of Fig. 10.44(a) = (a) (a) $= a^2$ square units.
Area of Fig. 10.44(b) = ($3a$) ($3a$) = $9a^2$ square units.

Hence Fig. 10.44(b) has an area $(3)^2$, i.e. 9 times the area of Fig. 10.44(a).

In general, **the areas of similar shapes are proportional to the squares of corresponding linear dimensions.** Thus if the sides of a square are doubled, for example, the area becomes $(2)^2$, i.e. 4 times its original area.

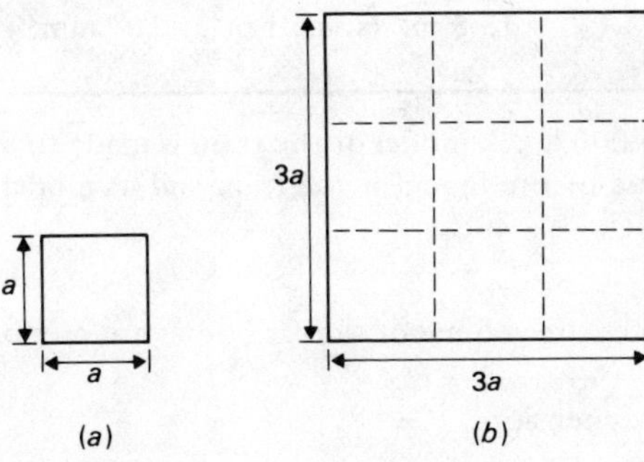

Fig. 10.44

Volumes

Figure 10.45 shows two cubes, one of which has sides three times as long as that of the other. Volume of Fig. 10.45(a) = (a) (a) (a) $= a^3$ cubic units.
Volume of Fig. 10.45(b) = ($3a$) ($3a$) ($3a$) = $27\,a^3$ cubic units.

Hence Fig. 10.45(b) has a volume $(3)^3$, i.e. 27 times the volume of Fig. 10.45(a).

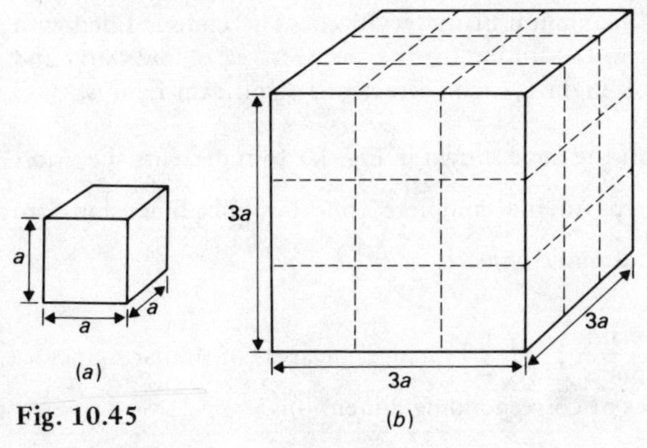

Fig. 10.45

In general, **the volumes of similar bodies are proportional to the cubes of corresponding linear dimensions.** Thus if the sides of a cube are doubled the volume becomes $(2)^3$, i.e. 8 times its original volume.

Worked problems on areas and volumes of similar shapes

Problem 1. A rectangular building is shown on a building plan having dimensions 10 mm by 15 mm. If the plan is drawn to a scale of 1 to 300, determine the true area of the building, in square metres.

Area of building on the plan = 10 mm × 15 mm = 150 mm^2.

Since the areas of similar shapes are proportional to the square of corresponding dimensions then:

$$\text{true area of building} = 150 \times (300)^2 = 13.5 \times 10^6 \text{ mm}^2$$
$$= \mathbf{13.5 \ m^2} \quad (\text{since } 1 \text{ m}^2 = 10^6 \text{ mm}^2)$$

Problem 2. A train has a mass of 20 000 kg. A model of the train is made to a scale of 1 to 50. Determine the mass of the model if the train and its model are made of the same material.

$\left(\dfrac{\text{Volume of model}}{\text{Volume of train}}\right) = \left(\dfrac{1}{50}\right)^3$, since the volume of similar bodies are proportional to the cube of corresponding dimensions.

Mass = density × volume, and since both train and model are made of the same material then $\left(\dfrac{\text{mass of model}}{\text{mass of train}}\right) = \left(\dfrac{1}{50}\right)^3$.

Hence mass of model = (mass of train) $\left(\dfrac{1}{50}\right)^3 = \dfrac{20\,000}{125 \times 10^3} = \mathbf{0.16 \ kg \ or \ 160 \ g.}$

Problem 3. A water container is in the form of an inverted cone of perpendicular height 80 cm and maximum diameter 40 cm. The cone is filled with water to a height of 20 cm. Determine (*a*) the surface area of the water and (*b*) the volume of the water in dm^3, each correct to 3 significant figures.

(*a*) The shaded portion of the cone shown in Fig. 10.46 represents the water and this is similar in shape to the complete cone but the dimensions are $\dfrac{20}{80}$, i.e. $\dfrac{1}{4}$ of the size of the whole cone.

Hence $\left(\dfrac{\text{surface area of water}}{\text{area of end of cone}}\right) = \left(\dfrac{1}{4}\right)^2$, since the areas of similar shapes are proportional to the squares of corresponding dimensions.

Thus surface area of water $= \left(\frac{1}{4}\right)^2 (\pi R^2) = \left(\frac{1}{4}\right)^2 \pi (20)^2 = 78.54\ \text{cm}^2$

$= \mathbf{78.5\ cm^2}$, correct to 3 significant figures.

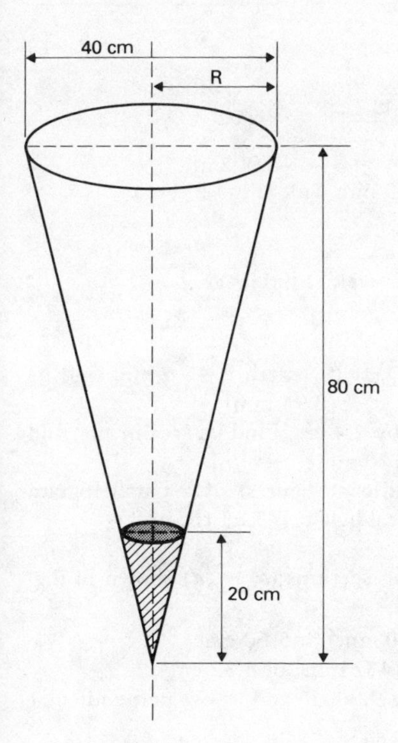

Fig. 10.46

(*b*) $\dfrac{\text{Volume of water}}{\text{volume of cone}} = \left(\frac{1}{4}\right)^3$, since the volume of similar bodies are proportional to the cube of corresponding dimensions.

Hence volume of water $= \left(\frac{1}{4}\right)^3$ (volume of cone) $= \left(\frac{1}{4}\right)^3 (\frac{1}{3}\pi R^2 h)$

$= \left(\frac{1}{4}\right)^3 (\frac{1}{3}\pi(20)^2\ 80)$

$= 523.6\ \text{cm}^3$

$= \dfrac{523.6}{1\,000}\ \text{dm}^3 = \mathbf{0.524\ dm^3}$, correct to 3 significant figures.

Further problems on areas and volumes of similar shapes may be found in the following Section (10.6) (Problems 47–52), page 307.

10.6 Further problems

Where necessary in the following problems assume that $\pi = 3.142$

Areas of plane figures

1. Convert 18.623 m into (*a*) mm (*b*) cm
   (*a*) [1 8623 mm] (*b*) [1862.3 cm]
2. Convert 72 460 mm into (*a*) cm (*b*) m
   (*a*) [7 246 cm] (*b*) [72.46 m]
3. Add 4.23 m, 92 cm, 143 mm and 1.609 m [6.902 m]
4. A metal plate is 91.4 mm long and 37.5 mm wide. Find its area (*a*) in mm^2 (*b*) in cm^2
   (*a*) [3 428 mm^2] (*b*) [34.28 cm^2]
5. A tray is 841.3 mm long and 412.7 mm wide. Find its area (*a*) in cm^2 (*b*) in m^2
   (*a*) [3 472 cm^2] (*b*) [0.347 2 m^2]
6. The area of a metal plate is 426 mm^2. If its length is 47.6 mm find its width in mm (to 3 significant figures). [8.95 mm]
7. A rectangular plate measures 36 mm by 2.4 cm. Find its area in cm^2 and its perimeter in mm. [8.64 cm^2; 120 mm]
8. Calculate the area (correct to 4 significant figures) of a parallelogram whose base is 9.41 cm and whose vertical height is 5.63 cm. [52.98 cm^2]
9. Find the area of each of the angle iron sections (*a*) – (*d*) shown in Fig. 10.47, all the dimensions being in mm.
   (*a*) [272 mm^2 or 2.72 cm^2] (*b*) [560 mm^2 or 5.60 cm^2]
   (*c*) [1 624 mm^2 or 16.24 cm^2] (*d*) [432 mm^2 or 4.32 cm^2]
10. Find the area of a triangle whose base is 9.80 cm and whose perpendicular height is 7.40 cm [36.26 cm^2]
11. Find the area (correct to 4 significant figures), of a triangle XYZ if
    (*a*) $x = 3.10$ cm, $z = 4.30$ cm and $\angle y = 90°$;
    (*b*) $x = 11.62$ cm, $y = 4.92$ cm and $\angle x = 90°$
    (*a*) [6.665 cm^2] (*b*) [25.90 cm^2]
12. A rectangular garden measures 50 m by 20 m. A 1.5 m flower border is made round the two longest sides and one short side. A circular fish pond of diameter 7 m is constructed in the middle of the garden. Find to the nearest square metre the area remaining. [786 m^2]
13. Find the area of an equilateral triangle of side 15.0 cm. [97.43 cm^2]
14. Find the area of a trapezium whose parallel sides measure 4.60 cm and 8.10 cm. The perpendicular distance between these two sides is 5.40 cm. [34.29 cm^2]
15. A square has an area of 79.23 cm^2. Calculate the length of a side (to 2 decimal places). [8.90 cm]
16. Find the length of the side of a square whose area is equal to that of a parallelogram with a 6.0 cm base and a perpendicular height to the base of 4.70 cm. [5.31 cm]
17. The area of a trapezium is 19.4 cm^2 and the perpendicular distance

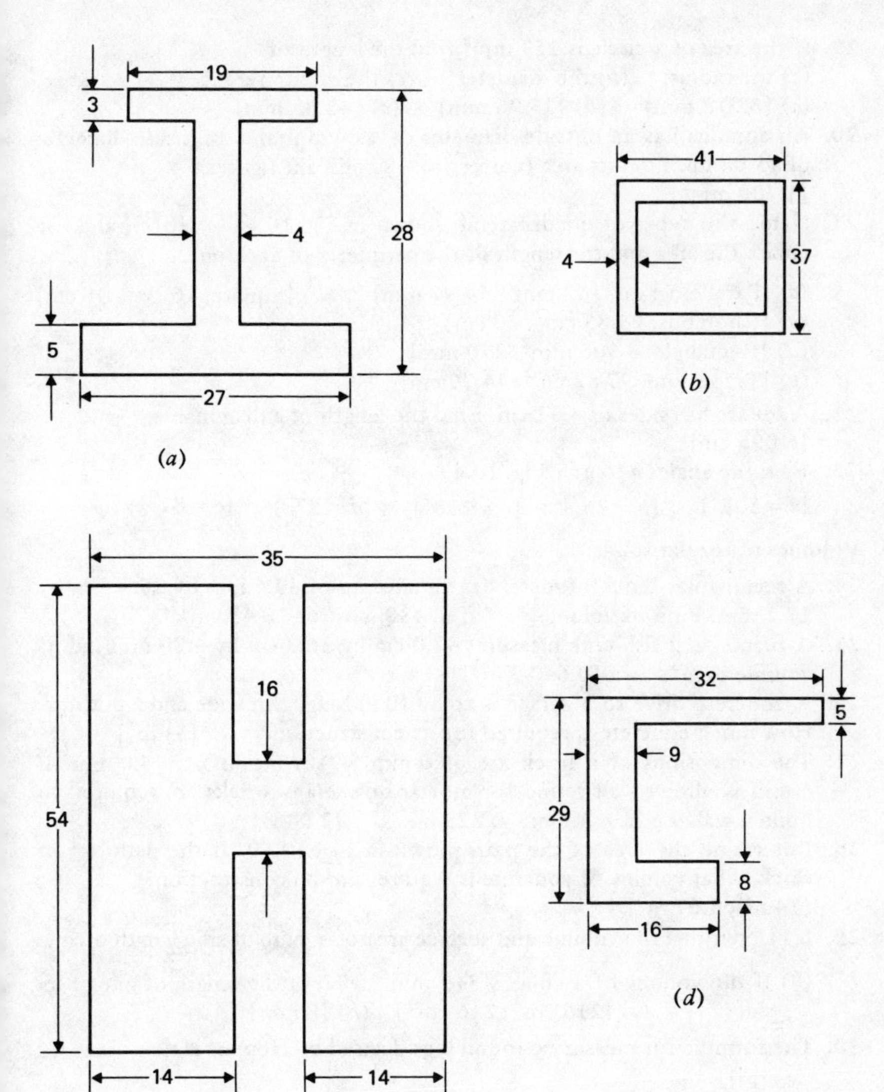

Fig. 10.47

between its parallel sides is 2.80 cm. If the length of one of the parallel sides is 4.23 cm find the length of the other parallel side. [9.63 cm]

18. Find the area of the following circles given that
(*a*) the radius is 15.0 mm in length
(*b*) the diameter is 2.80 cm in length
(*c*) the circumference is 42.4 cm in length.
(*a*) [707.0 mm^2] (*b*) [6.158 cm^2] (*c*) [143.0 cm^2]

19. If the area of a circle is 153 mm^2 find the length of
(*a*) the radius, (*b*) the diameter, (*c*) the circumference.
(*a*) [6.978 mm] (*b*) [13.96 mm] (*c*) [43.85 mm]
20. An annulus has an outside diameter of 49.0 mm and an inside diameter of 15.0 mm. Find its area (correct to 4 significant figures).
[1 709 mm^2]
21. Name the types of quadrilateral, shown in Fig. 10.48 (*a*) – (*e*), and calculate the area and the length of the perimeter of each one.
(*a*) [Parallelogram, 162 mm^2, 54.97 mm] (*b*) [Square, 16 cm^2, 16 cm]
(*c*) [Rhombus, 23.85 cm^2, 20 cm]
(*d*) [Rectangle, 3 906 mm^2, 270 mm]
(*e*) [Trapezium, 97.82 cm^2, 44.79 cm]
22. A square has sides of 4.31 cm. Find the length of a diagonal.
[6.095 cm]
23. Find the angles *a* to *e* in Fig. 10.49.
[$a = 108°$] [$b = 26°$] [$c = 138°$] [$d = 87°$] [$e = 81°$]

Volumes of regular solids

24. A rectangular block of metal has dimensions of 37.6 mm by 25.4 mm by 17.2 mm. Find its volume. [16 430 mm^3 or 16.43 cm^3]
25. A rectangular fish tank measures 42.0 cm by 36.3 cm by 4.20 m. Find its volume in m^3. [0.640 3 m^3]
26. A concrete drive to a garage is to be 10 m long, 2 m wide and $\frac{1}{6}$ m thick. How much concrete is required for its construction? [$3\frac{1}{3}$ m^3]
27. The dimensions of a brick are 65.0 mm × 102.5 mm × 215.0 mm. If 5 mm is allowed all round for mortar how many bricks are required to build a wall 9.6 m × 1.8 m× 0.225 m? [2 048]
28. Determine the area of the path shown in Fig. 10.50. If the path is $\frac{1}{6}$ m thick, what volume of concrete is required for its construction?
[34 m^2, 5.67 m^3]
29. (*a*) Determine the volume and surface area of a cube of side length 6 cm.
(*b*) If the volume of a cube is 512 mm^3 what is the length of one of its sides. (*a*) [216 cm^3 ; 216 cm^2] (*b*) [8 mm]
30. The formula for measuring round logs devised by Hoppus is
$$V = \frac{G^2L}{5\pi}$$
where V = volume, G = girth or circumference and L = length of log. How much allowance has been made for waste? [20%]
31. Find the volume of a metal tube whose outside diameter is 7.0 cm and whose internal diameter is 5.80 cm, if the length of the tube is 8.0 m.
[9 651 cm^3]
32. The volume of a cylinder is 348 cm^3. If its radius is 4.0 cm find its height.
[6.922 cm]
33. If the height of a cylindrical can is 5.50 cm and its volume is 252 cm^3 find its diameter. [7.637 cm]

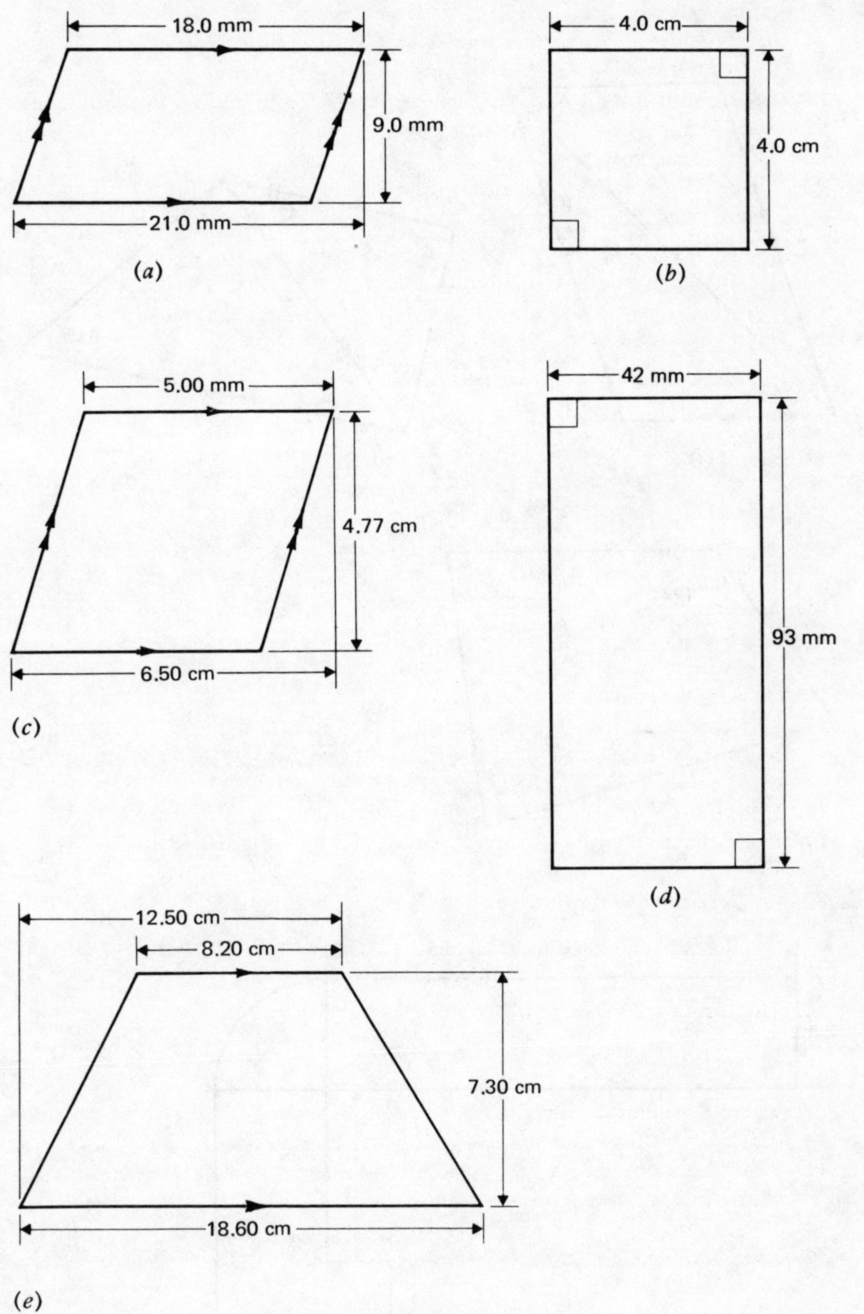
18.0 mm
9.0 mm
21.0 mm
(a)
4.0 cm
4.0 cm
(b)
5.00 mm
4.77 cm
6.50 cm
(c)
42 mm
93 mm
(d)
12.50 cm
8.20 cm
7.30 cm
18.60 cm
(e)

Fig. 10.48

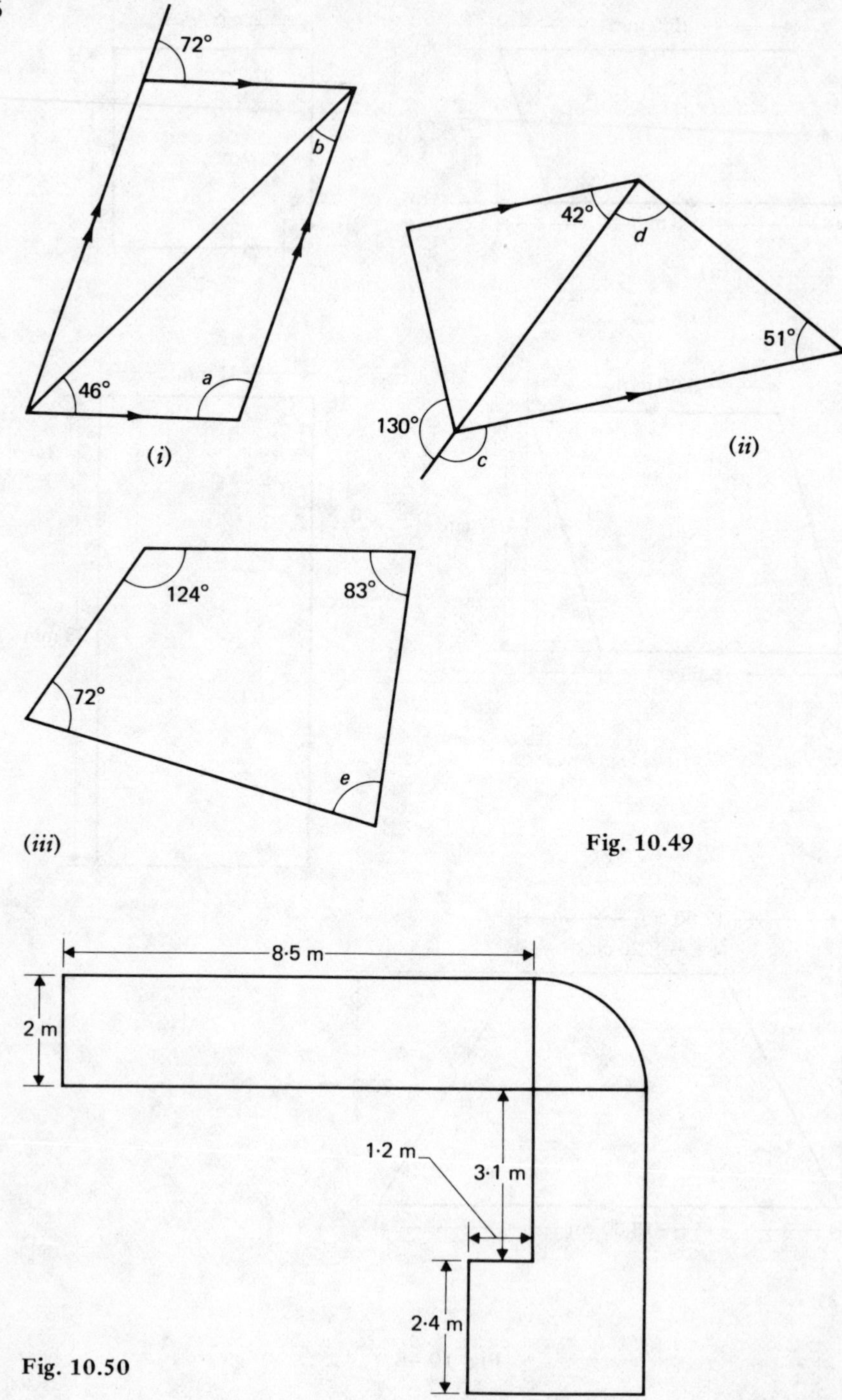

Fig. 10.49

Fig. 10.50

34. If a metal cone has a diameter of 92 mm and a perpendicular height of 124 mm find its volume in cm^3 (correct to 4 significant figures). [274.8 cm^3]
35. A cylinder is cast from a piece of alloy 4.60 cm by 7.30 cm by 12.60 cm. If the length of the cylinder is to be 80.0 cm, find its circumference. [8.153 cm]
36. If a cylindrical tin is to have a volume of 520 cm^3 find its height if the height and diameter are equal. [8.715 cm]
37. A square pyramid has a perpendicular height of 24.0 mm. If a side of the base is 3.20 cm, find the volume of the pyramid (to 4 significant figures). [8.192 cm^3]
38. A hemisphere has a diameter of 3.60 cm. Find its volume. [12.22 cm^3]
39. A trench is 12.0 m long. Its width at the top is 4.0 m and at the bottom 2.70 m and has a depth of 2.0 m. Find the volume of the trench, assuming its sides slant uniformly. [80.4 m^3]
40. Find the weight in kilograms of a hemispherical bowl of copper whose external and internal radii are 15.0 cm and 11.0 cm. Assume that 1 cm^3 of copper weighs 8.9 g. [38.11 kg]
41. A rivet consists of a cylindrical head, of diameter 1 cm and depth 5 mm, and a shaft of diameter 1.5 mm and length 1.5 cm. Find the volume of metal in 1 000 such rivets. [419.3 cm^3]
42. If the volume of a sphere is 476.0 cm^3 find its radius. [4.843 cm]
43. Calculate the volume of metal required to make a pipe of length 11.50 m with external diameter of 24.0 cm and thickness 4.0 mm. [34 110 cm^3]
44. A pyramid stands on a square base of side 15.0 cm. If the length of each sloping edge is 21.0 cm calculate the volume of the pyramid. [1 359 cm^3]
45. The solid shown in Fig. 10.51 comprises a hemisphere and a cone. The radius of the base of the cone is the same as the radius of the hemisphere, and the perpendicular height of the cone is twice its radius. If the volume of the solid is 382.0 cm^3 find the diameter and length l of the solid. [Diameter = 9.002 cm l = 13.50 cm]
46. A metal washer has an outside diameter of 2.0 cm and an inside diameter of 1.20 cm and is 3.0 mm thick. Find the weight in kilograms of 2 000 washers if 1 cm^3 weighs 9 grams. [10.86 kg]

Areas and volumes of similar shapes

47. A rectangular garage is shown on a building plan having dimensions 15 mm by 24 mm. If the plan is drawn to a scale of 1 to 200, determine the true area of the garage, in square metres. [14.4 m^2]
48. The area of a plot of land on a map is 400 mm^2. If the scale of the map is 1 to 50 000, determine the true area of the land in hectares (1 hectare = 10^4 m^2). [100 ha]

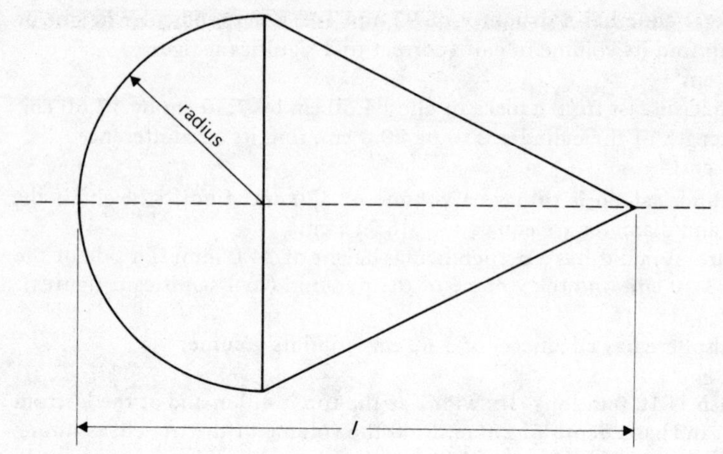

Fig. 10.51

49. A car has a mass of 800 kg. A model of the car is made to a scale of 1 to 20. Determine the mass of the model if the car and its model are made of the same material. [100 g]

50. The diameter of two spherical bearings are in the ratio of 3 : 5. What is the ratio of their volumes? [27 : 125]

51. A component has a mass of 600 g. If each of its dimensions are reduced by 25%, determine its new mass. [253.1 g]

52. A container is in the form of an inverted cone of perpendicular height 500 mm and maximum diameter 200 mm. The cone is filled with a liquid to a height of 100 mm. Determine (*a*) the surface area of the liquid, and (*b*) the volume of the liquid, each correct to 4 significant figures.
(*a*) [1 257 mm^2] (*b*) [41 890 mm^3]

Chapter 11

Trigonometry

11.1 Definition

Trigonometry is the branch of mathematics which deals with the measurement of sides and angles of triangles, and their relationships with each other.

11.2 Trigonometric ratios of acute angles

In Chapter 9 it was shown that the third side of a right-angled triangle could be calculated using the theorem of Pythagoras, provided that the other two sides were known. However the theorem does not provide a method for finding the angles of a triangle.

Given certain information, a triangle can be constructed by scale drawing and the unknown sides or angles measured. However, in the case of a right-angled triangle, a knowledge of the trigonometric ratios enables both sides and angles to be calcualted. Consider a right-angled triangle ABC shown in Fig. 11.1 and let the angle CAB be denoted by θ.
The three sides of the triangle are given special names.

The longest side, that is the side opposite the 90° angle, is called the **hypotenuse.** The side opposite the angle θ is called the **opposite side.** The third side is called the **adjacent side.** Hence, referring to angle θ in Fig. 11.1:

AC = hypotenuse
BC = opposite side
AB = adjacent side

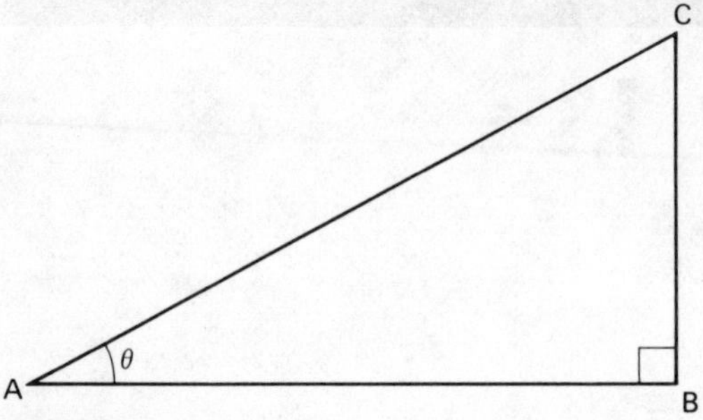

Fig. 11.1

Note that the hypotenuse is always the side opposite the right angle. However, the opposite and adjacent sides in a triangle will depend on which angle is being referred to. If the angle ACB in Fig. 11.1 is referred to, the opposite side would then be AB and the adjacent side, BC.

The following three definitions of trigonometric ratios are true for any right-angled triangle:

1. Natural sine

The ratio $\frac{\text{length of opposite side}}{\text{length of hypotenuse}}$ is called the **natural sine** of the angle θ. (The abbreviation 'sin' is usually used for sine.)
Thus,

$$\sin\theta = \frac{\textbf{opposite side}}{\textbf{hypotenuse}} \text{ or } \frac{\textbf{opposite}}{\textbf{hypotenuse}}$$

In Fig. 11.1, $\sin\theta = \frac{BC}{AC}$

2. Natural cosine

The ratio $\frac{\text{length of adjacent side}}{\text{length of hypotenuse}}$ is called the **natural cosine** of the angle θ. (The abbreviation 'cos' is usually used for cosine.)
Thus,

$$\cos\theta = \frac{\textbf{adjacent side}}{\textbf{hypotenuse}} \text{ or } \frac{\textbf{adjacent}}{\textbf{hypotenuse}}$$

In Fig. 11.1, $\cos\theta = \frac{AB}{AC}$

3. Natural tangent

The ratio $\frac{\text{length of opposite side}}{\text{length of adjacent side}}$ is called the **natural tangent** of the angle θ. (The abbreviation 'tan' is usually used for tangent.)

Thus,

$$\tan\theta = \frac{\textbf{opposite side}}{\textbf{adjacent side}} \textbf{ or } \frac{\textbf{opposite}}{\textbf{adjacent}}$$

In Fig. 11.1, $\tan\theta = \frac{BC}{AB}$

Validity of trigonometric ratios

Sine, cosine and tangent are the three fundamental trigonometric ratios used to determine the size of angles. As they are each a ratio of length divided by length, they have no units.

The definitions of sine, cosine and tangent are true **only** for a right-angled triangle. If a triangle is not right-angled the definitions must not be used.

Worked problems on trigonometric ratios of acute angles

Problem 1. Find $\sin\theta$, $\cos\theta$ and $\tan\theta$ in the triangle PQR shown in Fig. 11.2.

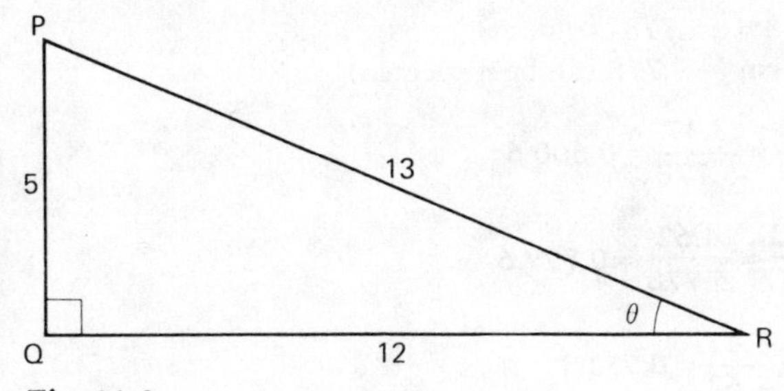

Fig. 11.2

$$\sin\theta = \frac{\text{opposite side}}{\text{hypotenuse}} = \frac{PQ}{PR} = \frac{5}{13} = 0.384\,6$$

$$\cos\theta = \frac{\text{adjacent side}}{\text{hypotenuse}} = \frac{QR}{PR} = \frac{12}{13} = 0.923\,1$$

$$\tan\theta = \frac{\text{opposite side}}{\text{adjacent side}} = \frac{PQ}{QR} = \frac{5}{12} = 0.416\,7$$

Hence, $\mathbf{\sin\theta = 0.384\,6}$, $\mathbf{\cos\theta = 0.923\,1}$, **and** $\mathbf{\tan\theta = 0.416\,7}$

Problem 2. In a triangle ABC, $\angle B = 90°$, $a = 4.62$ cm and $c = 3.47$ cm. Find sin C, cos C and tan C.

The triangle ABC is shown in Fig. 11.3.

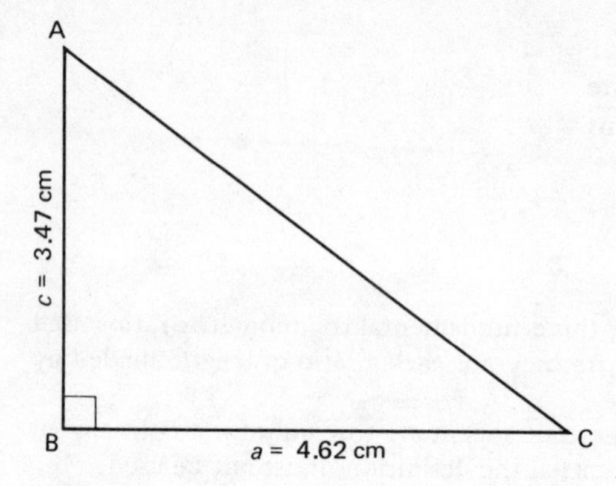

Fig. 11.3

Pythagoras's theorem is used to find the hypotenuse AC:

$$(AC)^2 = (AB)^2 + (BC)^2$$
$$= (3.47)^2 + (4.62)^2$$
$$= 12.04 + 21.34$$
$$= 33.38$$

Therefore $AC = \sqrt{33.38} = \pm 5.778$

AC = 5.778 cm (−5.778 can be neglected)

$$\sin C = \frac{\text{opposite}}{\text{hypotenuse}} = \frac{AB}{AC} = \frac{3.47}{5.778} = 0.600\ 6$$

$$\cos C = \frac{\text{adjacent}}{\text{hypotenuse}} = \frac{BC}{AC} = \frac{4.62}{5.778} = 0.799\ 6$$

$$\tan C = \frac{\text{opposite}}{\text{adjacent}} = \frac{AB}{BC} = \frac{3.47}{4.62} = 0.751\ 1$$

Thus, **sin C = 0.600 6, cos C = 0.799 6, and tan C = 0.751 1**

Problem 3.

(*a*) If $\cos A = \frac{11}{61}$, find sin A and tan A

(*b*) If $\tan B = \frac{8}{15}$, find sin B and cos B

(*a*) Figure 11.4 shows a right-angled triangle ABC with $\cos A = \frac{11}{61}$

Pythagoras's theorem is used to find the opposite side BC:

$$(AB)^2 = (AC)^2 + (BC)^2$$

Therefore $(61)^2 = (11)^2 + (BC)^2$

$$(BC)^2 = (61)^2 - (11)^2$$
$$= 3\ 721 - 121$$
$$= 3\ 600$$

Therefore $BC = \sqrt{3\ 600}$

$$= \pm 60$$

BC = 60 (−60 can be neglected)

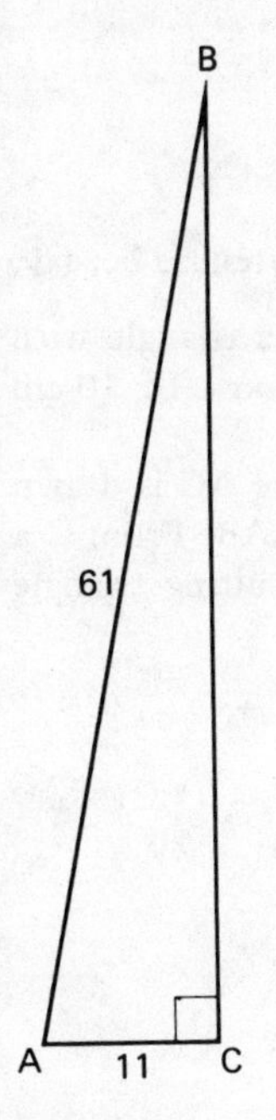

Fig. 11.4

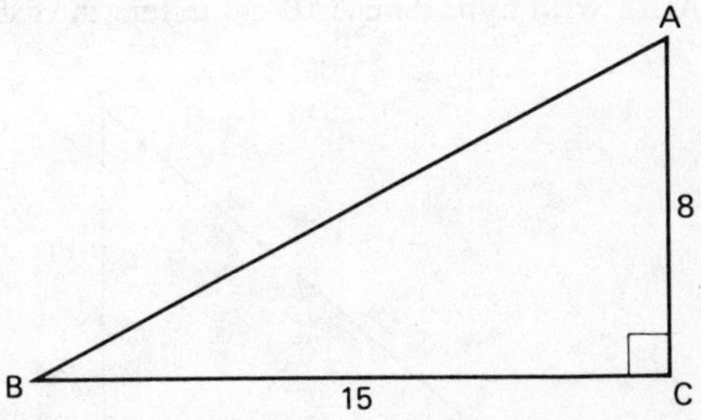

Fig. 11.5

Therefore $\sin A = \dfrac{BC}{AB} = \dfrac{60}{61} = 0.983\,6$

and $\qquad \tan A = \dfrac{BC}{AC} = \dfrac{60}{11} = 5\tfrac{5}{11} = 5.455$

Thus, sin A = 0.983 6, and tan A = 5.455

(*b*) Figure 11.5 shows a right-angled triangle ABC with $\tan B = \frac{8}{15}$

Pythagoras's theorem is used to find the hypotenuse AB:

$$
\begin{aligned}
(AB)^2 &= (AC)^2 + (BC)^2 \\
&= (8)^2 + (15)^2 \\
&= 64 + 225 \\
&= 289
\end{aligned}
$$

Therefore $\qquad AB = \sqrt{289}$
$\qquad\qquad = \pm 17$

Therefore $\qquad$ **AB = 17** (−17 can be neglected)

Therefore $\sin B = \dfrac{AC}{AB} = \dfrac{8}{17} = 0.470\,6$

and $\qquad \cos B = \dfrac{BC}{AB} = \dfrac{15}{17} = 0.882\,4$

Hence, sin B = 0.470 6, and cos B = 0.882 4

 Problem 4. Find, by drawing suitable triangles, the values of
(*a*) sin 40° (*b*) cos 53° (*c*) tan 24°

(*a*) *To find the value of sin 40° by construction*

Since the sine of an angle is the ratio $\frac{\text{opposite side}}{\text{hypotenuse}}$ let the hypotenuse be, say, 10 cm long. This will help to make calculations easier. Thus, a triangle with angles of 90°, 40° (and thus another of 50°) and a hypotenuse of 10 cm length has to be constructed using a set square or protractor.

Let any length AB be drawn horizontally. From A a line AC is drawn 10 cm long, the arm AC being drawn at an angle of 40° to AB. From C a perpendicular to AB is drawn to intersect AB at D. The resulting triangle ACD, with hypotenuse 10 cm in length is shown in Fig. 11.6.

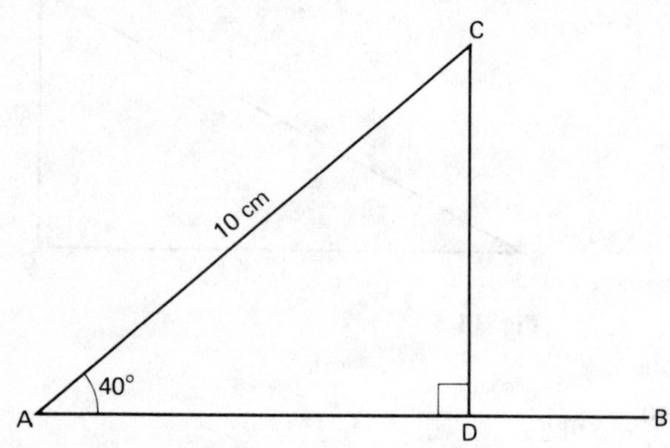

Fig. 11.6

$$\sin 40° = \frac{\text{opposite}}{\text{hypotenuse}} = \frac{CD}{AC}$$

Therefore by measurement, $\mathbf{\sin 40° = \frac{6.43}{10} = 0.643}$

(*b*) *To find the value of cos 53° by construction*

Since the cosine of an angle is the ratio $\frac{\text{adjacent side}}{\text{hypotenuse}}$ let the hypotenuse again be 10 cm long. Thus a triangle with angles of 90°, 53° (and thus another of 37°) and a hypotenuse of 10 cm length has to be constructed. This is done in a similar way to (*a*) above to produce a triangle PRS as shown in Fig. 11.7.

$$\cos 53° = \frac{\text{adjacent}}{\text{hypotenuse}} = \frac{PS}{PR}$$

Therefore by measurement, $\mathbf{\cos 53° = \frac{6.0}{10} = 0.60}$

(*c*) *To find the value of tan 24° by construction*

Since the tangent of an angle is the ratio $\frac{\text{opposite side}}{\text{adjacent side}}$ let the adjacent side

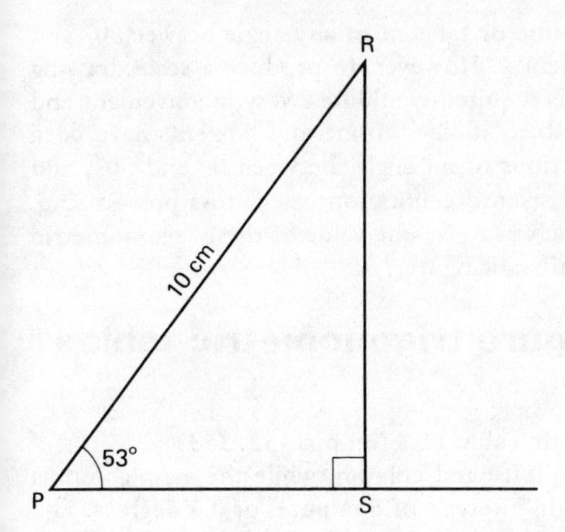

Fig. 11.7

be 10 cm long. Thus a triangle with angles of 90°, 24° (and thus another of 66°) and an adjacent side of 10 cm length has to be constructed.

Let the length XY be drawn horizontally 10 cm long to represent the adjacent side. At X an angle of 24° is drawn to XY. A perpendicular to XY is drawn at Y and the point where the opposite side and hypotenuse intersect is the vertex Z. The resulting triangle XYZ is shown in Fig. 11.8.

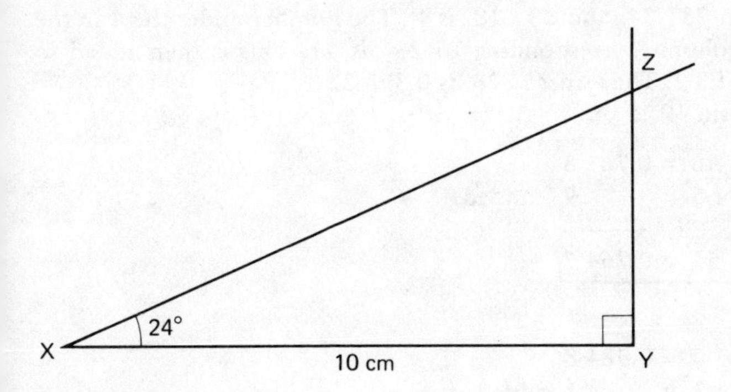

Fig. 11.8

$$\tan 24^\circ = \frac{\text{opposite}}{\text{adjacent}} = \frac{YZ}{XY}$$

Therefore by measurement, $\mathbf{\tan 24^\circ} = \frac{4.45}{10} = \mathbf{0.445}$

Further problems on trigonometric ratios of acute angles may be found in Section 11.9 (Problems 1–6), page 341.

 It is possible to find the sine, cosine or tangent of any angle between 0° and 90° by scale drawing as in Problem 4. However, to produce a scale drawing every time a trigonometric ratio is required would be a very inconvenient and time-consuming exercise. Thus tables of sine, cosine and tangents have been produced for all angles, and fractions of an angle, between 0° and 90°, and are readily available. Also, most scientific notation calculators possess sine, cosine and tangent functions which can give the value of these trigonometric ratios correct to as many as 9 significant figures.

11.3 Use of four-figure trigonometric tables

1. Natural sine tables

Values of natural sines are shown in Table 11.2 (on pp. 352, 353).
Angles in degrees are given in the left-hand column, while the main columns of the page subdivide the degrees in intervals of 6 minutes or 0.1 degrees. The mean difference column on the right gives the corrections for angles within the 6-minute intervals.

It can be seen that the value of a sine increase from 0 at 0° to a maximum of 1 at 90°.

For example, sin 18° is 0.309 0 and sin 76° is 0.970 3.

Similarly, sin 31° 18′ is 0.519 5 and sin 87° 42′ is 0.999 2.

In each of these examples the sine is read directly from the tables. However sin 23° 28′ is not as straightforward, since it lies between sin 23° 24′ and sin 23° 30′. Thus, sin 23° 28′ lies somewhere between 0.397 1 and 0.398 7. To find the exact value the mean difference column must be used. The difference between 23° 24′ and 23° 28′ is 4′. The number under the 4 in the mean difference column corresponding to 23° is 11. This is then added to 0.397 1 to give 0.398 2. Thus sin 23° 28′ is 0.398 2.

Similarly, for sin 49° 53′:

sin 49° 48′ = 0.763 8
mean difference for 5′ = 9 (added)

Therefore sin 49° 53′ = 0.764 7

For sin 80° 2′:

sin 80° 0′ = 0.984 8
mean difference for 2′ = 1 (added)

Therefore sin 80° 2′ = 0.984 9

The following values of sine may be checked:

sin 2° 47′ = 0.048 6
sin 15° 16′ = 0.263 3
sin 39° 58′ = 0.642 3
sin 62° 14′ = 0.884 9
sin 86° 33′ = 0.998 2

To find the angle whose sine is 0.734 5

The nearest number **less** than 0.734 5 in the main column is 0.733 7, which corresponds to $47^\circ\ 12'$. The number 7 337 is 8 less than 7 345. Eight in the mean difference column corresponds to $4'$. Therefore the angle whose sine is 0.734 5 is $47^\circ\ 12' + 4'$, that is, $\mathbf{47^\circ\ 16'}$.

To find the angle whose sine is 0.062 4

Given value of sine = 0.062 4
$\sin 3^\circ\ 30' = 0.061\ 0$ (subtracted)

$$\overline{\quad 14 \quad}$$

From the mean difference column 14 corresponds closest to an increase of $5'$. Hence the angle is $\mathbf{3^\circ\ 35'}$.

To find the angle whose sine is 0.264 8

Given value of sine = 0.264 8
$\sin 15^\circ\ 18' = 0.263\ 9$ (subtracted)

$$\overline{\quad 9 \quad}$$

From the mean difference column 9 corresponds to an increase of $3'$. Hence the angle is $\mathbf{15^\circ\ 21'}$.

[Note that '**arcsin θ**' is a short way of writing '**the angle whose sine is equal to** θ.]

Thus, from above, $\arcsin 0.734\ 5 = 47^\circ\ 16'$
$\arcsin 0.062\ 4 = 3^\circ\ 35'$
and $\arcsin 0.264\ 8 = 15^\circ\ 21'$.

Graph of $\sin\theta$ in the range $\theta = 0^\circ$ to $\theta = 90^\circ$

If θ denotes any angle between 0° and 90°, a graph can be plotted with θ degrees horizontal and $\sin\theta$ vertical.

From sine tables the following values are plotted:

θ degrees	0	10	20	30	40	50	60	70	80	90
$\sin\theta$	0	0.174	0.342	0.500	0.643	0.766	0.866	0.940	0.985	1.000

The curve of $\sin\theta$ is shown in Fig. 11.9.

2. Natural cosine tables

Values of natural cosines are shown in Table 11.3 (on pp. 354, 355).
It can be seen that the value of a cosine decreases from a maximum of 1 at 0°

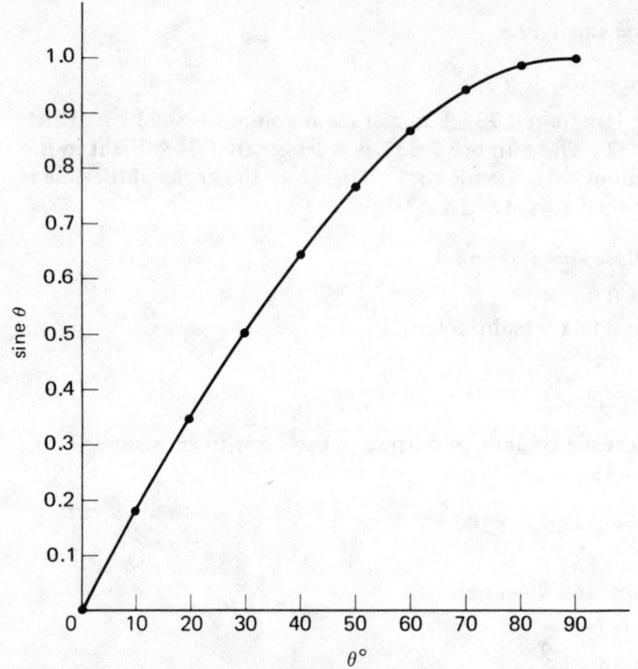

Fig. 11.9 Graph of sine θ/ θ degrees

to 0 at 90°. That is, as the angle increases the cosine decreases. (This is directly opposite to that of a sine.)

For example, cos 14° is 0.970 3 and cos 63° is 0.454 0.

Similarly, cos 25° 24′ is 0.903 3 and cos 82° 54′ is 0.123 6.

In each of these examples the value of the cosine is read directly from the tables. However cos 33° 10′ is not so straightforward, as it lies between 33° 6′ and 33° 12′, that is, cos 33° 10′ lies between 0.837 7 and 0.836 8. The difference between 33° 6′ and 33° 10′ is 4′. The number under the 4 in the mean difference column corresponding to 33° is 6. This is then **subtracted** from 0.837 7 to give 0.837 1, since the cosines are decreasing in value as the angles are increasing. Thus **cos 33° 10′ is 0.837 1.**

Similarly for cos 11° 57′:

cos 11° 54′ = 0.978 5
mean difference for 3′ = 2 (subtracted)

Therefore cos 11° 47′ = 0.978 3

For cos 77° 22′:

cos 77° 18′ = 0.219 8
mean difference for 4′ = 1 1 (subtracted)

Therefore cos 77° 22′ = 0.218 7

The following values of cosine may be checked:

$\cos 4^\circ 39' = 0.996\,7$
$\cos 11^\circ 27' = 0.980\,1$
$\cos 26^\circ 4' = 0.898\,3$
$\cos 58^\circ 59' = 0.515\,3$
$\cos 87^\circ 11' = 0.049\,1$

Reverse use of natural cosine tables

(*a*) *To find the angle whose cosine is 0.631 6*

The nearest number greater than 0.631 6 in the main column is 0.632 0 which corresponds to 50° 48′. The number 6 320 is 4 greater than 6 316. Four in the mean difference column corresponds to 2′. Therefore the angle whose cosine is 0.631 6 is 50° 48′ + 2′, i.e. **50° 50′**.

(*b*) *To find the angle whose cosine is 0.213 8*

Given value of cosine = 0.213 8
cosine 77° 36′ = 0.214 7 (subtracted)

−9

−9 in the mean difference column corresponds to an increase of 3′ in the angle. Hence the angle is **77° 39′**.

(*c*) *To find the angle whose cosine is 0.881 5*

Given value of cosine = 0.881 5
cos 28° 6′ = 0.882 1 (subtracted)

−6

−6 in the mean difference column corresponds to an increase of 4′ in the angle. Hence the angle is **28° 10′**.

[Note that '**arccos θ**' is a short way of writing '**the angle whose cosine is equal to θ**'.]

Thus, from above, arccos 0.631 6 = 50° 50′
arccos 0.213 8 = 77° 39′
and arccos 0.881 5 = 28° 10′

Graph of $\cos\theta$ in the range $\theta = 0^\circ$ to $\theta = 90^\circ$

If θ denotes any angle between 0° and 90°, a graph can be plotted with θ degrees horizontal and $\cos\theta$ vertical.

From cosine tables the following values are plotted:

θ degrees	0	10	20	30	40	50	60	70	80	90
$\cos\theta$	1.000	0.985	0.940	0.866	0.766	0.643	0.500	0.342	0.174	0

The curve of $\cos\theta$ is shown in Fig. 11.10.

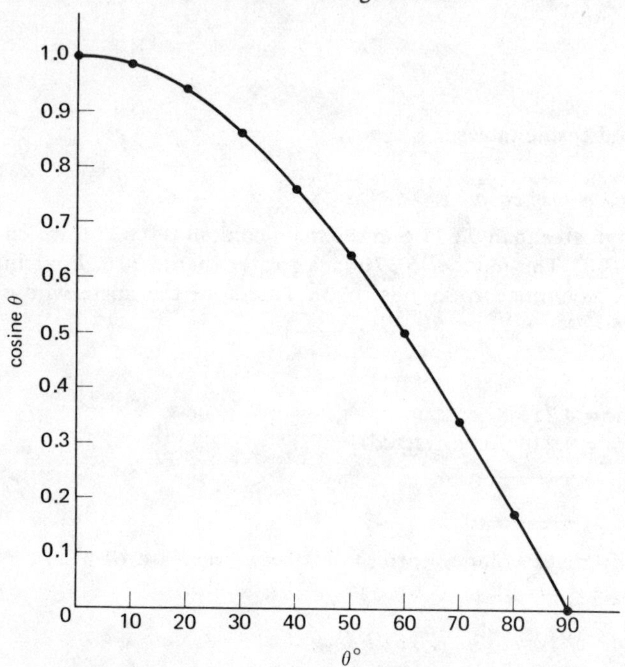

Fig. 11.10 Graph of cosine θ/ θ degrees

3. Natural tangent tables

Values of natural tangents are shown in Table 11.4 (on pp. 356, 357). It can be seen that the value of a tangent increases from 0 at 0° to 1 at 45° and then to infinity at 90°. The sign for infinity is ∞.

For example, tan 24° is 0.445 2, tan 69° is 2.605 1 and tan 86° is 14.30. Similarly, tan 15° 42′ is 0.281 1 and tan 56° 12′ is 1.493 8.

However tan 36° 16′ is not as straightforward as it lies between 36° 12′ and 36° 18′, that is, between 0.731 9 and 0.734 6. The mean difference of 4′ corresponding to 36° is 18. This is **added** to 0.731 9 to give 0.733 7. Thus **tan 36° 16′ is 0.733 7.**

Similarly for tan 47° 33′:

tan 47° 30′ = 1.091 3
mean difference for 3′ = 1 9 (added)

Therefore tan 47° 33′ = 1.093 2

For tan 82° 20′:

tan 82° 18′ = 7.396 2
tan 82° 24′ = 7.494 7

After 76° the mean differences cease to be sufficiently accurate. However if a linear relationship is assumed then a small error will be incurred.

The difference between 7.396 2 and 7.494 7 is 0.098 5. If a linear relationship is assumed then 0.098 5 represents 6′ in equal increments of $\frac{0.098\,5}{6}$. Hence 1′ represents 0.016 4, 2′ represents 2(0.016 4) or 0.032 8, and so on. Then

tan 82° 18′ = 7.396 2
mean difference for 2′ ≏ 0.032 8 (added)

Therefore **tan 82° 20′ ≏ 7.429 0**
[Note, the sign '≏' means 'is approximately equal to'.]

The following values of tangents may be checked:

tan 4° 11′ = 0.073 2
tan 42° 52′ = 0.928 1
tan 46° 26′ = 1.051 3
tan 72° 43′ = 3.213 8
tan 79° 14′ ≏ 5.258 9
tan 88° 27′ ≏ 37.00

Reverse use of natural tangent tables

(*a*) *To find the angle whose tangent is 1.423 8*

The nearest number less than 1.423 8 in the main columns is 1.422 9 which corresponds to 54° 54′. The number 1.422 9 is 9 less than 1.423 8 and 9 in the mean difference column corresponds to an increase of 1′. Thus the angle whose tangent is 1.423 8 is 54° 54′ + 1′, that is, 54° 55′.

(*b*) *To find the angle whose tangent is 0.235 4*

Given value of tangent = 0.235 4
tan 13° 12′ = 0.234 5 (subtracted)

9

From the mean difference column 9 corresponds to an increase of 3′.
Thus the angle whose tangent is 0.235 4 is 13° 15′.

(*c*) *To find the angle whose tangent is 6.637 7*

The given value of 6.637 7 lies between tan 81° 24′ and tan 81° 30′, that is, between 6.612 2 and 6.691 2.

The mean difference column is insufficiently accurate and therefore only an estimate of the angle can be made.

The difference between 6.691 2 and 6.612 2 is 0.079 0. If a linear relationship is assumed then 0.079 0 represents 6′ in equal increments of $\frac{0.0790}{6}$ or 0.013 2.

Hence, if	6.612 2	= tan 81° 24′
then	6.612 2 + 0.013 2	≏ tan 81° 25′
that is	6.625 4	≏ tan 81° 25′
Similarly	6.638 6	≏ tan 81° 26′
	6.651 8	≏ tan 81° 27′

and so on.

Hence, by interpolating, the angle whose tangent is 6.637 7 is found to be closest to **81° 26′**.

[Note that **'arctan θ'** is a short way of writing **'the angle whose tangent is θ'**.]

Thus from above, arctan 1.423 8 = 54° 55′
arctan 0.235 4 = 13° 15′
and arctan 6.637 7 ≏ 81° 26′

Graph of tan θ in the range $\theta = 0°$ to $\theta = 90°$

If θ denotes any angle between 0° and 90°, a graph can be plotted with θ degrees horizontal and tan θ vertical.

From tangent tables the following values are plotted:

θ degrees	0	10	20	30	40	50	60	70	80	90
tan θ	0	0.176	0.364	0.577	0.839	1.192	1.732	2.748	5.671	∞

The curve of tan θ is shown in Fig. 11.11.

Problems on the use of trigonometric tables may be found in Section 11.9 (Problems 7–16), page 343.

11.4 The fractional and surd form of trigonometric ratios for 30°, 45° and 60°

Trigonometric ratios for 30° and 60°

Consider an equilateral triangle ABC with each side equal to 2 units and each angle equal to 60°, as shown in Fig. 11.12.

Let ∠ A be bisected by the straight line AD as shown in Fig. 11.13. The side BC will also be bisected and AD is perpendicular to BC.

Thus, ∠ BAD = ∠ CAD = 30° and BD = DC = 1 unit.

If Pythagoras's theorem is applied to triangle ABD then:

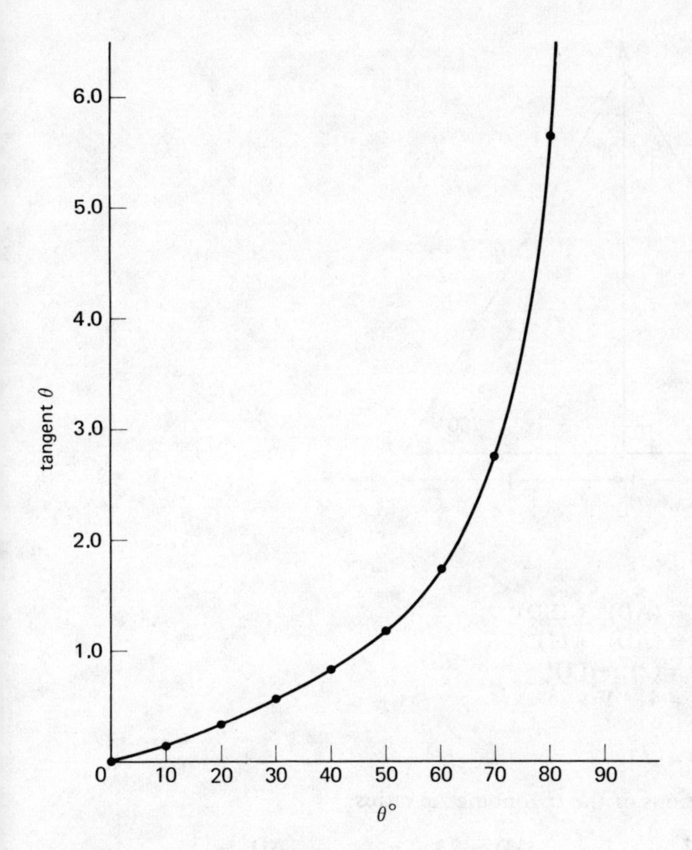

Fig. 11.11 Graph of tangent θ/ θ degrees

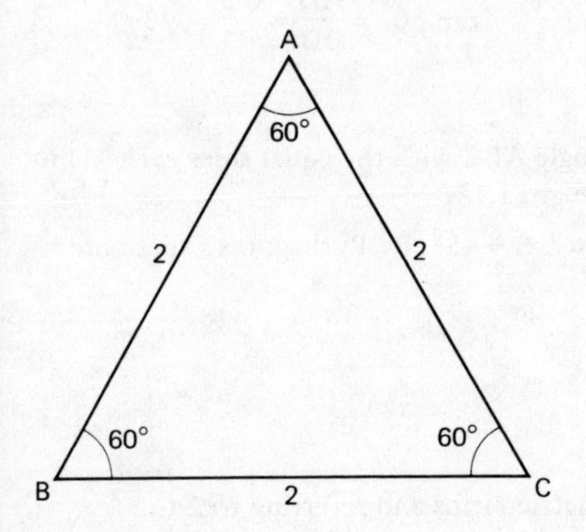

Fig. 11.12

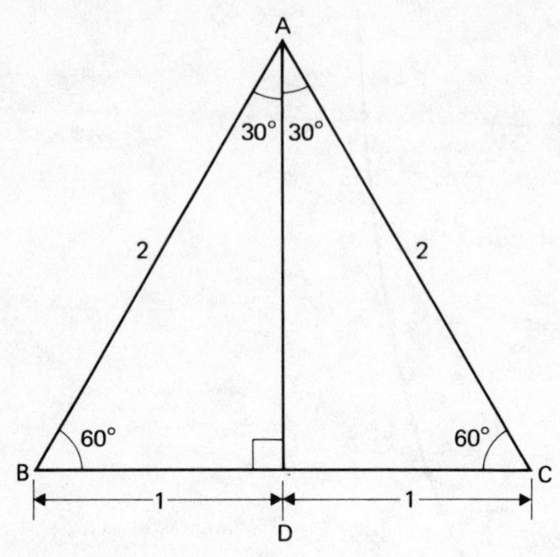

Fig. 11.13

$$(AB)^2 = (AD)^2 + (BD)^2$$

Therefore $(2)^2 = (AD)^2 + (1)^2$

$$AD^2 = (2)^2 - (1)^2$$
$$= 4 - 1$$
$$= 3$$

Therefore $\mathbf{AD} = \sqrt{3}$

Using the definitions of the trigonometric ratios:

$\sin 30° = \frac{BD}{AB} = \frac{1}{2}$ $\sin 60° = \frac{AD}{AB} = \frac{\sqrt{3}}{2}$ $\cos 30° = \frac{AD}{AB} = \frac{\sqrt{3}}{2}$

$\cos 60° = \frac{BD}{AB} = \frac{1}{2}$ $\tan 30° = \frac{BD}{AD} = \frac{1}{\sqrt{3}}$ $\tan 60° = \frac{AD}{BD} = \frac{\sqrt{3}}{1} = \sqrt{3}$

Trigonometric ratios for 45°

Consider a right-angled isosceles triangle ABC with the equal sides each 1 unit in length and ∠ B = 90° as shown in Fig. 11.14.

As the triangle ABC is isosceles, ∠ A = ∠ C = 45°. By Pythagoras's theorem:

$$(AC)^2 = (AB)^2 + (BC)^2$$

Therefore $(AC)^2 = (1)^2 + (1)^2$

$$= 1 + 1$$
$$= 2$$

Therefore $\mathbf{AC} = \sqrt{2}$

Using the definitions of the trigonometric ratios and referring to ∠ C:

$$\sin 45° = \frac{AB}{AC} = \frac{1}{\sqrt{2}}$$

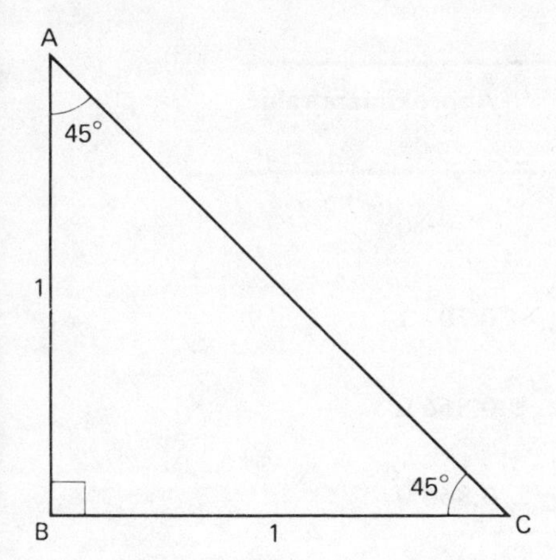

Fig. 11.14

$$\cos 45^\circ = \frac{BC}{AC} = \frac{1}{\sqrt{2}}$$

$$\tan 45^\circ = \frac{AB}{BC} = \frac{1}{1} = 1$$

A quantity which is not **exactly** expressible in figures is called a **surd.** For example, $\sqrt{3}$ or $\sqrt{2}$ are called surds because they cannot be expressed as a fraction and the decimal part may be continued indefinitely. Obviously an approximation can be made, such as letting $\sqrt{3}$ = 1.732 1 (instead of 1.732 050 8 . . .) and letting $\sqrt{2}$ = 1.414 (instead of 1.414 213 6 . . .). Surd form is used often in trigonometry.

A summary of 30°, 45° and 60° trigonometric ratios is shown in Table 11.1

It will be noticed that:

$$\sin 30^\circ = \cos 60^\circ$$
$$\sin 60^\circ = \cos 30^\circ$$
and $$\sin 45^\circ = \cos 45^\circ$$

It can be checked by reference to tables that such results are true for any pair of **complementary angles.** (That is, angles which add up to 90°.) If θ is any angle then:

$$\sin \theta = \cos (90 - \theta)$$
and $$\cos \theta = \sin (90 - \theta)$$

For example $\sin 20^\circ = \cos 70^\circ$
$\sin 36^\circ = \cos 54^\circ$
$\cos 41^\circ = \sin 49^\circ$
$\cos 72^\circ = \sin 18^\circ$ and so on.

Table 11.1

Trigonometric ratio	Exact value in surd form	Approximate value
Sine 30°	$\frac{1}{2}$	
Sine 45°	$\frac{1}{\sqrt{2}}$	0.707 1
Sine 60°	$\frac{\sqrt{3}}{2}$	0.866 0
Cosine 30°	$\frac{\sqrt{3}}{2}$	0.866 0
Cosine 45°	$\frac{1}{\sqrt{2}}$	0.707 1
Cosine 60°	$\frac{1}{2}$	
Tangent 30°	$\frac{1}{\sqrt{3}}$	0.577 4
Tangent 45°	1	
Tangent 60°	$\sqrt{3}$	1.732 1

Problems on trigonometrical ratios in surd form and complementary angles may be found in Section 11.9 (Problems 17–19), page 345.

11.5 Logarithms of trigonometric ratios

If logarithms are used for calculating problems in trigonometry, the tables of logarithms of sine, logarithms of cosine and logarithms of tangent are useful time saving aids. Such tables are readily available and are shown in Tables 11.5, 11.6, and 11.7 (on pp. 356, 363).

The method of reading such tables is exactly the same as for natural sines, cosines, and tangents. For example, from Table 11.5 (p. 358), log sin 43° = $\bar{1}$.833 8.

If log sine tables were not available then:

$$\sin 43^\circ = 0.682\,0 \quad \text{(from Table 11.2)}$$

and $\log_{10} (0.682\,0) = \bar{1}.833\,8$ (from tables of logarithms)

The same result has been achieved but two sets of tables have been referred to.

Similarly, the following may be checked:

log sin 76° 15′ = $\bar{1}$.987 4

From Table 11.6 (pp. 360, 361), log cos 57° = $\bar{1}$.736 1
and log cos 23° 52′ = $\bar{1}$.961 2

From Table 11.7 (pp. 362, 363), log tan 17° = $\bar{1}$.485 3
and log tan 52° 10′ = 0.109 8

Reverse use of such tables is possible in the same way as for natural sines, cosines and tangents.

For example, if log sin x = $\bar{1}$.462 3 then
x = 16° 51′

Similarly, if log cos x = $\bar{1}$.832 5 then
x = 47° 10′

Also, if log tan x = 0.474 3 then
x = 71° 27′.

Problems on logarithms of trigonometric ratios may be found in Section 11.9 (Problems 20–27), page 346.

11.6 Solution of right-angled triangles

To '**solve** a right-angled triangle' means 'to find the unknown sides and angles'.

In order to solve a right-angled triangle it is sufficient to know the length of one side and the value of one angle other than the right angle. Alternatively, if the lengths of two sides of a right-angled triangle are known then the triangle can be solved by using the theorem of Pythagoras in addition to trigonometric ratios.

Worked problems on the solution of right-angled triangles

Problem 1. In the triangle ABC shown in Fig. 11.15, find the lengths AC and AB.

There is usually more than one way in which such a problem can be solved.

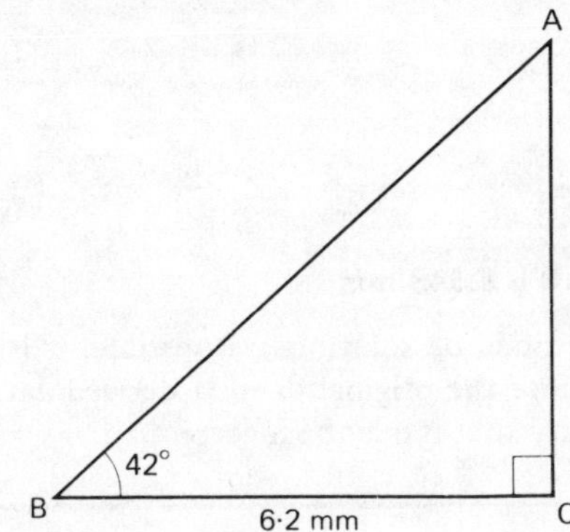

Fig. 11.15

Now $\tan 42^\circ = \dfrac{AC}{BC} = \dfrac{AC}{6.2}$

Therefore $AC = 6.2 \tan 42^\circ$

$= 6.2(0.900\,4)$

Therefore **AC = 5.582 mm**

If logarithms are used for this calculation, then logarithms of tangents can be used.

Number	**Logarithm**	
6.2	0.792 4	
log tan 42°	$\bar{1}.954\,4$	(to be added)
5.582	0.746 8	

Now $\cos 42^\circ = \dfrac{BC}{AB} = \dfrac{6.2}{AB}$

Therefore $AB = \dfrac{6.2}{\cos 42^\circ} = \dfrac{6.2}{0.743\,1}$

Therefore **AB = 8.343 mm**

Again if logarithms are used for this calculation then logarithms of cosine can instead be used.

Number	**Logarithm**	
6.2	0.792 4	
log cos 42°	1. 871 1	(to be subtracted)
8.343	0.921 3	

Alternatively, if BC = 6.2 mm and AC = 5.582 mm, then Pythagoras's theorem can be used to find AB. Thus:

$(AB)^2 = (AC)^2 + (BC)^2$

$= (5.582)^2 + (6.2)^2$

$= 31.16 + 38.44$

$= 69.60$

$AB = \sqrt{69.60}$

AB = 8.343 mm

Thus, **AC is 5.582 mm and AB is 8.343 mm**

[Note: when alternative methods of solution are possible, it is advisable to use those methods which utilise the original data. If derived data is used in a calculation there is a possibility that it may be incorrect.]

Problem 2. Solve the triangle DEF shown in Fig. 11.16.

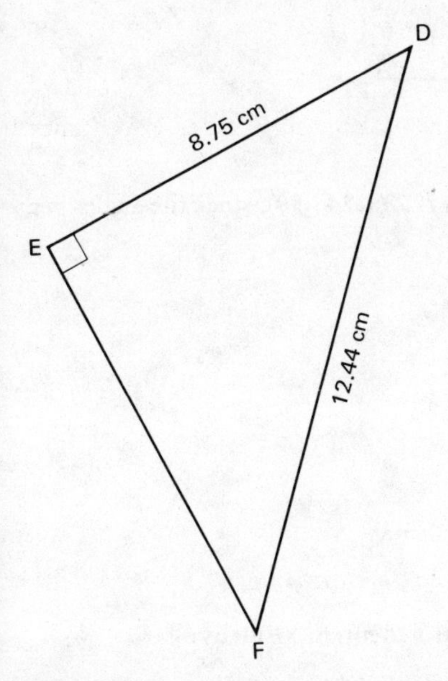

Fig. 11.16

To solve the triangle DEF the angles D and F and the length of the side EF must be found.

$\sin F = \dfrac{DE}{DF} = \dfrac{8.75}{12.44} = 0.703\,4$

Therefore $\angle$ **F** = arcsin 0.703 4 = **44° 42′**

Hence $\angle D = 180° - 90° - 44° \; 42'$

Therefore $\angle$ **D = 45° 18′**

Now $\cos F = \dfrac{EF}{12.44}$

Therefore EF = 12.44 cos 44° 42′

= 12.44 (0.710 8)

Therefore **EF = 8.842 cm**

(EF can be checked using Pythagoras's theorem.)

Hence $\angle$ **D is 45° 18′**, $\angle$ **F = 44° 42′ and EF = 8.842 cm**

Problem 3. Solve triangle XYZ given $\angle$ Y = 90°, $\angle$ X = 15° 21′ and XZ = 17.68 mm. Find also the area of the triangle.

The triangle XYZ is shown in Fig. 11.17.

Solution of the triangle XYZ requires the finding of $\angle$ Z and the lengths of sides XY and YZ.

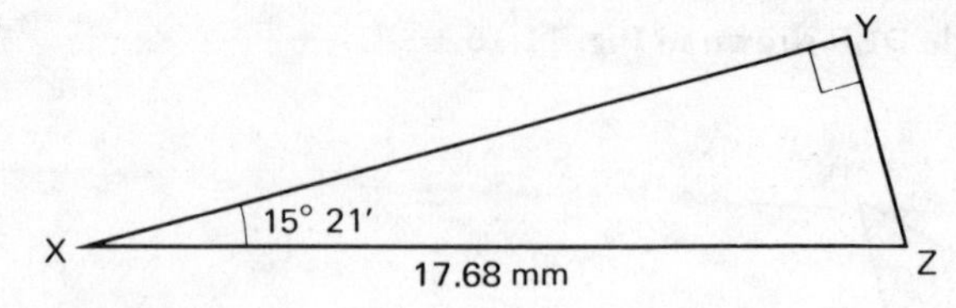

Fig. 11.17

If $\angle$ Y = 90° and $\angle$ X = 15° 21′ then $\angle$ **Z = 74° 39′**, since the angles in a triangle add up to 180°.

$$\sin 15^\circ\ 21' = \frac{YZ}{XZ} = \frac{YZ}{17.68}$$

Therefore YZ = 17.68 sin 15° 21′
= 17.68 (0.264 7)

Therefore **YZ = 4.680 mm**

$$\cos 15^\circ\ 21' = \frac{XY}{XZ} = \frac{XY}{17.68}$$

Therefore XY = 17.68 (cos 15° 21′)
= 17.68 (0.964 4)

Therefore **XY = 17.05 mm**

(YZ or XY can be checked using Pythagoras's theorem.) Finally:

Area of triangle XYZ = $\frac{1}{2}$ (base) (perpendicular height)
= $\frac{1}{2}$ (YZ) (XY)
= $\frac{1}{2}$ (4.68) (17.05)
= **39.90 mm²**

Hence $\angle$ Z is 74° 39′, YZ is 4.68 mm, XY is 17.05 mm and the area of triangle XYZ is 39.90 mm².

Further problems on the solution of right-angled triangles may be found in Section 11.9 (Problems 28–38), page 346.

11.7 Angles of elevation and depression

Angle of elevation

Consider a point A on horizontal ground 100 m away from a vertical flag-pole BC which is 30 m high, as shown in Fig. 11.18.

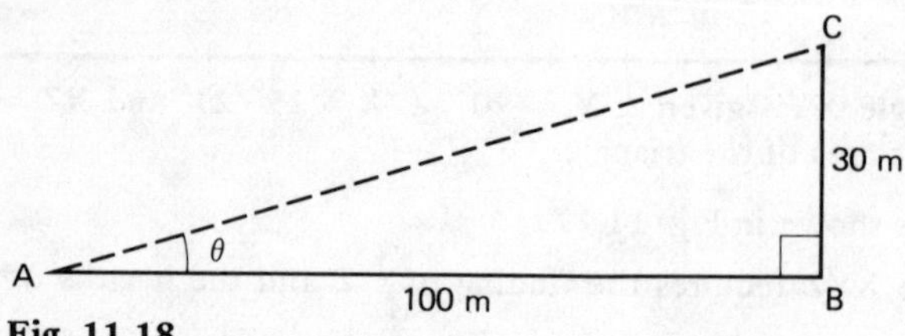

Fig. 11.18

The **angle of elevation** of the top of the flag-pole C is the angle that the imaginary straight line AC must be raised (or elevated) from the horizontal AB.

Hence the angle of elevation in Fig. 11.18 is θ.

$$\text{Now } \tan\theta = \frac{\text{opposite}}{\text{adjacent}} = \frac{BC}{AB} = \frac{30}{100} = 0.300\,0$$

Hence θ is the angle whose tangent is 0.300 0

that is, $\theta = \arctan 0.300\,0$
$= 16^\circ\,42'$.

Hence the angle of elevation, θ, is $16^\circ\,42'$.

Angle of depression

Consider a point D on the edge of a 50 m high vertical cliff DE looking out to sea. A ship, F, is situated 120 m from the base of the cliff. This is shown in Fig. 11.19.

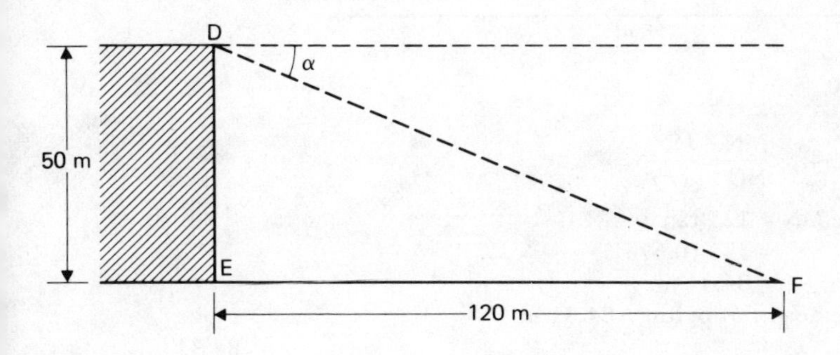

Fig. 11.19

The **angle of depression** of the ship from the point D is the angle through which the imaginary straight line DF must be lowered (or depressed) from the horizontal to the ship.

Hence the angle of depression in Fig. 11.19 is α.

Angle DFE is also equal to α (alternate angles between parallel lines).

$$\text{Now } \tan\alpha = \frac{\text{opposite}}{\text{adjacent}} = \frac{DE}{EF} = \frac{50}{120} = 0.416\,7$$

Therefore $\alpha = \arctan 0.416\,7 = 22^\circ\,37'$

Hence the angle of depression, α, is $22^\circ\,37'$.

Worked problems on angles of elevation and depression

Problem 1. An electricity pylon stands on level ground. At a point 125 m from the foot of the pylon the angle of elevation of the top of the pylon is 34°. Find the height of the pylon and the angle of elevation of its mid point.

Figure 11.20 shows the pylon MN and the angle of elevation of M from the point O as 34°.

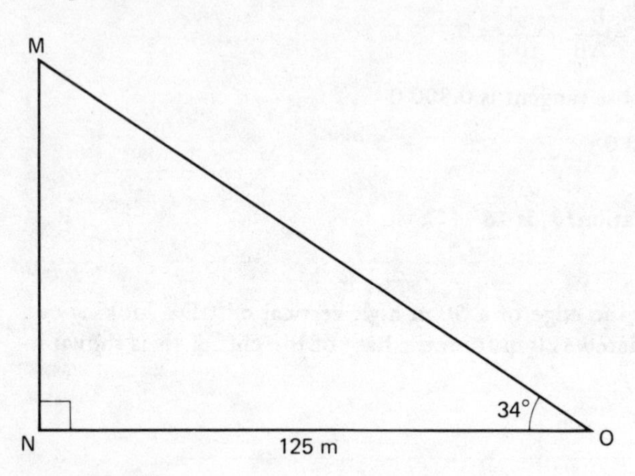

Fig. 11.20

$$\text{Now } \tan 34° = \frac{MN}{NO} = \frac{MN}{125}$$

Therefore MN = 125 tan 34°
= 125 (0.674 5)
= 84.31 m

Therefore **height of pylon = 84.31 m.**

The mid point, X, shown in Fig. 11.21 is at a height of $\frac{84.31}{2}$ m, that is, 42.155 m, or 42.16 m to 4 significant figures.

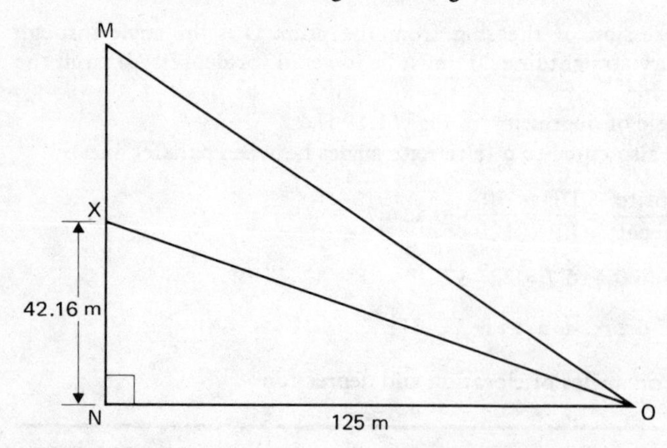

Fig. 11.21

Now tan XON $= \dfrac{42.16}{125} = 0.337\,3$

Therefore $\angle$ XON = arctan 0.337 3 = $18^\circ\ 38'$

Hence the angle of elevation of the mid point of the pylon is $18^\circ\ 38'$.

[Note the angle of elevation of the mid point is **not** half of the angle of elevation of the top of the pylon, that is, it is **not** 17°. This is because the graph of θ against tan θ is not linear.]

Problem 2. A theodolite is an instrument by which a surveyor measures angles. When a surveyor sets up a theodolite on horizontal ground some distance from a vertical tower the angle of elevation is 22°. The surveyor moves 120 m nearer to the tower and finds the angle of elevation is now 41°. Find the height of the tower.

The tower BC and the angles of elevation are shown in Fig. 11.22.

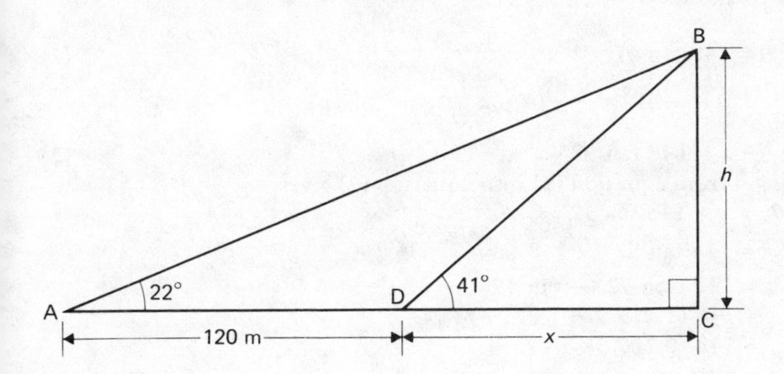

Fig. 11.22

Let DC = x m and BC = h m.

In triangle ABC, $\tan 22^\circ = \dfrac{h}{120 + x}$

Therefore $h = (\tan 22^\circ)(120 + x)$

$$h = (0.404\,0)(120 + x) \tag{1}$$

In triangle BDC, $\tan 41^\circ = \dfrac{h}{x}$

Therefore $h = (\tan 41^\circ)x$

$$h = 0.869\,3x \tag{2}$$

From equations (1) and (2)

$$(0.404\,0)(120 + x) = 0.869\,3x$$

Therefore

$$0.869\,3x = (0.404\,0)\,120 + (0.404\,0)x$$
$$0.869\,3x = 48.48 + 0.404\,0x$$
$$0.869\,3x - 0.404\,0x = 48.48$$
$$0.465\,3x = 48.48$$

$$\text{Therefore } x = \frac{48.48}{0.465\,3} = \mathbf{104.2\ m}$$

From equation (2) $h = (0.869\,3)\,(104.2)$

$= \mathbf{90.58\ m}$

Hence the height of the tower is **90.58 m.**

Problem 3. A vertical flagstaff stands on the edge of the top of a vertical building. A surveyor 148 m from the building measures the angles of elevation of the top and bottom of the flagstaff as 52° and 47°. Calculate the height of the flagstaff.

A flagstaff BC is shown in Fig. 11.23 on top of a building AB, with the surveyor standing at D.

Let the height of the building, AB = h m and the height of the flagstaff BC = x m.

In triangle ABD, $\tan 47^\circ = \dfrac{h}{148}$

Therefore $h = 148 \tan 47^\circ$ (1)

In triangle ACD, $\tan 52^\circ = \dfrac{h + x}{148}$

Therefore $h + x = 148 \tan 52^\circ$ (2)

Substituting h from equation (1) into equation (2) gives

$148 \tan 47^\circ + x = 148 \tan 52^\circ$

Therefore $x = 148 \tan 52^\circ - 148 \tan 47^\circ$

$= 148\,(\tan 52^\circ - \tan 47^\circ)$

$= 148\,(1.279\,9 - 1.072\,4)$

$= 148\,(0.207\,5)$

$x = \mathbf{30.71\ m}$

Hence the height of the flagstaff is **30.71 m.**

Problem 4. The angle of depression of a car viewed at a particular instant from the top of a 42 m high vertical building is 26° 32′. What is the angle of elevation of the top of the building viewed from the car? Find the distance of the car from the building at this instant. If the car is travelling away from the building at 10 km h^{-1}, how long will it take for the angle of depression from the top of the building to the car to become 10°?

Figure 11.24 shows the building AB with an angle of depression of 26° 32′ to the car C.

The angle of elevation of the top of the building B from the car C is 26° 32′, (alternate angles between parallel lines). That is, **∠ ACB is 26° 32′.**

To find AC, triangle ABC is used:

$$\tan 26^\circ\, 32' = \frac{AB}{AC} = \frac{42}{AC}$$

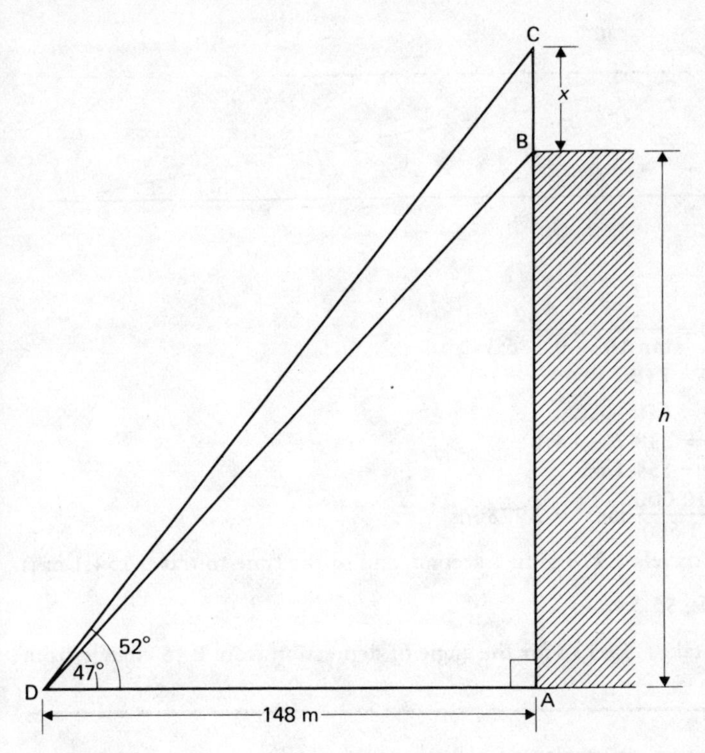

Fig. 11.23

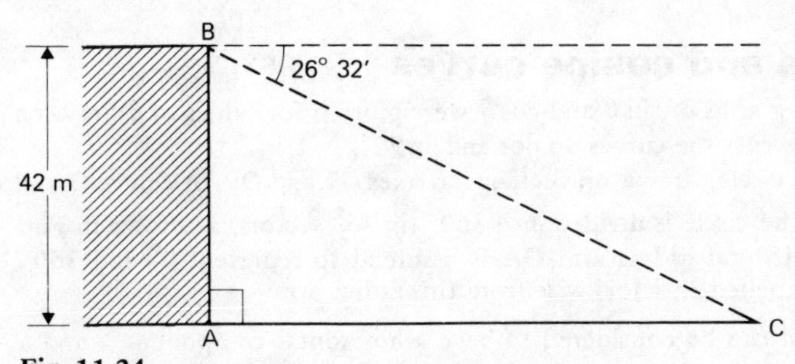

Fig. 11.24

Therefore $AC = \dfrac{42}{\tan 26° \; 32'} = \dfrac{42}{0.499\,3}$

$AC = 84.1$ m

Hence the distance of the car from the base of the building is 84.1 m.

Figure 11.25 shows the instant when the angle of depression from B to point D is 10°.

In triangle ABD, $\tan 10° = \dfrac{AB}{AD} = \dfrac{42}{AD}$

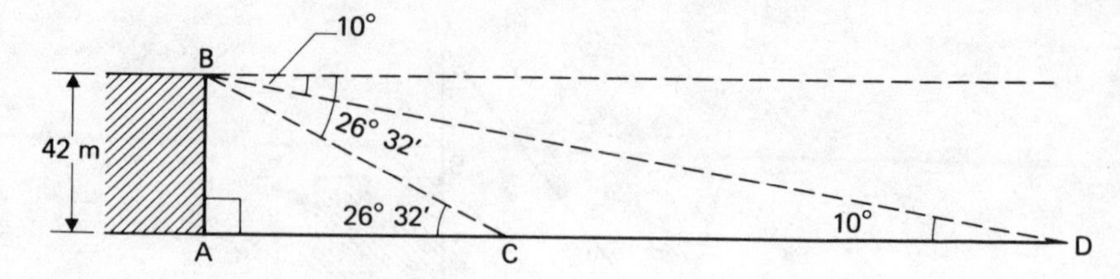

Fig. 11.25

Therefore $AD = \dfrac{42}{\tan 10^\circ} = \dfrac{42}{0.176\ 3}$

$AD = 238.2$ m

Now $CD = AD - AC$

$= 238.2 - 84.1$

Therefore **CD = 154.1 m.**

$10 \text{ km h}^{-1} = \dfrac{10\ 000}{3\ 600} \text{ ms}^{-1} = 2.778 \text{ ms}^{-1}$.

Thus the car travels 2.778 m in 1 second, and so the time to travel 154.1 m is $\dfrac{154.1}{2.778}$ s, that is, 55.5 s.

Thus the car takes 55.5 s **for the angle of depression from B to change from $26^\circ\ 32'$ to 10°.**

Further problems on angles of elevation and depression may be found in Section 11.9 (Problems 39–52), page 350.

11.8 Sine and cosine curves

In Section 11.3 graphs of $\sin\theta$ and $\cos\theta$ were plotted for values of θ between 0° and 90°. However the curves do not end at 90°.

Consider a circle, drawn on rectangular axes Ox and Oy, of centre O and radius 1 unit. The circle is divided into 360° (in 15° sectors) as shown in Fig. 11.26. The horizontal radius arm OA is assumed to represent 0° and 360° and angles are labelled anticlockwise from this radius arm.

Each radius arm can be considered to have a horizontal component x and a vertical component y. For example, for the 30° radius arm in Fig. 11.26 the horizontal component is shown by the length OC and the vertical component is shown by BC.

At 0°: the horizontal component, $x = OA = 1$ unit
the vertical component, $y = 0$

At 15°: the right-angled triangle shown in Fig. 11.27 shows the horizontal component x and the vertical component y.

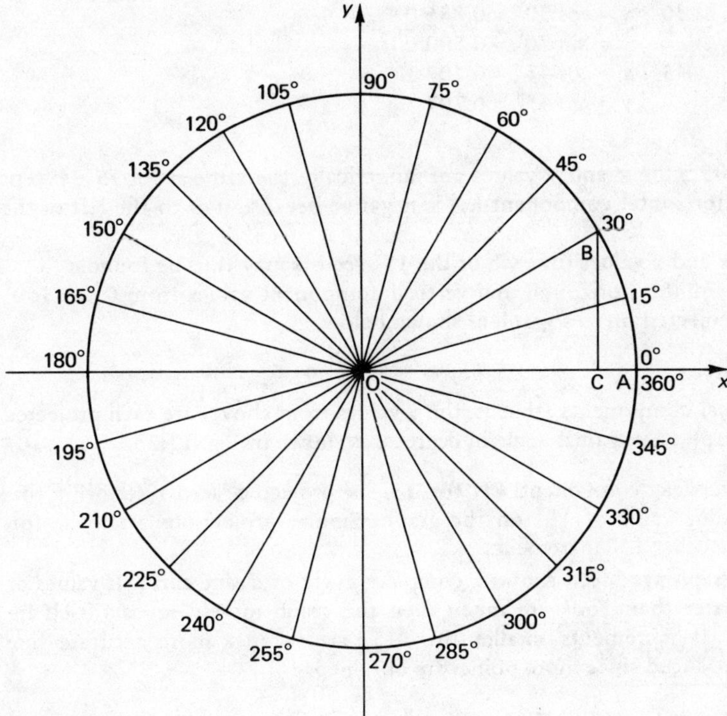

Fig. 11.26

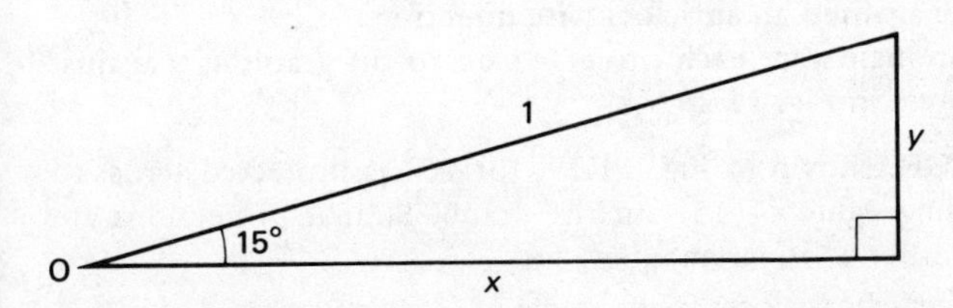

Fig. 11.27

The horizontal component x is calculated using the cosine:

$$\cos 15^\circ = \frac{x}{1}$$

Therefore $x = \cos 15^\circ = 0.965\,9$

The vertical component y is calculated using the sine:

$$\sin 15^\circ = \frac{y}{1}$$

Therefore $y = \sin 15^\circ = 0.258\,8$

Similarly, at 30°: $x = \cos 30° = 0.866\ 0$
$y = \sin 30° = 0.500\ 0$
and at 45°: $x = \cos 45° = 0.707\ 1$
$y = \sin 45° = 0.707\ 1$

and so on.

At 105°, the x and y values are numerically the same as for 75°, except that the horizontal component (x) is negative because it is to the left of the origin.

The x and y values for each of the 15° sectors may thus be found.

Each of the horizontal and vertical component values from 0° to 360° may be projected on to a graph as shown below.

1. Sine curve

The vertical components (that is, the y values from above) are each projected on to a graph of y against angle in degrees, as shown in Fig. 11.28.

The vertical component EF for 15° is projected across to E′F′, the corresponding value of 15° on the graph. Similar projections are made for each of the other 15° increments.

The graph produced shows a complete cycle of a **sine curve.** If values of angles greater than 360° are taken then the graph merely repeats itself indefinitely. If increments smaller than 15° are taken a more accurate sine curve is produced since more points are obtained.

2. Cosine curve

The circle is now redrawn with the radius arm OA in a vertical position and the 360° labelled once again in an anticlockwise direction.

The vertical components are each projected on to the graph of y against angle in degrees as shown in Fig. 11.29.

The component GH, shown in Fig. 11.29, for 15° is projected across to G′H′, the corresponding value of 15° on the graph. Similar projections are made for each of the other 15° increments.

The graph produced shows a complete cycle of a **cosine curve.**

If a sine curve, $y = \sin\theta$, and a cosine curve, $y = \cos\theta$, were sketched on the same axis over one complete cycle the result would be as in Fig. 11.30.

It may be seen that the cosine curve is of the same form as the sine curve, but displaced by 90°.

The sine curve representation has many important applications, including depicting the values obtained in simple harmonic motion, wave motion and in electrical alternating currents and voltages.

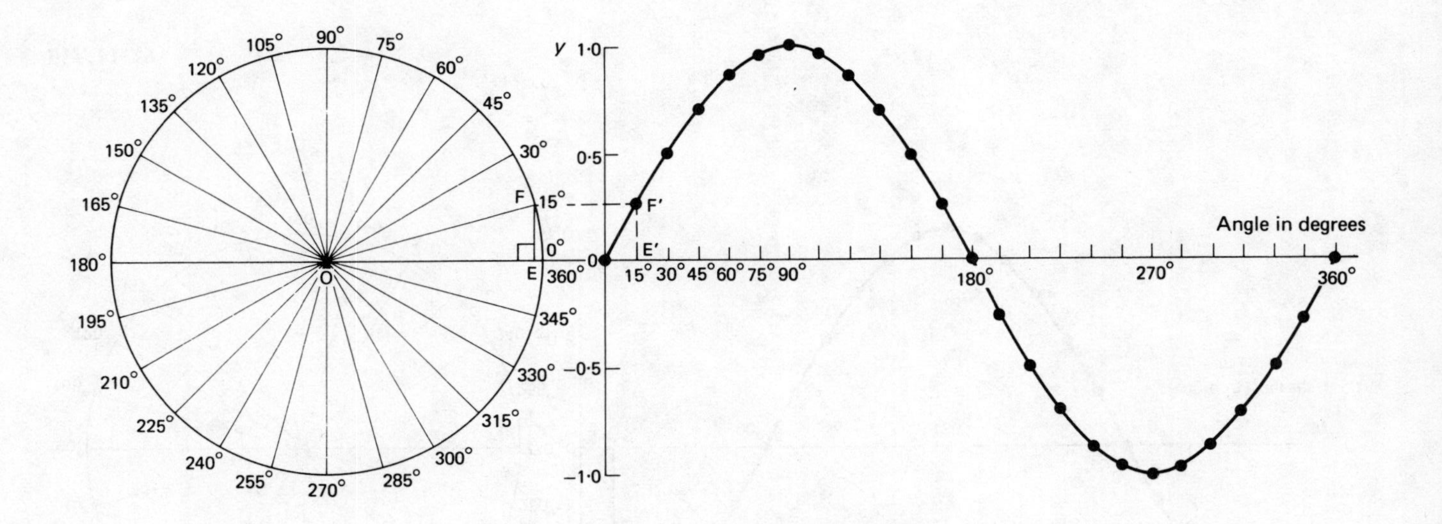

90°
75°
60°
45°
30°
15°
0°
360°
105°
120°
135°
150°
165°
180°
195°
210°
225°
240°
255°
270°
285°
300°
315°
330°
345°
O
E
F
y
1·0
0·5
0
−0·5
−1·0
F′
E′
15° 30° 45° 60° 75° 90°
180°
270°
360°
Angle in degrees

Fig. 11.28

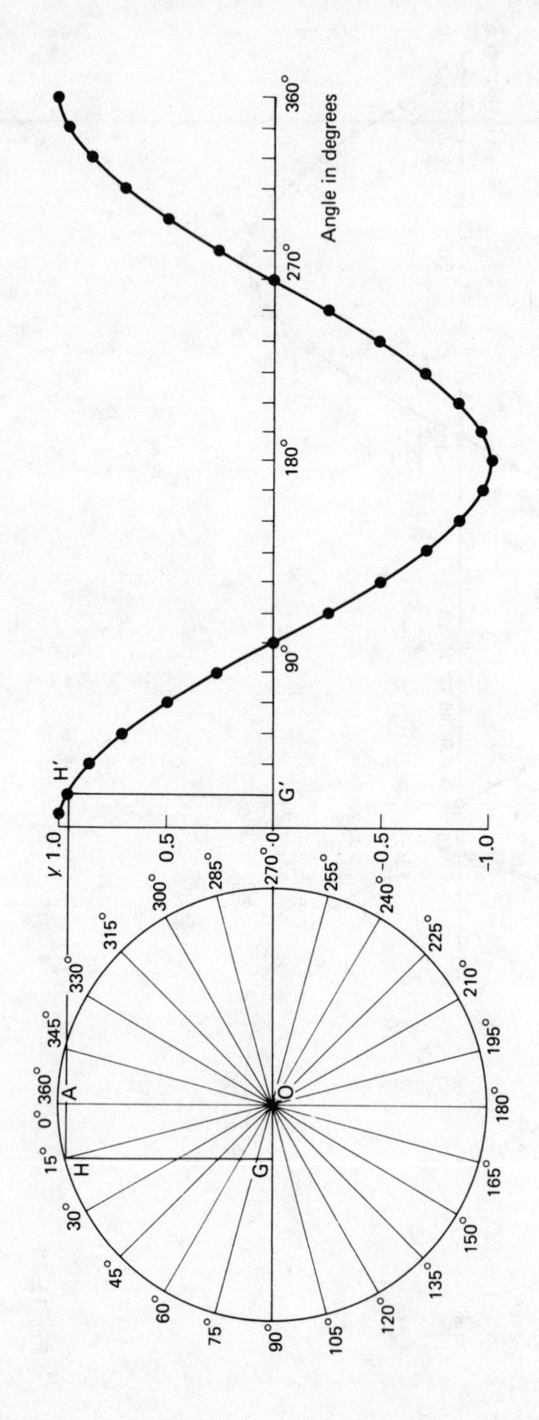

0° 360°
15°
30°
45°
60°
75°
90°
105°
120°
135°
150°
165°
180°
195°
210°
225°
240°
255°
270°
285°
300°
315°
330°
345°
A
H
G
O
y 1.0
0.5
0
-0.5
-1.0
H′
G′
90°
180°
270°
360°
Angle in degrees

Fig. 11.29

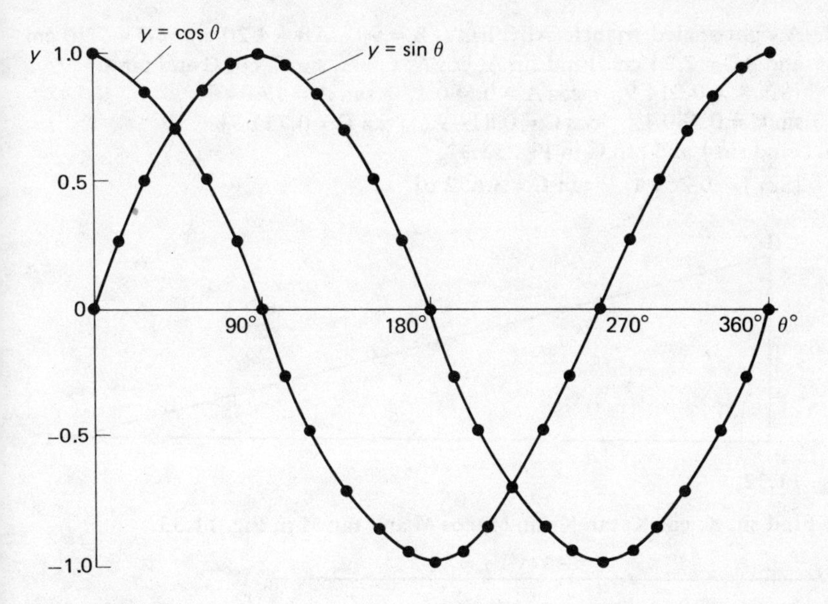

Fig. 11.30

11.9 Further problems

Trigonometric ratios

1. In Fig. 11.31 find sin E, cos E, tan E; sin F, cos F and tan F.

   [Sin E = $\frac{4}{5}$ or 0.8, cos E = $\frac{3}{5}$ or 0.6, tan E = $\frac{4}{3}$ or 1.333 3
   sin F = $\frac{3}{5}$ or 0.6, cos F = $\frac{4}{5}$ or 0.8, tan F = $\frac{3}{4}$ or 0.75]

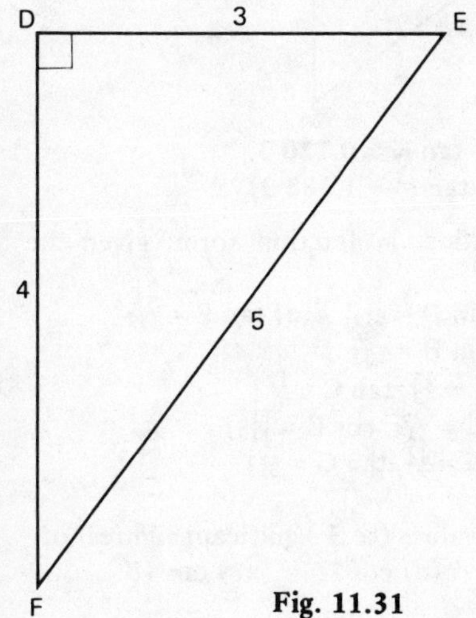

Fig. 11.31

2. A right-angled triangle ABC has $\angle B = 90^\circ$, AB = 4.20 cm, BC = 5.90 cm and AC = 7.24 cm. Find sin A, cos A, tan A; sin C, cos C and tan C.
[Sin A = 0.814 9, cos A = 0.580 1, tan A = 1.404 8;
sin C = 0.580 1, cos C = 0.814 9, tan C = 0.711 9]
3. Find sin J and tan G in Fig. 11.32.
[Sin J = 0.264 1, tan G = 3.652 6]

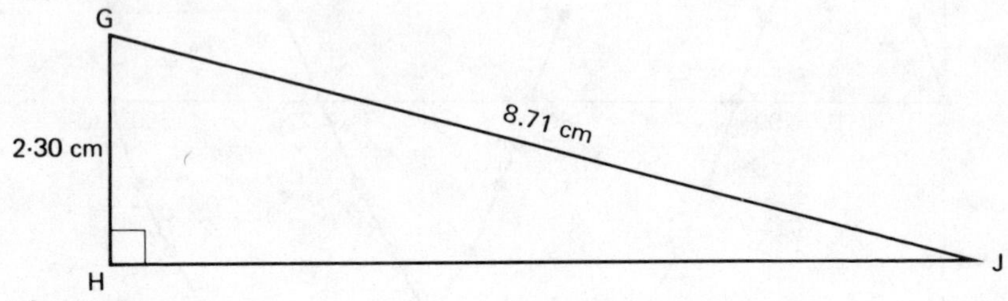

Fig. 11.32

4. Find sin K, cos K, tan K; sin M, cos M and tan M in Fig. 11.33.

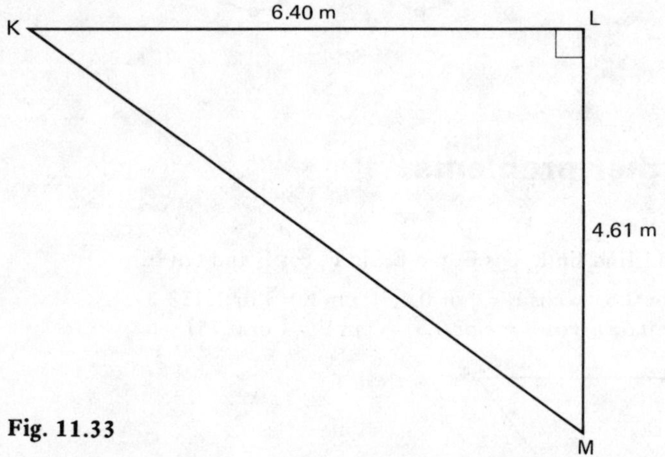

Fig. 11.33

[Sin K = 0.584 5, cos K = 0.811 5, tan K = 0.720 3,
sin M = 0.811 5, cos M = 0.584 5, tan M = 1.388 3]

5. Find the other two trigonometric ratios, in fraction form, given the following:
(*a*) tan A = $\frac{7}{24}$, (*b*) cos C = $\frac{5}{13}$, (*c*) sin D = $\frac{9}{41}$, (*d*) tan E = $\frac{15}{112}$,
(*e*) sin F = $\frac{13}{85}$, (*f*) cos G = $\frac{35}{37}$, (*g*) sin H = $\frac{19}{181}$
(*a*) [sin A = $\frac{7}{25}$, cos A = $\frac{24}{25}$] (*b*) [sin C = $\frac{12}{13}$, tan C = $\frac{12}{5}$]
(*c*) [cos D = $\frac{40}{41}$, tan D = $\frac{9}{40}$] (*d*) [sin E = $\frac{15}{113}$, cos E = $\frac{112}{113}$]
(*e*) [cos F = $\frac{84}{85}$, tan F = $\frac{13}{84}$] (*f*) [sin G = $\frac{12}{37}$, tan G = $\frac{12}{35}$]
(*g*) [cos H = $\frac{180}{181}$, tan H = $\frac{19}{180}$]
6. Find, by drawing suitable triangles, the values (to 3 significant figures) of:
(*a*) sin 33° (*b*) sin 84° (*c*) cos 25° (*d*) cos 76° (*e*) tan 18°
(*f*) tan 66°

(*a*) [0.545] (*b*) [0.995] (*c*) [0.906] (*d*) [0.242] (*e*) [0.325]
(*f*) [2.246]

Use of trigonometric tables

7. Use natural sine tables to find the values of the following:
(*a*) sin 5° (*b*) sin 12° 30′ (*c*) sin 20° 52′ (*d*) sin 37° 19′
(*e*) sin 49° 5′ (*f*) sin 64° 58′ (*g*) sin 82° 10′
(*a*) [0.087 2] (*b*) [0.216 4] (*c*) [0.356 2] (*d*) [0.606 2]
(*e*) [0.755 6] (*f*) [0.906 1] (*g*) [0.990 7]
8. Use natural cosine tables to find the values of the following:
(*a*) cos 12° (*b*) cos 16° 4′ (*c*) cos 29° 46′ (*d*) cos 42° 53′
(*e*) cos 51° 51′ (*f*) cos 73° 19′ (*g*) cos 86° 16′
(*a*) [0.978 1] (*b*) [0.961 0] (*c*) [0.868 0] (*d*) [0.732 7]
(*e*) [0.617 7] (*f*) [0.287 1] (*g*) [0.065 1]
9. Use natural tangent tables to find the values of the following:
(*a*) tan 31° (*b*) tan 49° (*c*) tan 83° (*d*) tan 12° 41′
(*e*) tan 24° 23′ (*f*) tan 42° 57′ (*g*) tan 63° 19′ (*h*) tan 86° 14′
(*a*) [0.600 9] (*b*) [1.150 4] (*c*) [8.144 3] (*d*) [0.225 0]
(*e*) [0.453 3] (*f*) [0.930 9] (*g*) [1.989 8] (*h*) [≏15.19]
10. Find the angles which have a sine of:
(*a*) 0.103 1 (*b*) 0.468 2 (*c*) 0.623 4 (*d*) 0.711 2 (*e*) 0.862 9
(*f*) 0.994 7
(*a*) [5° 55′] (*b*) [27° 55′] (*c*) [38° 34′] (*d*) [45° 20′]
(*e*) [59° 39′] (*f*) [84° 6′]
11. Find the angles which have a cosine of:
(*a*) 0.036 3 (*b*) 0.222 2 (*c*) 0.469 7 (*d*) 0.614 5 (*e*) 0.794 2
(*f*) 0.911 2
(*a*) [87° 55′] (*b*) [77° 10′] (*c*) [61° 59′] (*d*) [52° 5′]
(*e*) [37° 25′] (*f*) [24° 20′]
12. Find the angles which have a tangent of:
(*a*) 0.462 5 (*b*) 0.984 4 (*c*) 1.211 7 (*d*) 1.984 1 (*e*) 2.472 6
(*f*) 5.987 0
(*a*) [24° 49′] (*b*) [44° 33′] (*c*) [50° 28′] (*d*) [63° 15′]
(*e*) [67° 59′] (*f*) [≏80° 32′]
13. Find the dimension marked x in Fig. 11.34(*a*) to (*f*).

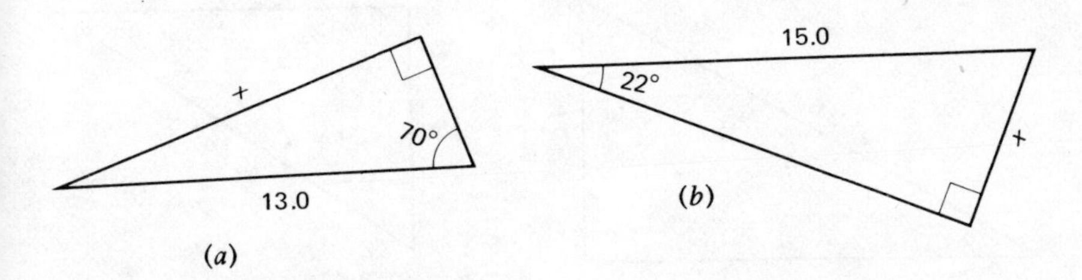

Fig. 11.34

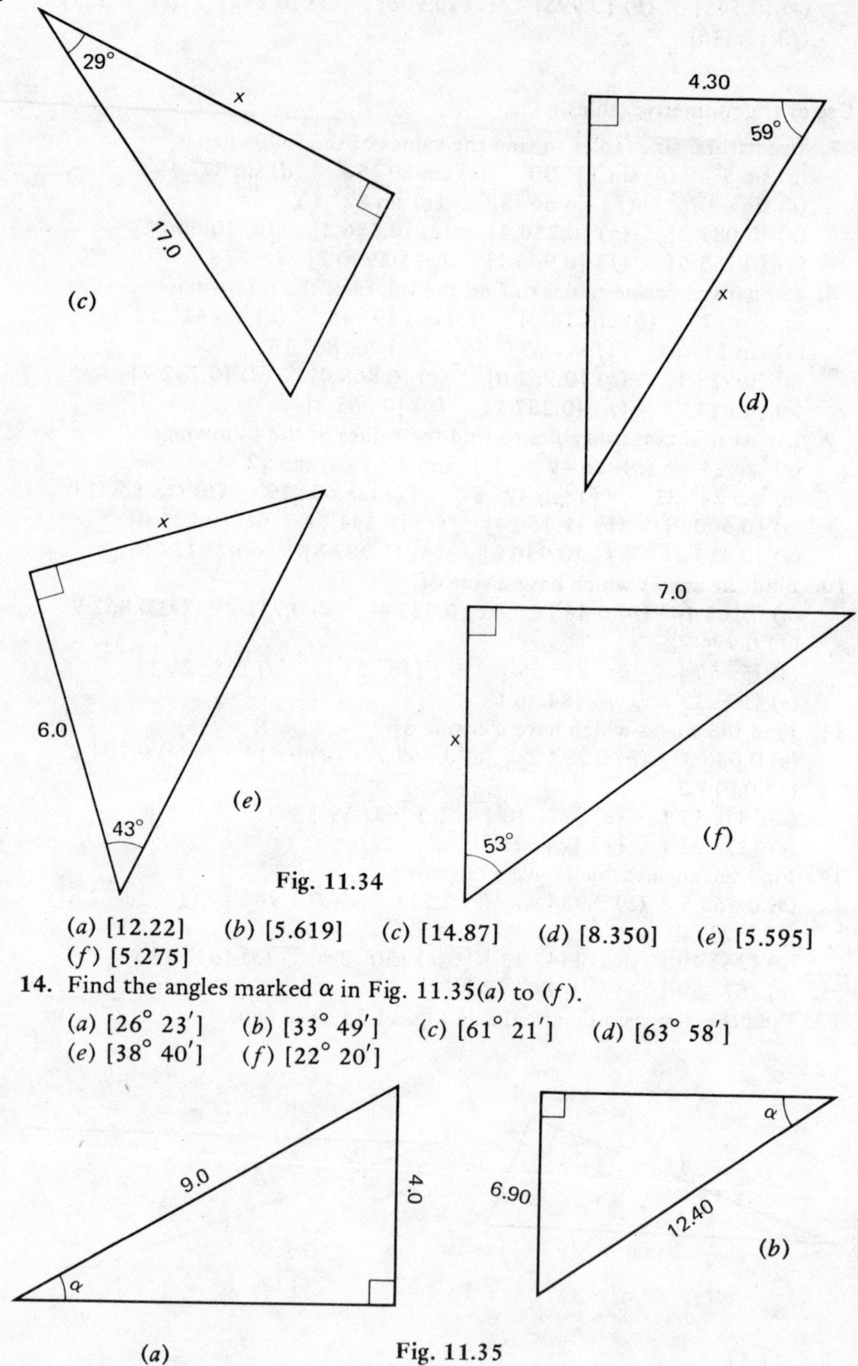

Fig. 11.34

(*a*) [12.22] (*b*) [5.619] (*c*) [14.87] (*d*) [8.350] (*e*) [5.595]
(*f*) [5.275]

14. Find the angles marked α in Fig. 11.35(*a*) to (*f*).

(*a*) [26° 23′] (*b*) [33° 49′] (*c*) [61° 21′] (*d*) [63° 58′]
(*e*) [38° 40′] (*f*) [22° 20′]

Fig. 11.35

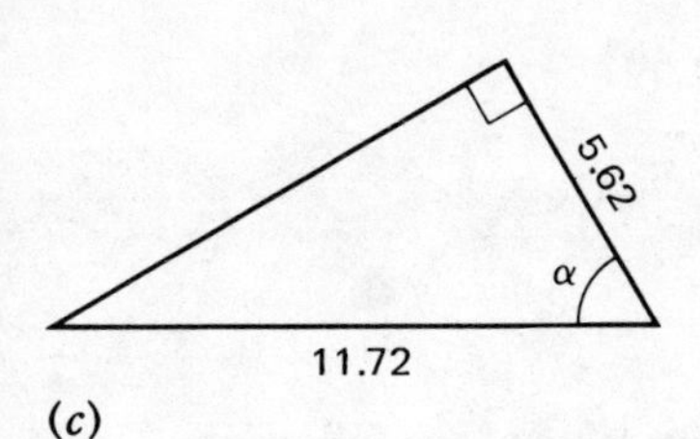

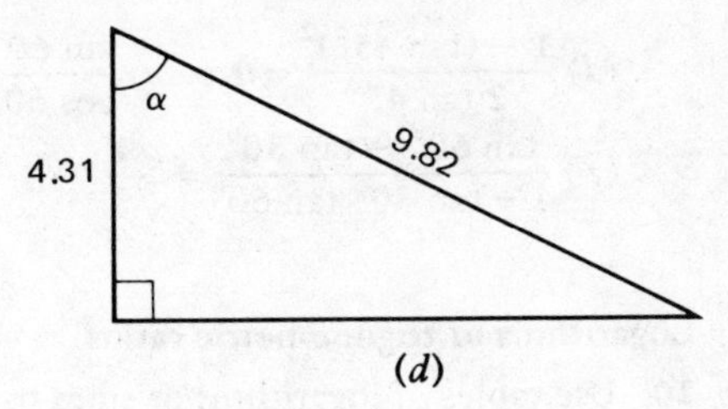

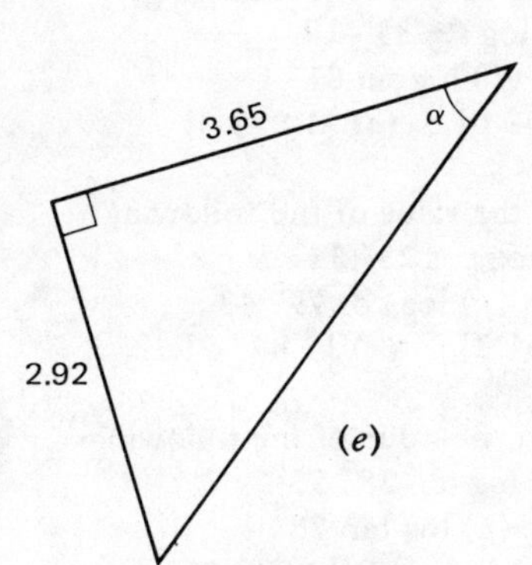

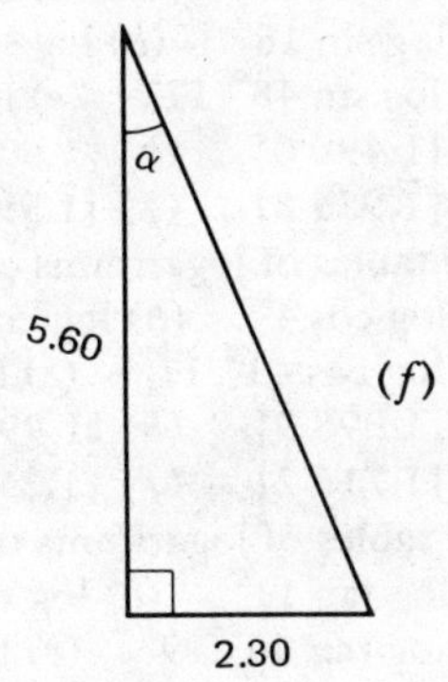

Fig. 11.35

15. Evaluate the following:

(*a*) $4 \cos 63^\circ + 2 \sin 27^\circ$ (*b*) $2.6 \cos 22^\circ - 4.1 \sin 67^\circ + 6.2 \tan 55^\circ$

(*c*) $\tan 82^\circ\, 6' - 3 \cos 12^\circ$ (*d*) $\dfrac{5.3 \tan 10^\circ}{4 \cos 17^\circ - 3 \sin 47^\circ}$

(*a*) [2.724] (*b*) [7.491] (*c*) [4.272] (*d*) [0.572 9]

16. If $\angle D = 32^\circ\, 17'$ and $\angle E = 73^\circ\, 52'$ evaluate the following (to 4 significant figures):

(*a*) $2 \sin D \cos E$ (*b*) $\dfrac{\tan D + \tan E}{1 - \tan D \tan E}$ (*c*) $\dfrac{3 \tan D}{2 \sin D \sin E}$

(*a*) [0.296 9] (*b*) [−3.455] (*c*) [1.847]

Surd forms and complementary angles

17. If $\cos 41^\circ = 0.766$ evaluate without using trigonometric tables:

$$2\left(\frac{5.4 \sin 49^\circ}{3 \cos 41^\circ} - \sin 90^\circ\right)$$ [1.6]

18. Evaluate the following without using tables and, if necessary, leave the answer in surd form:

(*a*) $2 \sin 30^\circ + 3 \cos 60^\circ$ (*b*) $3 \tan 60^\circ - 2 \sin 60^\circ$ (*c*) $\dfrac{3 \tan 30^\circ}{\tan 60^\circ}$

(*d*) $4 \cos 60^\circ - 2 \sin 60^\circ$ (*e*) $(\tan 45^\circ)(2 \cos 60^\circ - \sin 45^\circ \cos 45^\circ)$

(*a*) [2.5] (*b*) [$2\sqrt{3}$] (*c*) [1] (*d*) [$2 - \sqrt{3}$] (*e*) [0.5]

19. Show that:

(*a*) $\sin 30^\circ \cos 60^\circ + \cos 30^\circ \sin 60^\circ = 1$

(*b*) $\cos 60^\circ \cos 30^\circ - \sin 60^\circ \sin 30^\circ = 0$ (*c*) $(\cos 45^\circ)^2 = 1 - (\sin 45^\circ)^2$

(*d*) $\dfrac{1-(\tan 45^\circ)^2}{2\tan 45^\circ}=0$ (*e*) $\dfrac{\sin 60^\circ}{\cos 60^\circ}=\tan 60^\circ$

(*f*) $\dfrac{\tan 60^\circ - \tan 30^\circ}{1+\tan 30^\circ \tan 60^\circ}=\dfrac{1}{\sqrt{3}}$

Logarithms of trigonometric ratios

20. Use tables of logarithms of sines to find the value of the following:
(*a*) log sin 18° (*b*) log sin 23° 46′ (*c*) log s ·· 35° 19′
(*d*) log sin 48° 17′ (*e*) log sin 64° 35′ (*f*) log sin 81° 1
(*a*) [$\bar{1}$.490 0] (*b*) [$\bar{1}$.605 4] (*c*) [$\bar{1}$.762 0] (*d*) [$\bar{1}$.8 ,3 0]
(*e*) [$\bar{1}$.955 8] (*f*) [$\bar{1}$.994 9]

21. Use tables of logarithms of cosines to find the value of the following:
(*a*) log cos 4° (*b*) log cos 12° 15′ (*c*) log cos 23° 51′
(*d*) log cos 41° 17′ (*e*) log cos 58° 39′ (*f*) log cos 79° 43′
(*a*) [$\bar{1}$.998 9] (*b*) [$\bar{1}$.990 0] (*c*) [$\bar{1}$.961 2] (*d*) [$\bar{1}$.875 9]
(*e*) [$\bar{1}$.716 2] (*f*) [$\bar{1}$.251 7]

22. Use tables of logarithms of tangents to find the value of the following:
(*a*) log tan 11° (*b*) log tan 17° 49′ (*c*) log tan 28° 53′
(*d*) log tan 46° 19′ (*e*) log tan 63° 27′ (*f*) log tan 78° 4′
(*a*) [$\bar{1}$.288 7] (*b*) [$\bar{1}$.507 0] (*c*) [$\bar{1}$.741 7] (*d*) [0.020 0]
(*e*) [0.301 3] (*f*) [0.675 1]

23. Find the angles A to G given the following:
(*a*) log sin A = $\bar{1}$.923 6 (*b*) log sin B = $\bar{2}$.942 6
(*c*) log cos C = $\bar{1}$.824 7 (*d*) log cos D = $\bar{1}$.234 6
(*e*) log tan E = $\bar{1}$.823 5 (*f*) log tan F = 0.462 6
(*g*) log tan G = 1.133 7
(*a*) [57°] (*b*) [5° 2′] (*c*) [48° 6′] (*d*) [80° 7′] (*e*) [33° 40′]
(*f*) [70° 59′] (*g*) [85° 48′]

24. If $\cos X = \dfrac{9.46}{11.28}\sin 26^\circ$ find X. [68° 26′]

25. If $3\sin A = \dfrac{19.46}{22}\cos 73^\circ\ 15'$ find A. [4° 52′]

26. Evaluate ∠ B given $\dfrac{\cos 43^\circ\ 10'}{4}=\dfrac{3\sin 10^\circ\ 11'}{2.4\tan B}$ [50° 28′]

27. Calculate the value of $\dfrac{(\sin 56^\circ\ 52')(\cos 71^\circ\ 14')}{(3\tan 50^\circ\ 32')}$ (to 4 significant figures).
[0.073 94]

Solution of right-angled triangles

28. In an isosceles triangle ABC, AB = BC = 7 cm. If ∠ ABC is 76° find the length of AC and the value of angles BAC and ACB.
[8.619 cm, 52°, 52°]

29. Find the unknown sides and angles in the following right-angled triangles (shown in Fig. 11.36). The dimensions shown are centimetres.

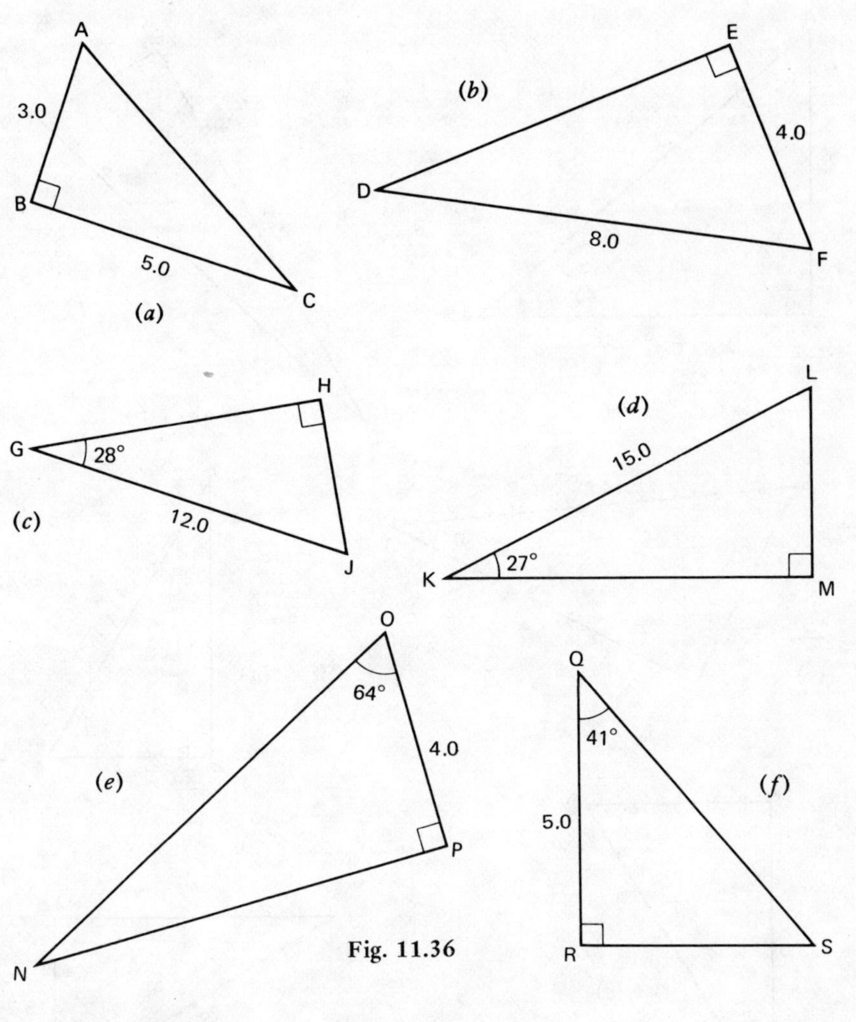

Fig. 11.36

(*a*) [AC = 5.831 cm, ∠ A = 59° 2′, ∠ C = 30° 58′]
(*b*) [DE = 6.928 cm, ∠ D = 30°, ∠ F = 60°]
(*c*) [∠ J = 62°, HJ = 5.634 cm, GH = 10.59 cm]
(*d*) [∠ L = 63°, LM = 6.810 cm, KM = 13.37 cm]
(*e*) [∠ N = 26°, ON = 9.125 cm, PM = 8.201 cm]
(*f*) [∠ S = 49°, RS = 4.346 cm, QS = 6.625 cm]

30. In a triangle XYZ, ∠ X = 90°, ∠ Y = 15° 43′ and XZ = 4.23 cm. Find the lengths of XY and YZ. [15.03 cm, 15.61 cm]

31. Solve the triangles shown in Fig. 11.37 and find their areas. The dimensions are in centimetres.

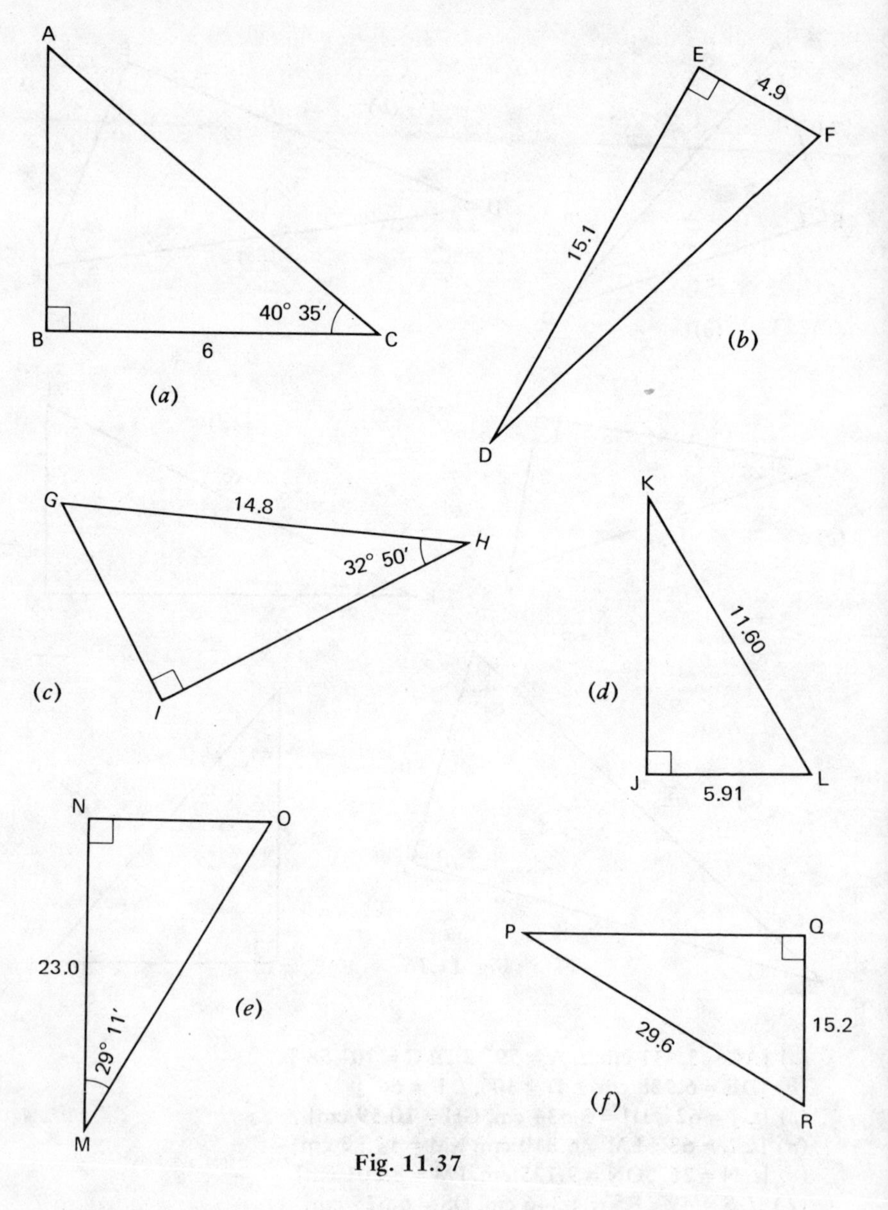

Fig. 11.37

(*a*) [AB = 5.140 cm, AC = 7.900 cm, ∠ A = 49° 25′, Area = 15.42 cm^2]
(*b*) [DF = 15.88 cm, ∠ D = 17° 59′, ∠ F = 72° 1′, Area = 37.00 cm^2]
(*c*) [∠ G = 57° 10′, GI = 8.025 cm, IH = 12.45 cm, Area = 49.96 cm^2]
(*d*) [KJ = 9.982 cm, ∠ K = 30° 38′, ∠ L = 59° 22′, Area = 29.50 cm^2]
(*e*) [∠ O = 60° 49′, NO = 12.85 cm, MO = 26.34 cm, Area = 147.8 cm^2]
(*f*) [PQ = 25.40 cm, ∠ P = 30° 54′, ∠ R = 59° 6′, Area = 193.0 cm^2]

32. A ladder rests against the top of the wall of a house and makes an angle of 65° with the ground. If the foot of the ladder is 16 m from the wall find the height of the house. [34.31 m]
33. A right-angled triangle ABC has an area of $16.8\ cm^2$. If BC is 5.2 cm and $\angle$ B is 90° find the length of AC and the angle BAC.
[8.293 cm, $38^\circ\ 50'$]
34. In the roof strut shown in Fig. 11.38 determine x and y and the angles β and θ.

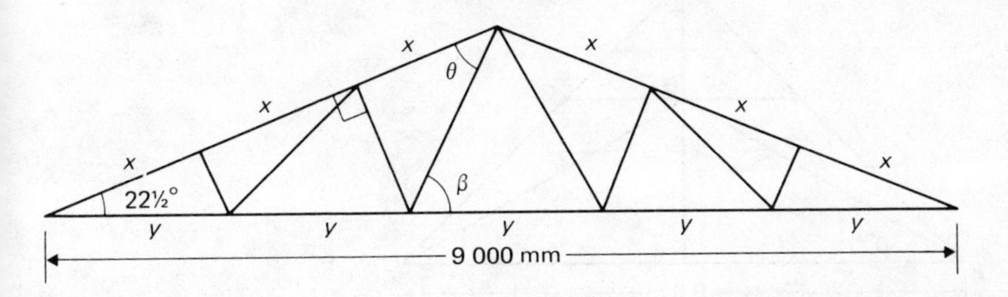

Fig. 11.38

[$x = 1\ 623.6$ mm, $y = 1\ 800$ mm, $\beta = 64^\circ\ 14'$, $\theta = 38^\circ\ 21'$]
35. Calculate the area of the symmetrical roof shown in Fig. 11.39.

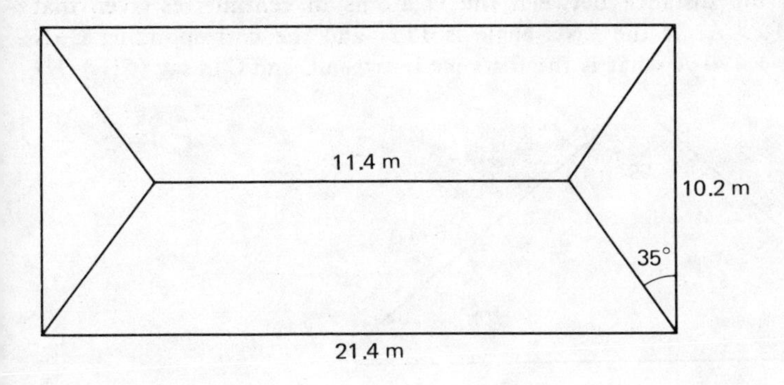

Fig. 11.39

[$158.1\ m^2$]
36. The cross-section of a service road to a housing estate is given in Fig. 11.40, the road being 35.6 m long. What volume of earth must be removed to form the cutting? Determine the angles α and β.
[$11\ 390\ m^3$, $\alpha = 30^\circ\ 46'$, $\beta = 33^\circ\ 20'$]
37. Figure 11.41 shows a symmetrical scissors truss with BC = 9.0 m and $\angle$ DBC = 30°. Determine the lengths of the struts and ties.

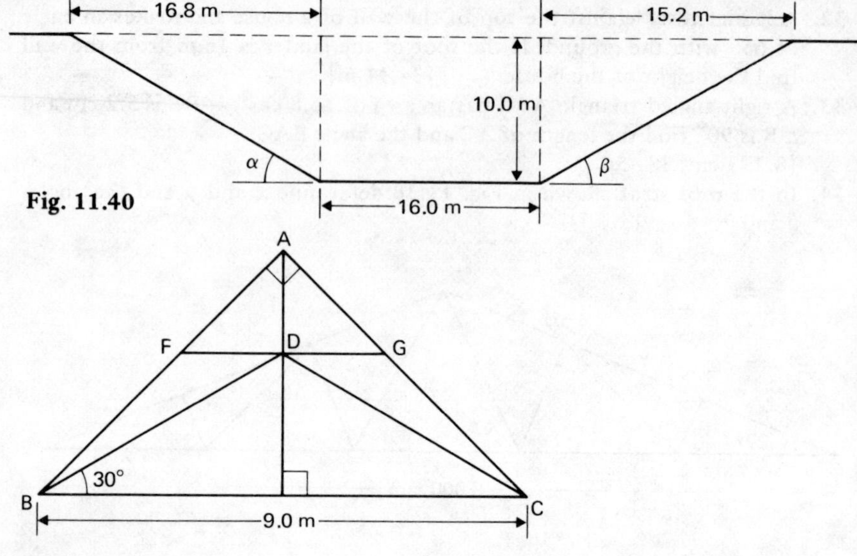

Fig. 11.40

Fig. 11.41

[AF = AG = 2.69 m, FB = GC = 3.67 m, FD = DG = 1.90 m, BE = EC = 4.50 m, BD = DC = 5.20 m, AD = 1.90 m, DE = 2.60 m]

38. In the molecule $(CH_3)_3$ P the C–P–C bond is represented by Fig. 11.42. What is the distance between the C atoms in centimetres given that 1 m = 10^{10} Å? If the CNC angle is 111° and the corresponding C–N distance 1.470 Å what is the distance between C and C in say $(CH_3)_3$ N?

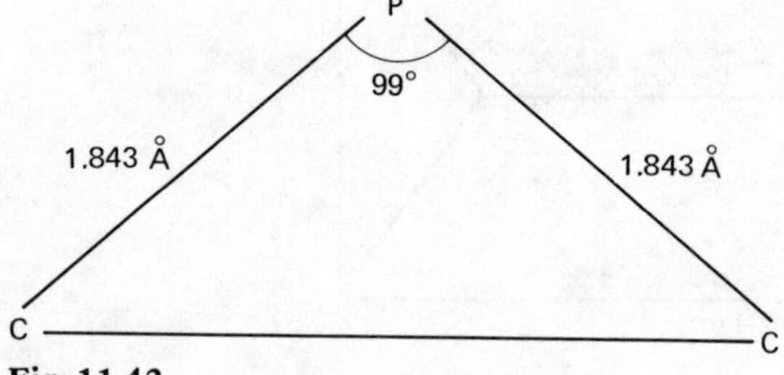

Fig. 11.42

[2.803×10^{-8} cm, 2.423×10^{-8} cm]

Angles of elevation and depression

39. A vertical tower stands on horizontal ground. At a point 120 m from the foot of the tower the angle of elevation of the top is 21°. Find the height of the tower. [46.06 m]

40. If the angle of elevation of the top of a vertical 52 m high building is 31° 15′, how far is it to the building? [85.69 m]

41. At a point 275 m from the foot of a vertical aerial, the angle of elevation of the top of the aerial is 16°. Find the height of the aerial. What is the angle of elevation of its mid point? [78.85 m, 8° $10'$]

42. From a particular point on horizontal ground a surveyor measures the angle of elevation of the top of a flagpole as 21° $21'$. He moves 75 m nearer to the flagpole and measures the angle of elevation as 32° $49'$. Find the height of the flagpole. [74.44 m]
43. A vertical aerial stands on the edge of the top of a building. At a point 230 m from the building the angles of elevation of the top and bottom of the aerial are 37° and 34°. Calculate the height of the aerial. [18.18 m]
44. From the window of a house 110 m away from a tower, the angle of elevation of the top of the tower is 28° and the angle of depression of the bottom is 10°. Find the height of the tower. [77.88 m]
45. From the top of a vertical cliff 125 m high the angle of depression of a boat is 18° $50'$. Find the distance of the boat from the cliff. [366.6 m]
46. From the top of a vertical cliff 82 m high the angles of depression of two boats lying due south of the cliff are 25° and 19°. How far are the boats apart? [62.30 m]
47. At a particular instant the angle of depression of a car, viewed from the top of a vertical building, is 21° $15'$. The building is 55 m high and the car is moving perpendicularly away from the building. If, 10 seconds later, the angle of depression is measured as 15° $10'$, find the speed of the car in km h^{-1}. [22.12 km h^{-1}]
48. A vertical flagstaff BE, 10 m high, stands at the corner B of a horizontal rectangular parade ground ABCD. The angles of elevation of the top of the flagstaff from A and D are 15° $11'$ and 13° $16'$, respectively. Calculate (*a*) the length of AB; (*b*) the length of BD; (*c*) the length of BC; and (*d*) the angle of elevation of E from C.
(*a*) [36.85 m] (*b*) [42.43 m] (*c*) [21.03 m] (*d*) [25° $26'$]
49. From a boat at sea, the angles of elevation of the top and bottom of a vertical lighthouse standing on the edge of a cliff are 33° and 27° respectively. If the lighthouse is 23 m high, calculate the height of the cliff. [83.76 m]
50. From the top of a cliff 143 m high, the angles of depression of the top and bottom of a lighthouse which is at sea level are observed to be 35° and 42°. Find the height of the lighthouse. [31.80 m]
51. From a window 5 m above the ground a man notices that the angle of depression of the foot of a building across the road is 26°, and the angle of elevation of the top of the same building is 32°. Find the height of the building and the width of the road (assuming the road to be horizontal). [11.41 m, 10.25 m]
52. The elevation of a tower from two points, one due south of the tower and the other due north of it are 16° and 21° respectively and the two points of observation are 400 m apart. Find the height of the tower. [65.65 m]

Sine and cosine curves

53. Draw to scale a sine curve and a cosine curve on the same axis over one complete cycle. Make the maximum value 4 cm.

Degrees	0′ 0°·0	6′ 0°·1	12′ 0°·2	18′ 0°·3	24′ 0°·4	30′ 0°·5	36′ 0°·6	42′ 0°·7	48′ 0°·8	54′ 0°·9	Mean Differences 1	2	3	4	5
0	·0000	0017	0035	0052	0070	0087	0105	0122	0140	0157	3	6	9	12	15
1	·0175	0192	0209	0227	0244	0262	0279	0297	0314	0332	3	6	9	12	15
2	·0349	0366	0384	0401	0419	0436	0454	0471	0488	0506	3	6	9	12	15
3	·0523	0541	0558	0576	0593	0610	0628	0645	0663	0680	3	6	9	12	15
4	·0698	0715	0732	0750	0767	0785	0802	0819	0837	0854	3	6	9	12	15
5	·0872	0889	0906	0924	0941	0958	0976	0993	1011	1028	3	6	9	12	14
6	·1045	1063	1080	1097	1115	1132	1149	1167	1184	1201	3	6	9	12	14
7	·1219	1236	1253	1271	1288	1305	1323	1340	1357	1374	3	6	9	12	14
8	·1392	1409	1426	1444	1461	1478	1495	1513	1530	1547	3	6	9	12	14
9	·1564	1582	1599	1616	1633	1650	1668	1685	1702	1719	3	6	9	12	14
10	·1736	1754	1771	1788	1805	1822	1840	1857	1874	1891	3	6	9	12	14
11	·1908	1925	1942	1959	1977	1994	2011	2028	2045	2062	3	6	9	11	14
12	·2079	2096	2113	2130	2147	2164	2181	2198	2215	2232	3	6	9	11	14
13	·2250	2267	2284	2300	2317	2334	2351	2368	2385	2402	3	6	8	11	14
14	·2419	2436	2453	2470	2487	2504	2521	2538	2554	2571	3	6	8	11	14
15	·2588	2605	2622	2639	2656	2672	2689	2706	2723	2740	3	6	8	11	14
16	·2756	2773	2790	2807	2823	2840	2857	2874	2890	2907	3	6	8	11	14
17	·2924	2940	2957	2974	2990	3007	3024	3040	3057	3074	3	6	8	11	14
18	·3090	3107	3123	3140	3156	3173	3190	3206	3223	3239	3	6	8	11	14
19	·3256	3272	3289	3305	3322	3338	3355	3371	3387	3404	3	5	8	11	14
20	·3420	3437	3453	3469	3486	3502	3518	3535	3551	3567	3	5	8	11	14
21	·3584	3600	3616	3633	3649	3665	3681	3697	3714	3730	3	5	8	11	14
22	·3746	3762	3778	3795	3811	3827	3843	3859	3875	3891	3	5	8	11	14
23	·3907	3923	3939	3955	3971	3987	4003	4019	4035	4051	3	5	8	11	14
24	·4067	4083	4099	4115	4131	4147	4163	4179	4195	4210	3	5	8	11	13
25	·4226	4242	4258	4274	4289	4305	4321	4337	4352	4368	3	5	8	11	13
26	·4384	4399	4415	4431	4446	4462	4478	4493	4509	4524	3	5	8	10	13
27	·4540	4555	4571	4586	4602	4617	4633	4648	4664	4679	3	5	8	10	13
28	·4695	4710	4726	4741	4756	4772	4787	4802	4818	4833	3	5	8	10	13
29	·4848	4863	4879	4894	4909	4924	4939	4955	4970	4985	3	5	8	10	13
30	·5000	5015	5030	5045	5060	5075	5090	5105	5120	5135	3	5	8	10	13
31	·5150	5165	5180	5195	5210	5225	5240	5255	5270	5284	2	5	7	10	12
32	·5299	5314	5329	5344	5358	5373	5388	5402	5417	5432	2	5	7	10	12
33	·5446	5461	5476	5490	5505	5519	5534	5548	5563	5577	2	5	7	10	12
34	·5592	5606	5621	5635	5650	5664	5678	5693	5707	5721	2	5	7	10	12
35	·5736	5750	5764	5779	5793	5807	5821	5835	5850	5864	2	5	7	10	12
36	·5878	5892	5906	5920	5934	5948	5962	5976	5990	6004	2	5	7	9	12
37	·6018	6032	6046	6060	6074	6088	6101	6115	6129	6143	2	5	7	9	12
38	·6157	6170	6184	6198	6211	6225	6239	6252	6266	6280	2	5	7	9	11
39	·6293	6307	6320	6334	6347	6361	6374	6388	6401	6414	2	4	7	9	11
40	·6428	6441	6455	6468	6481	6494	6508	6521	6534	6547	2	4	7	9	11
41	·6561	6574	6587	6600	6613	6626	6639	6652	6665	6678	2	4	7	9	11
42	·6691	6704	6717	6730	6743	6756	6769	6782	6794	6807	2	4	6	9	11
43	·6820	6833	6845	6858	6871	6884	6896	6909	6921	6934	2	4	6	8	11
44	·6947	6959	6972	6984	6997	7009	7022	7034	7046	7059	2	4	6	8	10

Degrees	0′ 0°·0	6′ 0°·1	12′ 0°·2	18′ 0°·3	24′ 0°·4	30′ 0°·5	36′ 0°·6	42′ 0°·7	48′ 0°·8	54′ 0°·9	Mean Differences 1	2	3	4	5
45	·7071	7083	7096	7108	7120	7133	7145	7157	7169	7181	2	4	6	8	10
46	·7193	7206	7218	7230	7242	7254	7266	7278	7290	7302	2	4	6	8	10
47	·7314	7325	7337	7349	7361	7373	7385	7396	7408	7420	2	4	6	8	10
48	·7431	7443	7455	7466	7478	7490	7501	7513	7524	7536	2	4	6	8	10
49	·7547	7558	7570	7581	7593	7604	7615	7627	7638	7649	2	4	6	8	9
50	·7660	7672	7683	7694	7705	7716	7727	7738	7749	7760	2	4	6	7	9
51	·7771	7782	7793	7804	7815	7826	7837	7848	7859	7869	2	4	5	7	9
52	·7880	7891	7902	7912	7923	7934	7944	7955	7965	7976	2	4	5	7	9
53	·7986	7997	8007	8018	8028	8039	8049	8059	8070	8080	2	3	5	7	9
54	·8090	8100	8111	8121	8131	8141	8151	8161	8171	8181	2	3	5	7	8
55	·8192	8202	8211	8221	8231	8241	8251	8261	8271	8281	2	3	5	7	8
56	·8290	8300	8310	8320	8329	8339	8348	8358	8368	8377	2	3	5	6	8
57	·8387	8396	8406	8415	8425	8434	8443	8453	8462	8471	2	3	5	6	8
58	·8480	8490	8499	8508	8517	8526	8536	8545	8554	8563	2	3	5	6	8
59	·8572	8581	8590	8599	8607	8616	8625	8634	8643	8652	1	3	4	6	7
60	·8660	8669	8678	8686	8695	8704	8712	8721	8729	8738	1	3	4	6	7
61	·8746	8755	8763	8771	8780	8788	8796	8805	8813	8821	1	3	4	6	7
62	·8829	8838	8846	8854	8862	8870	8878	8886	8894	8902	1	3	4	5	7
63	·8910	8918	8926	8934	8942	8949	8957	8965	8973	8980	1	3	4	5	6
64	·8988	8996	9003	9011	9018	9026	9033	9041	9048	9056	1	3	4	5	6
65	·9063	9070	9078	9085	9092	9100	9107	9114	9121	9128	1	2	4	5	6
66	·9135	9143	9150	9157	9164	9171	9178	9184	9191	9198	1	2	3	5	6
67	·9205	9212	9219	9225	9232	9239	9245	9252	9259	9265	1	2	3	4	6
68	·9272	9278	9285	9291	9298	9304	9311	9317	9323	9330	1	2	3	4	5
69	·9336	9342	9348	9354	9361	9367	9373	9379	9385	9391	1	2	3	4	5
70	·9397	9403	9409	9415	9421	9426	9432	9438	9444	9449	1	2	3	4	5
71	·9455	9461	9466	9472	9478	9483	9489	9494	9500	9505	1	2	3	4	5
72	·9511	9516	9521	9527	9532	9537	9542	9548	9553	9558	1	2	3	3	4
73	·9563	9568	9573	9578	9583	9588	9593	9598	9603	9608	1	2	2	3	4
74	·9613	9617	9622	9627	9632	9636	9641	9646	9650	9655	1	2	2	3	4
75	·9659	9664	9668	9673	9677	9681	9686	9690	9694	9699	1	1	2	3	4
76	·9703	9707	9711	9715	9720	9724	9728	9732	9736	9740	1	1	2	3	3
77	·9744	9748	9751	9755	9759	9763	9767	9770	9774	9778	1	1	2	3	3
78	·9781	9785	9789	9792	9796	9799	9803	9806	9810	9813	1	1	2	2	3
79	·9816	9820	9823	9826	9829	9833	9836	9839	9842	9845	1	1	2	2	3
80	·9848	9851	9854	9857	9860	9863	9866	9869	9871	9874	0	1	1	2	2
81	·9877	9880	9882	9885	9888	9890	9893	9895	9898	9900	0	1	1	2	2
82	·9903	9905	9907	9910	9912	9914	9917	9919	9921	9923	0	1	1	2	2
83	·9925	9928	9930	9932	9934	9936	9938	9940	9942	9943	0	1	1	1	2
84	·9945	9947	9949	9951	9954	9954	9956	9957	9959	9960	0	1	1	1	2
85	·9962	9963	9965	9966	9968	9969	9971	9972	9973	9974	0	0	1	1	1
86	·9976	9977	9978	9979	9980	9981	9982	9983	9984	9985	0	0	1	1	1
87	·9986	9987	9988	9989	9990	9990	991	9992	9993	9993	0	0	0	1	1
88	·9994	9995	9995	9996	9996	9997	9997	9997	9998	9998	0	0	0	0	0
89	·9998	9999	9999	9999	9999	1·000	1·000	1·000	1·000	1·000	0	0	0	0	0
90	1·000														

Numbers in difference columns to be subtracted, not added

Degrees	0′ 0°·0	6′ 0°·1	12′ 0°·2	18′ 0°·3	24′ 0°·4	30′ 0°·5	36′ 0°·6	42′ 0°·7	48′ 0°·8	54′ 0°·9	Mean Differences 1	2	3	4	5
0	1·000	1·000	1·000	1·000	1·000	1·000	·9999	9999	9999	9999	0	0	0	0	0
1	·9998	9998	9998	9997	9997	9997	9996	9996	9995	9995	0	0	0	0	0
2	·9994	9993	9993	9992	9991	9990	9990	9989	9988	9987	0	0	0	1	1
3	·9986	9985	9984	9983	9982	9981	9980	9979	9978	9977	0	0	1	1	1
4	·9976	9974	9973	9972	9971	9969	9968	9966	9965	9963	0	0	1	1	1
5	·9962	9960	9959	9957	9956	9954	9952	9951	9949	9947	0	1	1	1	2
6	·9945	9943	9942	9940	9938	9936	9934	9932	9930	9928	0	1	1	1	2
7	·9925	9923	9921	9919	9917	9914	9912	9910	9907	9905	0	1	1	2	2
8	·9903	9900	9898	9895	9893	9890	9888	9885	9882	9880	0	1	1	2	2
9	·9877	9874	9871	9869	9866	9863	9860	9857	9854	9851	0	1	1	2	2
10	·9848	9845	9842	9839	9836	9833	9829	9826	9823	9820	1	1	2	2	3
11	·9816	9813	9810	9806	9803	9799	9796	9792	9789	9785	1	1	2	2	3
12	·9781	9778	9774	9770	9767	9763	9759	9755	9751	9748	1	1	2	3	3
13	·9744	9740	9736	9732	9728	9724	9720	9715	9711	9707	1	1	2	3	3
14	·9703	9699	9694	9690	9686	9681	9677	9673	9668	9664	1	1	2	3	4
15	·9659	9655	9650	9646	9641	9636	9632	9627	9622	9617	1	2	2	3	4
16	·9613	9608	9603	9598	9593	9588	9583	9578	9573	9568	1	2	2	3	4
17	·9563	9558	9553	9548	9542	9537	9532	9527	9521	9516	1	2	3	3	4
18	·9511	9505	9500	9494	9489	9483	9478	9472	9466	9461	1	2	3	4	5
19	·9455	9449	9444	9438	9432	9426	9421	9415	9409	9403	1	2	3	4	5
20	·9397	9391	9385	9379	9373	9367	9361	9354	9348	9342	1	2	3	4	5
21	·9336	9330	9323	9317	9311	9304	9298	9291	9285	9278	1	2	3	4	5
22	·9272	9265	9259	9252	9245	9239	9232	9225	9219	9212	1	2	3	4	6
23	·9205	9198	9191	9184	9178	9171	9164	9157	9150	9143	1	2	3	5	6
24	·9135	9128	9121	9114	9107	9100	9092	9085	9078	9070	1	2	4	5	6
25	·9063	9056	9048	9041	9033	9026	9018	9011	9003	8996	1	3	4	5	6
26	·8988	8980	8973	8965	8957	8949	8942	8934	8926	8918	1	3	4	5	6
27	·8910	8902	8894	8886	8878	8870	8862	8854	8846	8838	1	3	4	5	7
28	·8829	8821	8813	8805	8796	8788	8780	8771	8763	8755	1	3	4	6	7
29	·8746	8738	8729	8721	8712	8704	8695	8686	8678	8669	1	3	4	6	7
30	·8660	8652	8643	8634	8625	8616	8607	8599	8590	8581	1	3	4	6	7
31	·8572	8563	8554	8545	8536	8526	8517	8508	8499	8490	2	3	5	6	8
32	·8480	8471	8462	8453	8443	8434	8425	8415	8406	8396	2	3	5	6	8
33	·8387	8377	8368	8358	8348	8339	8329	8320	8310	8300	2	3	5	6	8
34	·8290	8281	8271	8261	8251	8241	8231	8221	8211	8202	2	3	5	7	8
35	·8192	8181	8171	8161	8151	8141	8131	8121	8111	8100	2	3	5	7	8
36	·8090	8080	8070	8059	8049	8039	8028	8018	8007	7997	2	3	5	7	9
37	·7986	7976	7965	7955	7944	7934	7923	7912	7902	7891	2	4	5	7	9
38	·7880	7869	7859	7848	7837	7826	7815	7804	7793	7782	2	4	5	7	9
39	·7771	7760	7749	7738	7727	7716	7705	7694	7683	7672	2	4	6	7	9
40	·7660	7649	7638	7627	7615	7604	7593	7581	7570	7559	2	4	6	8	9
41	·7547	7536	7524	7513	7501	7490	7478	7466	7455	7443	2	4	6	8	10
42	·7431	7420	7408	7396	7385	7373	7361	7349	7337	7325	2	4	6	8	10
43	·7314	7302	7290	7278	7266	7254	7242	7230	7218	7206	2	4	6	8	10
44	·7193	7181	7169	7157	7145	7133	7120	7108	7096	7083	2	4	6	8	10

Degrees	0′ 0°·0	6′ 0°·1	12′ 0°·2	18′ 0°·3	24′ 0°·4	30′ 0°·5	36′ 0°·6	42′ 0°·7	48′ 0°·8	54′ 0°·9	Mean Differences 1	2	3	4	5
45	·7071	7059	7046	7034	7022	7009	6997	6984	6972	6959	2	4	6	8	10
46	·6947	6934	6921	6909	6896	6884	6871	6858	6845	6833	2	4	6	8	11
47	·6820	6807	6794	6782	6769	6756	6743	6730	6717	6704	2	4	6	9	11
48	·6691	6678	6665	6652	6639	6626	6613	6600	6587	6574	2	4	7	9	11
49	·6561	6547	6534	6521	6508	6494	6481	6468	6455	6441	2	4	7	9	11
50	·6428	6414	6401	6388	6374	6361	6347	6334	6320	6307	2	4	7	9	11
51	·6293	6280	6266	6252	6239	6225	6211	6198	6184	6170	2	5	7	9	11
52	·6157	6143	6129	6115	6101	6088	6074	6060	6046	6032	2	5	7	9	12
53	·6018	6004	5990	5976	5962	5948	5934	5920	5906	5892	2	5	7	9	12
54	·5878	5864	5850	5835	5821	5807	5793	5779	5764	5750	2	5	7	9	12
55	·5736	5721	5707	5693	5678	5664	5650	5635	5621	5606	2	5	7	10	12
56	·5592	5577	5563	5548	5534	5519	5505	5490	5476	5461	2	5	7	10	12
57	·5446	5432	5417	5402	5388	5373	5358	5344	5329	5314	2	5	7	10	12
58	·5299	5284	5270	5255	5240	5225	5210	5195	5180	5165	2	5	7	10	12
59	·5150	5135	5120	5105	5090	5075	5060	5045	5030	5015	3	5	8	10	13
60	·5000	4985	4970	4955	4939	4924	4909	4894	4879	4863	3	5	8	10	13
61	·4848	4833	4818	4802	4787	4772	4756	4741	4726	4710	3	5	8	10	13
62	·4695	4679	4664	4648	4633	4617	4602	4586	4571	4555	3	5	8	10	13
63	·4540	4524	4509	4493	4478	4462	4446	4431	4415	4399	3	5	8	10	13
64	·4384	4368	4352	4337	4321	4305	4289	4274	4258	4242	3	5	8	11	13
65	·4226	4210	4195	4179	4163	4147	4131	4115	4099	4083	3	5	8	11	13
66	·4067	4051	4035	4019	4003	3987	3971	3955	3939	3923	3	5	8	11	14
67	·3907	3891	3875	3859	3843	3827	3811	3795	3778	3762	3	5	8	11	14
68	·3746	3730	3714	3697	3681	3665	3649	3633	3616	3600	3	5	8	11	14
69	·3584	3567	3551	3535	3518	3502	3486	3469	3453	3437	3	5	8	11	14
70	·3420	3404	3387	3371	3355	3338	3322	3305	3289	3272	3	5	8	11	14
71	·3256	3239	3223	3206	3190	3173	3156	3140	3123	3107	3	6	8	11	14
72	·3090	3074	3057	3040	3024	3007	2990	2974	2957	2940	3	6	8	11	14
73	·2924	2907	2890	2874	2857	2840	2823	2807	2790	2773	3	6	8	11	14
74	·2756	2740	2723	2706	2689	2672	2656	2639	2622	2605	3	6	8	11	14
75	·2588	2571	2554	2538	2521	2504	2487	2470	2453	2436	3	6	8	11	14
76	·2419	2402	2385	2368	2351	2334	2317	2300	2284	2267	3	6	8	11	14
77	·2250	2233	2215	2198	2181	2164	2147	2130	2113	2096	3	6	9	11	14
78	·2079	2062	2045	2028	2011	1994	1977	1959	1942	1925	3	6	9	11	14
79	·1908	1891	1874	1857	1840	1822	1805	1788	1771	1754	3	6	9	11	14
80	·1736	1719	1702	1685	1668	1650	1633	1616	1599	1582	3	6	9	12	14
81	·1564	1547	1530	1513	1495	1478	1461	1444	1426	1409	3	6	9	12	14
82	·1392	1374	1357	1340	1323	1305	1228	1271	1253	1236	3	6	9	12	14
83	·1219	1201	1184	1167	1149	1132	1115	1097	1080	1063	3	6	9	12	14
84	·1045	1028	1011	0993	0976	0958	0941	0924	0906	0889	3	6	9	12	14
85	·0872	0854	0837	0819	0802	0785	0767	0750	0732	0715	3	6	9	12	15
86	·0698	0680	0663	0645	0628	0610	0593	0576	0558	0541	3	6	9	12	15
87	·0523	0506	0488	0471	0454	0436	0419	0401	0384	0366	3	6	9	12	15
88	·0349	0332	0314	0297	0279	0262	0244	0227	0209	0192	3	6	9	12	15
89	·0175	0157	0140	0122	0105	0087	0070	0052	0035	0017	3	6	9	12	15
90	·0000														

Degrees	0′ 0°·0	6′ 0°·1	12′ 0°·2	18′ 0°·3	24′ 0°·4	30′ 0°·5	36′ 0°·6	42′ 0°·7	48′ 0°·8	54′ 0°·9	Mean Differences 1	2	3	4	5
0	·0000	0017	0035	0052	0070	0087	0105	0122	0140	0157	3	6	9	12	15
1	·0175	0192	0209	0227	0244	0262	0279	0297	0314	0332	3	6	9	12	15
2	·0349	0367	0384	0402	0419	0437	0454	0472	0489	0507	3	6	9	12	15
3	·0524	0542	0559	0577	0594	0612	0629	0647	0664	0682	3	6	9	12	15
4	·0699	0717	0734	0752	0769	0787	0805	0822	0840	0857	3	6	9	12	15
5	·0875	0892	0910	0928	0945	0963	0981	0998	1016	1033	3	6	9	12	15
6	·1051	1069	1086	1104	1122	1139	1157	1175	1192	1210	3	6	9	12	15
7	·1228	1246	1263	1281	1299	1317	1334	1352	1370	1388	3	6	9	12	15
8	·1405	1423	1441	1459	1477	1495	1512	1530	1548	1566	3	6	9	12	15
9	·1584	1602	1620	1638	1655	1673	1691	1709	1727	1745	3	6	9	12	15
10	·1763	1781	1799	1817	1835	1853	1871	1890	1908	1926	3	6	9	12	15
11	·1944	1962	1980	1998	2016	2035	2053	2071	2089	2107	3	6	9	12	15
12	·2126	2144	2162	2180	2199	2217	2235	2254	2272	2290	3	6	9	12	15
13	·2309	2327	2345	2364	2382	2401	2419	2338	2456	2475	3	6	9	12	15
14	·2493	2512	2530	2549	2568	2586	2605	2623	2642	2661	3	6	9	12	16
15	·2679	2698	2717	2736	2754	2773	2792	2811	2830	2849	3	6	9	13	16
16	·2867	2886	2905	2924	2943	2962	2981	3000	3019	3038	3	6	9	13	16
17	·3057	3076	3096	3115	3134	3153	3172	3191	3211	3230	3	6	10	13	16
18	·3249	3269	3288	3307	3327	3346	3365	3385	3404	3424	3	6	10	13	16
19	·3443	3463	3482	3502	3522	3541	3561	3581	3600	3620	3	7	10	13	16
20	·3640	3659	3679	3699	3719	3739	3759	3779	3799	3819	3	7	10	13	17
21	·3839	3895	3879	3899	3919	3939	3959	3979	4000	4020	3	7	10	13	17
22	·4040	4061	4081	4101	4122	4142	4163	4183	4204	4224	3	7	10	14	17
23	·4245	4265	4286	4307	4327	4348	4369	4390	4411	4431	3	7	10	14	17
24	·4452	4473	4494	4515	4536	4557	4578	4599	4621	4642	4	7	11	14	18
25	·4663	4684	4706	4727	4748	4770	4791	4813	4834	4856	4	7	11	14	18
26	·4877	4899	4921	4942	4964	4986	5008	5029	5051	5073	4	7	11	15	18
27	·5095	5117	5139	5161	5184	5206	5228	5250	5272	5295	4	7	11	15	18
28	·5317	5340	5362	5384	5407	5430	5452	5475	5498	5520	4	8	11	15	19
29	·5543	5566	5589	5612	5635	5658	5681	5704	5727	5750	4	8	12	15	19
30	·5774	5797	5820	5844	5867	5890	5914	5938	5961	5985	4	8	12	16	20
31	·6009	6032	6056	6080	6104	6128	6152	6176	6200	6224	4	8	12	16	20
32	·6249	6273	6297	6322	6346	6371	6395	6420	6445	6469	4	8	12	16	20
33	·6494	6519	6544	6569	6594	6619	6644	6669	6694	6720	4	8	13	17	21
34	·6745	6771	6796	6822	6847	6873	6899	6924	6950	6976	4	9	13	17	21
35	·7002	7028	7054	7080	7107	7133	7159	7186	7212	7239	4	9	13	18	22
36	·7265	7292	7319	7346	7373	7400	7427	7454	7481	7508	5	9	14	18	23
37	·7536	7563	7590	7618	7646	7673	7701	7729	7757	7785	5	9	14	18	23
38	·7813	7841	7869	7898	7926	7954	7983	8012	8040	8069	5	9	14	19	24
39	·8098	8127	8156	8185	8214	8243	8273	8302	8332	8361	5	10	15	20	24
40	·8391	8421	8451	8481	8511	8541	8571	8601	8632	8662	5	10	15	20	25
41	·8693	8724	8754	8785	8816	8847	8878	8910	8941	8972	5	10	16	21	26
42	·9004	9036	9067	9099	9131	9163	9195	9228	9260	9293	5	11	16	21	27
43	·9325	9358	9391	9424	9457	9490	9523	9556	9590	9623	6	11	17	22	28
44	·9657	9691	9725	9759	9793	9827	9861	9896	9930	9965	6	11	17	23	29

Degrees	0′ 0°·0	6′ 0°·1	12′ 0°·2	18′ 0°·3	24′ 0°·4	30′ 0°·5	36′ 0°·6	42′ 0°·7	48′ 0°·8	54′ 0°·9	Mean Differences				
											1	2	3	4	5
45	1·0000	0035	0070	0105	0141	0176	0212	0247	0283	0319	6	12	18	24	30
46	1·0355	0392	0428	0464	0501	0538	0575	0612	0649	0686	6	12	18	25	31
47	1·0724	0761	0799	0837	0875	0913	0951	0990	1028	1067	6	13	19	25	32
48	1·1106	1145	1184	1224	1263	1303	1343	1383	1423	1463	7	13	20	27	33
49	1·1504	1544	1585	1626	1667	1708	1750	1792	1833	1875	7	14	21	28	34
50	1·1918	1960	2002	2045	2088	2131	2174	2218	2261	2305	7	14	22	29	36
51	1·2349	2393	2437	2482	2527	2572	2617	2662	2708	2753	8	15	23	30	38
52	1·2799	2846	2892	2938	2985	3032	3079	3127	3175	3222	8	16	24	31	39
53	1·3270	3319	3367	3416	3465	3514	3564	3613	3663	3713	8	16	25	33	41
54	1·3764	3814	3865	3916	3968	4019	4071	4124	4176	4229	9	17	26	34	43
55	1·4281	4335	4388	4442	4496	4550	4605	4659	4715	4770	9	18	27	36	45
56	1·4826	4882	4938	4994	5051	5108	5166	5224	5282	5340	10	19	29	38	48
57	1·5399	5458	5517	5577	5637	5697	5757	5818	5880	5941	10	20	30	40	50
58	1·6003	6066	6128	6191	6255	6319	6383	6447	6512	6577	11	21	32	43	53
59	1·6643	6709	6775	6842	6909	6977	7045	7113	7182	7251	11	23	34	45	56
60	1·7321	7391	7461	7532	7603	7675	7747	7820	7893	7966	12	24	36	48	60
61	1·8040	8115	8190	8265	8341	8418	8495	8572	8650	8728	13	26	38	51	64
62	1·8807	8887	8967	9047	9128	9210	9292	9375	9458	9542	14	27	41	55	68
63	1·9626	9711	9797	9883	9970	2·0057	2·0145	2·0233	2·0323	2·0413	15	29	44	58	73
64	2·0503	0594	0686	0778	0872	0965	1060	1155	1251	1348	16	31	47	63	78
65	2·1445	1543	1642	1742	1842	1943	2045	2148	2251	2355	17	34	51	68	85
66	2·2460	2566	2673	2781	2889	2998	3109	3220	3332	3445	18	37	55	73	92
67	2·3559	3673	3789	3906	4023	4142	4262	4383	4504	4627	20	40	60	79	99
68	2·4751	4876	5002	5129	5257	5386	5517	5649	5782	5916	22	43	65	87	108
69	2·6051	6187	6325	6464	6605	6746	6889	7034	7179	7326	24	47	71	95	119
70	2·7475	7625	7776	7929	8083	8239	8397	8556	8716	8878	26	52	78	104	131
71	2·9042	9208	9375	9544	9714	9887	3·0061	3·0237	3·0415	3·0595	29	58	87	116	145
72	3·0777	9961	1146	1334	1524	1716	1910	2106	2305	2506	32	64	96	129	161
73	3·2709	2914	3122	3332	3544	3759	3977	4197	4420	4646	36	72	108	144	180
74	3·4874	5105	5339	5576	5816	6059	6305	6554	6806	7062	41	81	122	163	204
75	3·7321	7583	7848	8118	8391	8667	8947	9232	9520	9812	46	93	139	186	232
76	4·0108	0408	0713	1022	1335	1653	1976	2303	2635	2972	53	107	160	213	267
77	4·3315	3662	4015	4374	4737	5107	5483	5864	6252	6646					
78	4·7046	7453	7867	8288	8716	9152	9594	5·0045	5·0504	5·0970	Mean differences cease to be sufficiently accurate				
79	5·1446	1929	2422	2924	3435	3955	4486	5026	5578	6140					
80	5·6713	7297	7894	8502	9124	9758	6·0405	6·1066	6·1742	6·2432					
81	6·3138	3859	4596	5350	6122	6912	7720	8548	9395	7·0264					
82	7·1154	2066	3002	3962	4947	5958	6996	8062	9158	8·0285					
83	8·1443	2636	3863	5126	6427	7769	9152	9·0579	9·2052	9·3572					
84	9·5144	9·677	9·845	10·02	10·20	10·39	10·58	10·78	10·99	11·20					
85	11·43	11·66	11·91	12·16	12·43	12·71	13·00	13·30	13·62	13·95					
86	14·30	14·67	15·06	15·46	15·89	16·35	16·83	17·34	17·89	18·46					
87	19·08	19·74	20·45	21·20	22·02	22·90	23·86	24·90	26·03	27·27					
88	28·64	30·14	31·82	33·69	35·80	38·19	40·92	44·07	47·74	52·08					
89	57·29	63·66	71·62	81·85	95·49	114·6	143·2	191·0	286·5	573·0					
90	∞														

Degrees	0′ 0°.0	6′ 0°.1	12′ 0°.2	18′ 0°.3	24′ 0°.4	30′ 0°.5	36′ 0°.6	42′ 0°.7	48′ 0°.8	54′ 0°.9	Mean Differences				
											1	2	3	4	5
0	$-\infty$	$\bar{3}.2419$	$\bar{3}.5429$	7190	8439	9408	$\bar{2}.0200$	$\bar{2}.0870$	$\bar{2}.1450$	$\bar{2}.1961$					
1	$\bar{2}.2419$	2832	3210	3558	3880	4179	4459	4723	4971	5206					
2	$\bar{2}.5428$	5640	5842	6035	6220	6397	6567	6731	6889	7041					
3	$\bar{2}.7188$	7330	7468	7602	7731	7857	7979	8098	8213	8326					
4	$\bar{2}.8436$	8543	8647	8749	8849	8946	9042	9135	9226	9315	16	32	48	64	[illegible]
5	$\bar{2}.9403$	9489	9573	9655	9736	9816	9894	9970	$\bar{1}.0046$	$\bar{1}.0120$	13	26	39	52	[illegible]
6	$\bar{1}.0192$	0264	0334	0403	0472	0539	0605	0670	0734	0797	11	22	33	44	[illegible]
7	$\bar{1}.0859$	0920	0981	1040	1099	1157	1214	1271	1326	1381	10	19	29	38	[illegible]
8	$\bar{1}.1436$	1489	1542	1594	1646	1697	1747	1797	1847	1895	8	17	25	34	[illegible]
9	$\bar{1}.1943$	1991	2038	2085	2131	2176	2221	2266	2310	2353	8	15	23	30	[illegible]
10	$\bar{1}.2397$	2439	2482	2524	2565	2606	2647	2687	2727	2767	7	14	20	27	[illegible]
11	$\bar{1}.2806$	2845	2883	2921	2959	2997	3034	3070	3107	3143	6	12	19	25	[illegible]
12	$\bar{1}.3179$	3214	3250	3284	3319	3353	3387	3421	3455	3488	6	11	17	23	[illegible]
13	$\bar{1}.3521$	3554	3586	3618	3650	3682	3713	3745	3775	3806	5	11	16	21	[illegible]
14	$\bar{1}.3837$	3867	3897	3927	3957	3986	4015	4044	4073	4102	5	10	15	20	[illegible]
15	$\bar{1}.4130$	4158	4186	4214	4242	4269	4296	4323	4350	4377	5	9	14	18	[illegible]
16	$\bar{1}.4403$	4430	4456	4482	4508	4533	4559	4584	4609	4634	4	9	13	17	[illegible]
17	$\bar{1}.4659$	4684	4709	4733	4757	4781	4805	4829	4853	4876	4	8	12	16	[illegible]
18	$\bar{1}.4900$	4923	4946	4969	4992	5015	5037	5060	5082	5104	4	8	11	15	[illegible]
19	$\bar{1}.5126$	5148	5170	5192	5213	5235	5256	5278	5299	5320	4	7	11	14	[illegible]
20	$\bar{1}.5341$	5361	5382	5402	5423	5443	5463	5484	5504	5523	3	7	10	14	[illegible]
21	$\bar{1}.5543$	5563	5583	5602	5621	5641	5660	5679	5698	5717	3	6	10	13	[illegible]
22	$\bar{1}.5736$	5754	5773	5792	5810	5828	5847	5865	5883	5901	3	6	9	12	[illegible]
23	$\bar{1}.5919$	5937	5954	5972	5990	6007	6024	6042	6059	6076	3	6	9	12	[illegible]
24	$\bar{1}.6093$	6110	6127	6144	6161	6177	6194	6210	6227	6243	3	6	8	11	[illegible]
25	$\bar{1}.6259$	6276	6292	6308	6324	6340	6356	6371	6387	6403	3	5	8	11	[illegible]
26	$\bar{1}.6418$	6434	6449	6465	6480	6495	6510	6526	6541	6556	3	5	8	10	[illegible]
27	$\bar{1}.6570$	6585	6600	6615	6629	6644	6659	6673	6687	6702	2	5	7	10	[illegible]
28	$\bar{1}.6716$	6730	6744	6759	6773	6787	6801	6814	6828	6842	2	5	7	9	[illegible]
29	$\bar{1}.6856$	6869	6883	6896	6910	6923	6937	6950	6963	6977	2	4	7	9	[illegible]
30	$\bar{1}.6990$	7003	7016	7029	7042	7055	7068	7080	7093	7106	2	4	6	9	[illegible]
31	$\bar{1}.7118$	7131	7144	7156	7168	7181	7193	7205	7218	7230	2	4	6	8	[illegible]
32	$\bar{1}.7242$	7254	7266	7278	7290	7302	7314	7326	7338	7349	2	4	6	8	[illegible]
33	$\bar{1}.7361$	7373	7384	7396	7407	7419	7430	7442	7453	7464	2	4	6	8	[illegible]
34	$\bar{1}.7476$	7487	7498	7509	7520	7531	7542	7553	7564	7575	2	4	6	7	[illegible]
35	$\bar{1}.7586$	7597	7607	7618	7629	7640	7650	7661	7671	7682	2	4	5	7	[illegible]
36	$\bar{1}.7692$	7703	7713	7723	7734	7744	7754	7764	7774	7785	2	3	5	7	[illegible]
37	$\bar{1}.7795$	7805	7815	7825	7835	7844	7854	7864	7874	7884	2	3	5	7	[illegible]
38	$\bar{1}.7893$	7903	7913	7922	7932	7941	7951	7960	7970	7979	2	3	5	6	[illegible]
39	$\bar{1}.7989$	7998	8007	8017	8026	8035	8044	8053	8063	8072	2	3	5	6	[illegible]
40	$\bar{1}.8081$	8090	8099	8108	8117	8125	8134	8143	8152	8161	1	3	4	6	[illegible]
41	$\bar{1}.8169$	8178	8187	8195	8204	8213	8221	8230	8238	8247	1	3	4	6	[illegible]
42	$\bar{1}.8255$	8264	8272	8280	8289	8297	8305	8313	8322	8330	1	3	4	6	[illegible]
43	$\bar{1}.8338$	8346	8354	8362	8370	8378	8386	8394	8402	8410	1	3	4	5	[illegible]
44	$\bar{1}.8418$	8426	8433	8441	8449	8457	8464	8472	8480	8487	1	3	4	5	[illegible]

Table 11.5 (cont'd) Logarithms of sines

Degrees	0′ 0°.0	6′ 0°.1	12′ 0°.2	18′ 0°.3	24′ 0°.4	30′ 0°.5	36′ 0°.6	42′ 0°.7	48′ 0°.8	54′ 0°.9	Mean Differences 1	2	3	4	5
45	$\bar{1}$.8495	8502	8510	8517	8525	8532	8540	8547	8555	8562	1	2	4	5	6
46	$\bar{1}$.8569	8577	8584	8591	8598	8606	8613	8620	8627	8634	1	2	4	5	6
47	$\bar{1}$.8641	8648	8655	8662	8669	8676	8683	8690	8697	8704	1	2	3	5	6
48	$\bar{1}$.8711	8718	8724	8731	8738	8745	8751	8758	8765	8771	1	2	3	4	6
49	$\bar{1}$.8778	8784	8791	8797	8804	8810	8817	8823	8830	8836	1	2	3	4	5
50	$\bar{1}$.8843	8849	8855	8862	8868	8874	8880	8887	8893	8899	1	2	3	4	5
51	$\bar{1}$.8905	8911	8917	8923	8929	8935	8941	8947	8953	8959	1	2	3	4	5
52	$\bar{1}$.8965	8971	8977	8983	8989	8995	9000	9006	9012	9018	1	2	3	4	5
53	$\bar{1}$.9023	9029	9035	9041	9046	9052	9057	9063	9069	9074	1	2	3	4	5
54	$\bar{1}$.9080	9085	9091	9096	9101	9107	9112	9118	9123	9128	1	2	3	4	5
55	$\bar{1}$.9134	9139	9144	9149	9155	9160	9165	9170	9175	9181	1	2	3	3	4
56	$\bar{1}$.9186	9191	9196	9201	9206	9211	9216	9221	9226	9231	1	2	3	3	4
57	$\bar{1}$.9236	9241	9246	9251	9255	9260	9265	9270	9275	9279	1	2	2	3	4
58	$\bar{1}$.9284	9289	9294	9298	9303	9308	9312	9317	9322	9326	1	2	2	3	4
59	$\bar{1}$.9331	9335	9340	9344	9349	9353	9358	9362	9367	9371	1	1	2	3	4
60	$\bar{1}$.9375	9380	9384	9388	9393	9397	9401	9406	9410	9414	1	1	2	3	4
61	$\bar{1}$.9418	9422	9427	9431	9435	9439	9443	9447	9451	9455	1	1	2	3	3
62	$\bar{1}$.9459	9463	9467	9471	9475	9479	9483	9487	9491	9495	1	1	2	3	3
63	$\bar{1}$.9499	9503	9506	9510	9514	9518	9522	9525	9529	9533	1	1	2	3	3
64	$\bar{1}$.9537	9540	9544	9548	9551	9555	9558	9562	9566	9569	1	1	2	2	3
65	$\bar{1}$.9573	9576	9580	9583	9587	9590	9594	9597	9601	9604	1	1	2	2	3
66	$\bar{1}$.9607	9611	9614	9617	9621	9624	9627	9631	9634	9637	1	1	2	2	3
67	$\bar{1}$.9640	9643	9647	9650	9653	9656	9659	9662	9666	9669	1	1	2	2	3
68	$\bar{1}$.9672	9675	9678	9681	9684	9687	9690	9693	9696	9699	0	1	1	2	2
69	$\bar{1}$.9702	9704	9707	9710	9713	9716	9719	9722	9724	9727	0	1	1	2	2
70	$\bar{1}$.9730	9733	9735	9738	9741	9743	9746	9749	9751	9754	0	1	1	2	2
71	$\bar{1}$.9757	9759	9762	9764	9767	9770	9772	9775	9777	9780	0	1	1	2	2
72	$\bar{1}$.9782	9785	9787	9789	9792	9794	9797	9799	9801	9804	0	1	1	2	2
73	$\bar{1}$.9806	9808	9811	9813	9815	9817	9820	9822	9824	9826	0	1	1	2	2
74	$\bar{1}$.9828	9831	9833	9835	9837	9839	9841	9843	9845	9847	0	1	1	1	2
75	$\bar{1}$.9849	9851	9853	9855	9857	9859	9861	9863	9865	9867	0	1	1	1	2
76	$\bar{1}$.9869	9871	9873	9875	9876	9878	9880	9882	9884	9885	0	1	1	1	2
77	$\bar{1}$.9887	9889	9891	9892	9894	9896	9897	9899	9901	9902	0	1	1	1	1
78	$\bar{1}$.9904	9906	9907	9909	9910	9912	9913	9915	9916	9918	0	1	1	1	1
79	$\bar{1}$.9919	9921	9922	9924	9925	9927	9928	9929	9931	9932	0	0	1	1	1
80	$\bar{1}$.9934	9935	9936	9937	9939	9940	9941	9943	9944	9945	0	0	1	1	1
81	$\bar{1}$.9946	9947	9949	9950	9951	9952	9953	9954	9955	9956	0	0	1	1	1
82	$\bar{1}$.9958	9959	9960	9961	9962	9963	9964	9965	9966	9967	0	0	1	1	1
83	$\bar{1}$.9968	9968	9969	9970	9971	9972	9973	9974	9975	9975	0	0	0	1	1
84	$\bar{1}$.9976	9977	9978	9978	9979	9980	9981	9981	9982	9983	0	0	0	0	1
85	$\bar{1}$.9983	9984	9985	9985	9986	9987	9987	9988	9988	9989	0	0	0	0	0
86	$\bar{1}$.9989	9990	9990	9991	9991	9992	9992	9993	9993	9994	0	0	0	0	0
87	$\bar{1}$.9994	9994	9995	9995	9996	9996	9996	9996	9997	9997	0	0	0	0	0
88	$\bar{1}$.9997	9998	9998	9998	9998	9999	9999	9999	9999	9999	0	0	0	0	0
89	$\bar{1}$.9999	9999	0.0000	0000	0000	0000	0000	0000	0000	0000					
90	0.0000														

Numbers in difference columns to be subtracted, not added

Degrees	0′ 0°.0	6′ 0°.1	12′ 0°.2	18′ 0°.3	24′ 0°.4	30′ 0°.5	36′ 0°.6	42′ 0°.7	48′ 0°.8	54′ 0°.9	Mean Differences 1	2	3	4	5
0	0.0000	0000	0000	0000	0000	0000	0000	0000	0000	$\bar{1}$.9999	0	0	0	0	0
1	$\bar{1}$.9999	9999	9999	9999	9999	9999	9998	9998	9998	9998	0	0	0	0	0
2	$\bar{1}$.9997	9997	9997	9996	9996	9996	9996	9995	9995	9994	0	0	0	0	0
3	$\bar{1}$.9994	9994	9993	9993	9992	9992	9991	9991	9990	9990	0	0	0	0	0
4	$\bar{1}$.9989	9989	9988	9988	9987	9987	9986	9985	9985	9984	0	0	0	0	0
5	$\bar{1}$.9983	9983	9982	9981	9981	9980	9979	9978	9978	9977	0	0	0	0	1
6	$\bar{1}$.9976	9975	9975	9974	9973	9972	9971	9970	9969	9968	0	0	0	1	1
7	$\bar{1}$.9968	9967	9966	9965	9964	9963	9962	9961	9960	9959	0	0	1	1	1
8	$\bar{1}$.9958	9956	9955	9954	9953	9952	9951	9950	9949	9947	0	0	1	1	1
9	$\bar{1}$.9946	9945	9944	9943	9941	9940	9939	9937	9936	9935	0	0	1	1	1
10	$\bar{1}$.9934	9932	9931	9929	9928	9927	9925	9924	9922	9921	0	0	1	1	1
11	$\bar{1}$.9919	9918	9916	9915	9913	9912	9910	9909	9907	9906	0	1	1	1	1
12	$\bar{1}$.9904	9902	9901	9899	9897	9896	9894	9892	9891	9889	0	1	1	1	1
13	$\bar{1}$.9887	9885	9884	9882	9880	9878	9876	9875	9873	9871	0	1	1	1	2
14	$\bar{1}$.9869	9867	9865	9863	9861	9859	9857	9855	9853	9851	0	1	1	1	2
15	$\bar{1}$.9849	9847	9845	9843	9841	9839	9837	9835	9833	9831	0	1	1	1	2
16	$\bar{1}$.9828	9826	9824	9822	9820	9817	9815	9813	9811	9808	0	1	1	2	2
17	$\bar{1}$.9806	9804	9801	9799	9797	9794	9792	9789	9787	9785	0	1	1	2	2
18	$\bar{1}$.9782	9780	9777	9775	9772	9770	9767	9764	9762	9759	0	1	1	2	2
19	$\bar{1}$.9757	9754	9751	9749	9746	9743	9741	9738	9735	9733	0	1	1	2	2
20	$\bar{1}$.9730	9727	9724	9722	9719	9716	9713	9710	9707	9704	0	1	1	2	2
21	$\bar{1}$.9702	9699	9696	9693	9690	9687	9684	9681	9678	9675	0	1	1	2	2
22	$\bar{1}$.9672	9669	9666	9662	9659	9656	9653	9650	9647	9643	1	1	2	2	3
23	$\bar{1}$.9640	9637	9634	9631	9627	9624	9621	9617	9614	9611	1	1	2	2	3
24	$\bar{1}$.9607	9604	9601	9597	9594	9590	9587	9583	9580	9576	1	1	2	2	3
25	$\bar{1}$.9573	9569	9566	9562	9558	9555	9551	9548	9544	9540	1	1	2	2	3
26	$\bar{1}$.9537	9533	9529	9525	9522	9518	9514	9510	9506	9503	1	1	2	3	3
27	$\bar{1}$.9499	9495	9491	9487	9483	9479	9475	9471	9467	9463	1	1	2	3	3
28	$\bar{1}$.9459	9455	9451	9447	9443	9439	9435	9431	9427	9422	1	1	2	3	3
29	$\bar{1}$.9418	9414	9410	9406	9401	9397	9393	9388	9384	9380	1	1	2	3	4
30	$\bar{1}$.9375	9371	9367	9362	9358	9353	9349	9344	9340	9335	1	1	2	3	4
31	$\bar{1}$.9331	9326	9322	9317	9312	9308	9303	9298	9294	9289	1	2	2	3	4
32	$\bar{1}$.9284	9279	9275	9270	9265	9260	9255	9251	9246	9241	1	2	2	3	4
33	$\bar{1}$.9236	9231	9226	9221	9216	9211	9206	9201	9196	9191	1	2	3	3	4
34	$\bar{1}$.9186	9181	9175	9170	9165	9160	9155	9149	9144	9139	1	2	3	3	4
35	$\bar{1}$.9134	9128	9123	9118	9112	9107	9101	9096	9091	9085	1	2	3	4	5
36	$\bar{1}$.9080	9074	9069	9063	9057	9052	9046	9041	9035	9029	1	2	3	4	5
37	$\bar{1}$.9023	9018	9012	9006	9000	8995	8989	8983	8977	8971	1	2	3	4	5
38	$\bar{1}$.8965	8959	8953	8947	8941	8935	8929	8923	8917	8911	1	2	3	4	5
39	$\bar{1}$.8905	8899	8893	8887	8880	8874	8868	8862	8855	8849	1	2	3	4	5
40	$\bar{1}$.8843	8836	8830	8823	8817	8810	8804	8797	8791	8784	1	2	3	4	5
41	$\bar{1}$.8778	8771	8765	8758	8751	8745	8738	8731	8724	8718	1	2	3	5	6
42	$\bar{1}$.8711	8704	8697	8690	8683	8676	8669	8662	8655	8648	1	2	3	5	6
43	$\bar{1}$.8641	8634	8627	8620	8613	8606	8598	8591	8584	8577	1	2	4	5	6
44	$\bar{1}$.8569	8562	8555	8547	8540	8532	8525	8517	8510	8502	1	2	4	5	6

Table 11.6 (cont'd) Logarithms of cosines

umbers in difference columns to be subtracted, not added

	0′ 0°.0	6′ 0°.1	12′ 0°.2	18′ 0°.3	24′ 0°.4	30′ 0°.5	36′ 0°.6	42′ 0°.7	48′ 0°.8	54′ 0°.9	Mean Differences 1	2	3	4	5
5	$\bar{1}.8495$	8487	8480	8472	8464	8457	8449	8441	8433	8426	1	3	4	5	6
6	$\bar{1}.8418$	8410	8402	8394	8386	8378	8370	8362	8354	8346	1	3	4	5	7
7	$\bar{1}.8338$	8330	8322	8313	8305	8297	8289	8280	8272	8264	1	3	4	6	7
8	$\bar{1}.8255$	8247	8238	8230	8221	8213	8204	8195	8187	8178	1	3	4	6	7
9	$\bar{1}.8169$	8161	8152	8143	8134	8125	8117	8108	8099	8090	1	3	4	6	7
0	$\bar{1}.8081$	8072	8063	8053	8044	8035	8026	8017	8007	7998	2	3	5	6	8
1	$\bar{1}.7989$	7979	7970	7960	7951	7941	7932	7922	7913	7903	2	3	5	6	8
2	$\bar{1}.7893$	7884	7874	7864	7854	7844	7835	7825	7815	7805	2	3	5	7	8
3	$\bar{1}.7795$	7785	7774	7764	7754	7744	7734	7723	7713	7703	2	3	5	7	9
4	$\bar{1}.7692$	7682	7671	7661	7650	7640	7629	7618	7607	7597	2	4	5	7	9
5	$\bar{1}.7586$	7575	7564	7553	7542	7531	7520	7509	7498	7487	2	4	6	7	9
6	$\bar{1}.7476$	7464	7453	7442	7430	7419	7407	7396	7384	7373	2	4	6	8	10
7	$\bar{1}.7361$	7349	7338	7326	7314	7302	7290	7278	7266	7254	2	4	6	8	10
8	$\bar{1}.7242$	7230	7218	7205	7193	7181	7168	7156	7144	7131	2	4	6	8	10
9	$\bar{1}.7118$	7106	7093	7080	7068	7055	7042	7029	7016	7003	2	4	6	9	11
0	$\bar{1}.6990$	6977	6963	6950	6937	6923	6910	6896	6883	6869	2	4	7	9	11
1	$\bar{1}.6856$	6842	6828	6814	6801	6787	6773	6759	6744	6730	2	5	7	9	12
2	$\bar{1}.6716$	6702	6687	6673	6659	6644	6629	6615	6600	6585	2	5	7	10	12
3	$\bar{1}.6570$	6556	6541	6526	6510	6495	6480	6465	6449	6434	3	5	8	10	13
4	$\bar{1}.6418$	6403	6387	6371	6356	6340	6324	6308	6292	6276	3	5	8	11	13
5	$\bar{1}.6259$	6243	6227	6210	6194	6177	6161	6144	6127	6110	3	6	8	11	14
6	$\bar{1}.6093$	6076	6059	6042	6024	6007	5990	5972	5954	5937	3	6	9	12	15
7	$\bar{1}.5919$	5901	5883	5865	5847	5828	5810	5792	5773	5754	3	6	9	12	15
8	$\bar{1}.5736$	5717	5698	5679	5660	5641	5621	5602	5583	5563	3	6	10	13	16
9	$\bar{1}.5543$	5523	5504	5484	5463	5443	5423	5402	5382	5361	3	7	10	14	17
0	$\bar{1}.5341$	5320	5299	5278	5256	5235	5213	5192	5170	5148	4	7	11	14	18
1	$\bar{1}.5126$	5104	5082	5060	5037	5015	4992	4969	4946	4923	4	8	11	15	19
2	$\bar{1}.4900$	4876	4853	4829	4805	4781	4757	4733	4709	4684	4	8	12	16	20
3	$\bar{1}.4659$	4634	4609	4584	4559	4533	4508	4482	4456	4430	4	9	13	17	21
4	$\bar{1}.4403$	4377	4350	4323	4296	4269	4242	4214	4186	4158	5	9	14	18	23
5	$\bar{1}.4130$	4102	4073	4044	4015	3986	3957	3927	3897	3867	5	10	15	20	24
6	$\bar{1}.3837$	3806	3775	3745	3713	3682	3650	3618	3586	3554	5	11	16	21	26
7	$\bar{1}.3521$	3488	3455	3421	3387	3353	3319	3284	3250	3214	6	11	17	23	28
8	$\bar{1}.3179$	3143	3107	3070	3034	2997	2959	2921	2883	2845	6	12	19	25	31
9	$\bar{1}.2806$	2767	2727	2687	2647	2606	2565	2524	2482	2439	7	14	20	27	34
0	$\bar{1}.2397$	2353	2310	2266	2221	2176	2131	2085	2038	1991	8	15	23	30	38
1	$\bar{1}.1943$	1895	1847	1797	1747	1697	1646	1594	1542	1489	8	17	25	34	42
2	$\bar{1}.1436$	1381	1326	1271	1214	1157	1099	1040	0981	0920	10	19	29	38	48
3	$\bar{1}.0859$	0797	0734	0670	0605	0539	0472	0403	0334	0264	11	22	33	44	55
4	$\bar{1}.0192$	0120	0046	$\bar{2}.9970$	$\bar{2}.9894$	$\bar{2}.9816$	$\bar{2}.9736$	$\bar{2}.9655$	$\bar{2}.9573$	$\bar{2}.9489$	13	26	39	52	65
5	$\bar{2}.9403$	9315	9226	9135	9042	8946	8849	8749	8647	8543	16	32	48	64	80
6	$\bar{2}.8436$	8326	8213	8098	7979	7857	7731	7602	7468	7330					
7	$\bar{2}.7188$	7041	6889	6731	6567	6397	6220	6035	5842	5640					
8	$\bar{2}.5428$	5206	4971	4723	4459	4179	3880	3558	3210	2832					
9	$\bar{2}.2419$	1961	1450	0870	0200	$\bar{3}.9408$	$\bar{3}.8439$	$\bar{3}.7190$	$\bar{3}.5429$	$\bar{3}.2419$					
0	∞														

 Table 11.7 Logarithms of tangents

Degrees	0′ 0°.0	6′ 0°.1	12′ 0°.2	18′ 0°.3	24′ 0°.4	30′ 0°.5	36′ 0°.6	42′ 0°.7	48′ 0°.8	54′ 0°.9	Mean Differences				
											1	2	3	4	
0	$-\infty$	$\bar{3}.2419$	$\bar{3}.5429$	$\bar{3}.7190$	$\bar{3}.8439$	$\bar{3}.9409$	$\bar{2}.0200$	$\bar{2}.0870$	$\bar{2}.1450$	$\bar{2}.1962$					
1	$\bar{2}.2419$	2833	3211	3559	3881	4181	4461	4725	4973	5208					
2	$\bar{2}.5431$	5643	5845	6038	6223	6401	6571	6736	6894	7046					
3	$\bar{2}.7194$	7337	7475	7609	7739	7865	7988	8107	8223	8336					
4	$\bar{2}.8446$	8554	8659	8762	8862	8960	9056	9150	9241	9331	16	32	48	64	[illegible]
5	$\bar{2}.9420$	9506	9591	9674	9756	9836	9915	9992	$\bar{1}.0068$	$\bar{1}.0143$	13	26	40	53	[illegible]
6	$\bar{1}.0216$	0289	0360	0430	0499	0567	0633	0699	0764	0828	11	22	34	45	[illegible]
7	$\bar{1}.0891$	0954	1015	1076	1135	1194	1252	1310	1367	1423	10	20	29	39	[illegible]
8	$\bar{1}.1478$	1533	1587	1640	1693	1745	1797	1848	1898	1948	9	17	26	35	[illegible]
9	$\bar{1}.1997$	2046	2094	2142	2189	2236	2282	2328	2374	2419	8	16	23	31	[illegible]
10	$\bar{1}.2463$	2507	2551	2594	2637	2680	2722	2764	2805	2846	7	14	21	28	[illegible]
11	$\bar{1}.2887$	2927	2967	3006	3046	3085	3123	3162	3200	3237	6	13	19	26	[illegible]
12	$\bar{1}.3275$	3312	3349	3385	3422	3458	3493	3529	3564	3599	6	12	18	24	[illegible]
13	$\bar{1}.3634$	3668	3702	3736	3770	3804	3837	3870	3903	3935	6	11	17	22	[illegible]
14	$\bar{1}.3968$	4000	4032	4064	4095	4127	4158	4189	4220	4250	5	10	16	21	[illegible]
15	$\bar{1}.4281$	4311	4341	4371	4400	4430	4459	4488	4517	4546	5	10	15	20	[illegible]
16	$\bar{1}.4575$	4603	4632	4660	4688	4716	4744	4771	4799	4826	5	9	14	19	[illegible]
17	$\bar{1}.4853$	4880	4907	4934	4961	4987	5014	5040	5066	5092	4	9	13	18	[illegible]
18	$\bar{1}.5118$	5143	5169	5195	5220	5245	5270	5295	5320	5345	4	8	13	17	21
19	$\bar{1}.5370$	5394	5419	5443	5467	5491	5516	5539	5563	5587	4	8	12	16	20
20	$\bar{1}.5611$	5634	5658	5681	5704	5727	5750	5773	5796	5819	4	8	12	15	19
21	$\bar{1}.5842$	5864	5887	5909	5932	5954	5976	5998	6020	6042	4	7	11	15	19
22	$\bar{1}.6064$	6086	6108	6129	6151	6172	6194	6215	6236	6257	4	7	11	14	18
23	$\bar{1}.6279$	6300	6321	6341	6362	6383	6404	6424	6445	6465	3	7	10	14	17
24	$\bar{1}.6486$	6506	6527	6547	6567	6587	6607	6627	6647	6667	3	7	10	13	17
25	$\bar{1}.6687$	6706	6726	6746	6765	6785	6804	6824	6843	6863	3	7	10	13	16
26	$\bar{1}.6882$	6901	6920	6939	6958	6977	6996	7015	7034	7053	3	6	9	13	16
27	$\bar{1}.7072$	7090	7109	7128	7146	7165	7183	7202	7220	7238	3	6	9	12	15
28	$\bar{1}.7257$	7275	7293	7311	7330	7348	7366	7384	7402	7420	3	6	9	12	15
29	$\bar{1}.7438$	7455	7473	7491	7509	7526	7544	7562	7579	7597	3	6	9	12	15
30	$\bar{1}.7614$	7632	7649	7667	7684	7701	7719	7736	7753	7771	3	6	9	12	14
31	$\bar{1}.7788$	7805	7822	7839	7856	7873	7890	7907	7924	7941	3	6	9	11	14
32	$\bar{1}.7958$	7975	7992	8008	8025	8042	8059	8075	8092	8109	3	6	8	11	14
33	$\bar{1}.8125$	8142	8158	8175	8191	8208	8224	8241	8257	8274	3	5	8	11	14
34	$\bar{1}.8290$	8306	8323	8339	8355	8371	8388	8404	8420	8436	3	5	8	11	14
35	$\bar{1}.8452$	8468	8484	8501	8517	8533	8549	8565	8581	8597	3	5	8	11	13
36	$\bar{1}.8613$	8629	8644	8660	8676	8692	8708	8724	8740	8755	3	5	8	11	13
37	$\bar{1}.8771$	8787	8803	8818	8834	8850	8865	8881	8897	8912	3	5	8	10	13
38	$\bar{1}.8928$	8944	8959	8975	8990	9006	9022	9037	9053	9068	3	5	8	10	13
39	$\bar{1}.9084$	9099	9115	9130	9146	9161	9176	9192	9207	9223	3	5	8	10	13
40	$\bar{1}.9238$	9254	9269	9284	9300	9315	9330	9346	9361	9376	3	5	8	10	13
41	$\bar{1}.9392$	9407	9422	9438	9453	9468	9483	9499	9514	9529	3	5	8	10	13
42	$\bar{1}.9544$	9560	9575	9590	9605	9621	9636	9651	9666	9681	3	5	8	10	13
43	$\bar{1}.9697$	9712	9727	9742	9757	9772	9788	9803	9818	9833	3	5	8	10	13
44	$\bar{1}.9848$	9864	9879	9894	9909	9924	9939	9955	9970	9985	3	5	8	10	13

Table 11.7 (cont'd) Logarithms of tangents

Degrees	0′ 0°.0	6′ 0°.1	12′ 0°.2	18′ 0°.3	24′ 0°.4	30′ 0°.5	36′ 0°.6	42′ 0°.7	48′ 0°.8	54′ 0°.9	Mean Differences 1	2	3	4	5
45	.0000	0015	0030	0045	0061	0076	0091	0106	0121	0136	3	5	8	10	13
46	.0152	0167	0182	0197	0212	0228	0243	0258	0273	0288	3	5	8	10	13
47	.0303	0319	0334	0349	0364	0379	0395	0410	0425	0440	3	5	8	10	13
48	.0456	0471	0486	0501	0517	0532	0547	0562	0578	0593	3	5	8	10	13
49	.0608	0624	0639	0654	0670	0685	0700	0716	0731	0746	3	5	8	10	13
50	.0762	0777	0793	0808	0824	0839	0854	0870	0885	0901	3	5	8	10	13
51	.0916	0932	0947	0963	0978	0994	1010	1025	1041	1056	3	5	8	10	13
52	.1072	1088	1103	1119	1135	1150	1166	1182	1197	1213	3	5	8	10	13
53	.1229	1245	1260	1276	1292	1308	1324	1340	1356	1371	3	5	8	11	13
54	.1387	1403	1419	1435	1451	1467	1483	1499	1516	1532	3	5	8	11	13
55	.1548	1564	1580	1596	1612	1629	1645	1661	1677	1694	3	5	8	11	14
56	.1710	1726	1743	1759	1776	1792	1809	1825	1842	1858	3	5	8	11	14
57	.1875	1891	1908	1925	1941	1958	1975	1992	2008	2025	3	6	8	11	14
58	.2042	2059	2076	2093	2110	2127	2144	2161	2178	2195	3	6	9	11	14
59	.2212	2229	2247	2264	2281	2299	2316	2333	2351	2368	3	6	9	12	14
60	.2386	2403	2421	2438	2456	2474	2491	2509	2527	2545	3	6	9	12	15
61	.2562	2580	2598	2616	2634	2652	2670	2689	2707	2725	3	6	9	12	15
62	.2743	2762	2780	2798	2817	2835	2854	2872	2891	2910	3	6	9	12	15
63	.2928	2947	2966	2985	3004	3023	3042	3061	3080	3099	3	6	9	13	16
64	.3118	3137	3157	3176	3196	3215	3235	3254	3274	3294	3	6	10	13	16
65	.3313	3333	3353	3373	3393	3413	3433	3453	3473	3494	3	7	10	13	17
66	.3514	3535	3555	3576	3596	3617	3638	3659	3679	3700	3	7	10	14	17
67	.3721	3743	3764	3785	3806	3828	3849	3871	3892	3914	4	7	11	14	18
68	.3936	3958	3980	4002	4024	4046	4068	4091	4113	4136	4	7	11	15	19
69	.4158	4181	4204	4227	4250	4273	4296	4319	4342	4366	4	8	12	15	19
70	.4389	4413	4437	4461	4484	4509	4533	4557	4581	4606	4	8	12	16	20
71	.4630	4655	4680	4705	4730	4755	4780	4805	4831	4857	4	8	13	17	21
72	.4882	4908	4934	4960	4986	5013	5039	5066	5093	5120	4	9	13	18	22
73	.5147	5174	5201	5229	5256	5284	5312	5340	5368	5397	5	9	14	19	23
74	.5425	5454	5483	5512	5541	5570	5600	5629	5659	5689	5	10	15	20	25
75	.5719	5750	5780	5811	5842	5873	5905	5936	5968	6000	5	10	16	21	26
76	.6032	6065	6097	6130	6163	6196	6230	6264	6298	6332	6	11	17	22	28
77	.6366	6401	6436	6471	6507	6542	6578	6615	6651	6688	6	12	18	24	30
78	.6725	6763	6800	6838	6877	6915	6954	6994	7033	7073	6	13	19	26	32
79	.7113	7154	7195	7236	7278	7320	7363	7406	7449	7493	7	14	21	28	35
80	.7537	7581	7626	7672	7718	7764	7811	7858	7906	7954	8	16	23	31	39
81	.8003	8052	8102	8152	8203	8255	8307	8360	8413	8467	9	17	26	35	43
82	.8522	8577	8633	8690	8748	8806	8865	8924	8985	9046	10	20	29	39	49
83	.9109	9172	9236	9301	9367	9433	9501	9570	9640	9711	11	22	34	45	56
84	.9784	9857	9932	1.0008	1.0085	1.0164	1.0244	1.0326	1.0409	1.0494	13	26	40	53	66
85	1.0580	0669	0759	0850	0944	1040	1138	1238	1341	1446	16	32	48	64	81
86	1.1554	1664	1777	1893	2012	2135	2261	2391	2525	2663					
87	1.2806	2954	3106	3264	3429	3599	3777	3962	4155	4357					
88	1.4569	4792	5027	5275	5539	5819	6119	6441	6789	7167					
89	1.7581	8038	8550	9130	9800	2.0591	2.1561	2.2810	2.4571	2.7581					

Section IV

Statistics

Chapter 12

An introduction to statistics

12.1 Introduction

Although statistics has been little more than a form of State Arithmetic in the past, as used today it is a scientific approach to collecting, organising, summarising and analysing numerical data and presenting it in a form such that reasonable predictions and decisions can be based on the analysis.

12.2 Data classification

Quantities which vary, such as the masses or heights of a group of people, are called **variables.** When variables can have any value at all within certain limits they are called **continuous** and since, for example, the mass of a child passes through all possible values between certain limits, then a set of data based on masses is continuous. Other variables can only have certain specific values within limits; for example, the number of children in a family can only be integer values, one or two or three and so on, and not, for example, two and a quarter. Data in a set of this sort are called **discontinuous** or **discrete.**

In statistics, it is often the method of analysis which is adopted which determines whether a variable is discrete or continuous. When data which are continuous are formed into a frequency distribution (see section 12.4) – for example, heights of a group of people measured correct to the nearest centimetre – then it becomes discontinuous. Similarly, discrete data with only a relatively small difference in values compared with the size of the variable are often treated as if they are continuous; for example, analysis of the population of a country. In general, measurements give rise to continuous data and counting gives rise to discrete data.

A **set** is a group of data and an individual value within the set is called a **member** of the set. Thus if the heights of five objects are measured and are found to be 10 cm, 13 cm, 15 cm, 6 cm and 9 cm, correct to the nearest centimetre, then the set of the heights of the five objects is [10, 13, 15, 6, 9] and one of the members of the set is 13. A set containing all the members is called a **population,** thus, for example, **all** the telephone numbers for a given area form a population. However, if 30 telephone numbers are selected at random from a directory for the given area, the set of those 30 numbers is called a **sample.** The number of times that the value of a member occurs in a set is called the **frequency** of that member. Thus, for the set:

2, 3, 7, 4, 2, 5, 3, 4, 2, 5, 11, 7, 4

members 2 and 4 each have a frequency of three, since they each occur three times, members 3, 5 and 7 each have a frequency of two and member 11 has a frequency of one. The **relative frequency** with which a member of a set occurs is given by:

$$\text{relative frequency} = \frac{\text{frequency of member}}{\text{total number of members in the set}} \times 100\%.$$

Thus, the relative frequency of the member 2 is $\frac{3}{13} \times 100\%$, that is, approximately 23%.

Worked problems on data classification

Problem 1. State whether data obtained on the following topics is likely to be discrete or continuous:

(*a*) the number of premium bonds sold by a Post Office,
(*b*) temperature of air recorded every hour at a weather station,
(*c*) the time to failure of an electric light bulb, and
(*d*) the annual income of a group of workers.

(*a*) Discrete; since when counting the numbers sold, only integer values can result.

(*b*) Continuous; the air temperature is measured and can be any value within certain limits.

(*c*) Continuous; the amount of time is measured and can be any value within certain limits.

(*d*) Discrete; the annual income is likely to be to the nearest pound or even if an annual income is based on pounds and pence it will still be discrete.

Problem 2. For the set of data given, determine (*a*) the frequency of each member of the set and (*b*) the percentage relative frequency of each member.

$$\begin{Bmatrix} 19 & 14 & 15 & 17 & 15 \\ 17 & 15 & 17 & 15 & 14 \\ 15 & 19 & 15 & 14 & 15 \\ 12 & 15 & 17 & 14 & 17 \end{Bmatrix}$$

One method of determining the frequencies is to use a 'tally'. An inspection of the set shows that it contains the numbers 12, 14, 15, 17 and 19 only. These are listed in the left-hand column of Table 12.1. Each member of the

Table 12.1

Number	Tally	Frequency	Relative Frequency %
12	1	1	5
14	1111	4	20
15	~~1111~~ 111	8	40
17	~~1111~~	5	25
19	11	2	10
	Total:	20	

set is inspected in turn and allocated to a particular row, depending on its value, by placing a mark in the tally column. Each group of 4 marks is 'barred' when the fifth mark is allocated to a particular row, to assist in the counting process. The frequency values are as shown in Table 12.1, obtained by adding the appropriate tally marks. The relative frequency values are shown in the right-hand column of Table 12.1, and are obtained from

$$\text{relative frequency} = \frac{\text{frequency of member}}{\text{total frequency of all members}} \times 100\%.$$

For the first row, relative frequency $= \dfrac{1 \times 100}{20} = 5\%$. For the second row,

relative frequency $= \dfrac{4 \times 100}{20} = 20\%$, and so on.

Further problems on data classification may be found in Section 12.5 (Problems 1–4), page 383.

12.3 Presentation of data

A column of numbers can be uninspiring and do not really indicate trends at a glance. Graphs can be used to present data visually, but there are also many other ways of doing so. Some of these are dealt with in this section, for sets having a few members only.

Pictograms

A pictogram uses pictures to represent data, the usual method being to have a group of pictorial units, all the same size, the numerical value being represented by the number of pictures shown. For example, a survey was carried out in a firm to find out the mode of travel used by employees to get to work. It showed that 35% used a bus, 25% used a car or motor cycle, 20% came on a bicycle, 10% used a train and 10% walked. This data could be depicted as shown in the pictogram in Fig. 12.1.

A glance at the pictogram shows that the majority of employees arrive at work either by bus or in their own vehicles. Relatively few come by train or live near enough to walk.

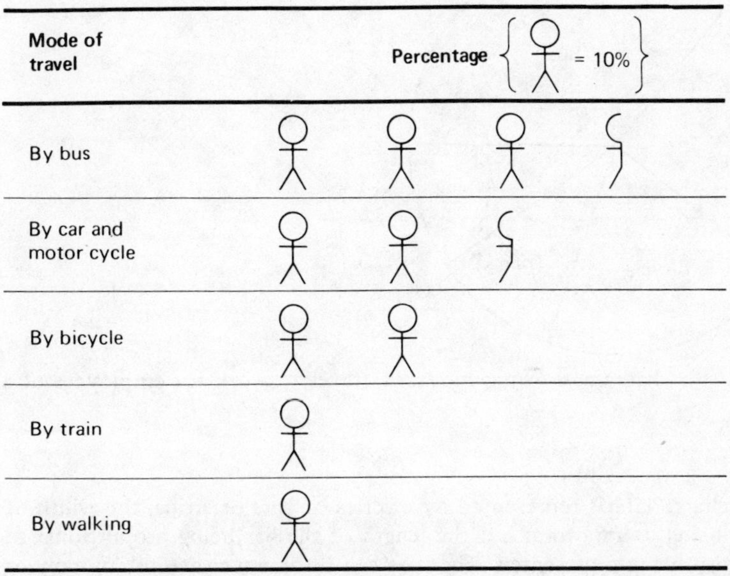

Mode of travel	Percentage { = 10% }
By bus	
By car and motor cycle	
By bicycle	
By train	
By walking	

Fig. 12.1 Pictogram showing mode of travel to work for people employed by a firm

Pie diagrams

A pie diagram is constructed by drawing a circle of any radius and subdividing it into sectors, so that the area of each sector is proportional to the quantity it is representing. To achieve this, the angles at the centre of the circle must also be in proportion to the quantities that are being represented. The information about the mode of travel depicted in Fig. 12.1 by a pictogram could equally well have been shown on a pie diagram. Since

100% corresponds to 360°
10% corresponds to 36°
20% corresponds to $36 \times 2 = 72^\circ$
25% corresponds to $36 \times \frac{5}{2} = 90^\circ$

and 35% corresponds to $36 \times \frac{7}{2} = 126°$,

by drawing a circle of any radius and using a protractor to divide the angle at the centre of the circle into angles of 126°, 90°, 72°, 36° and 36°, these angles being proportional to the percentages being represented, Fig. 12.2 is produced.

An examination of the pie diagram shows that the majority travel to work by bus or in their own vehicles and relatively few come by train or walk.

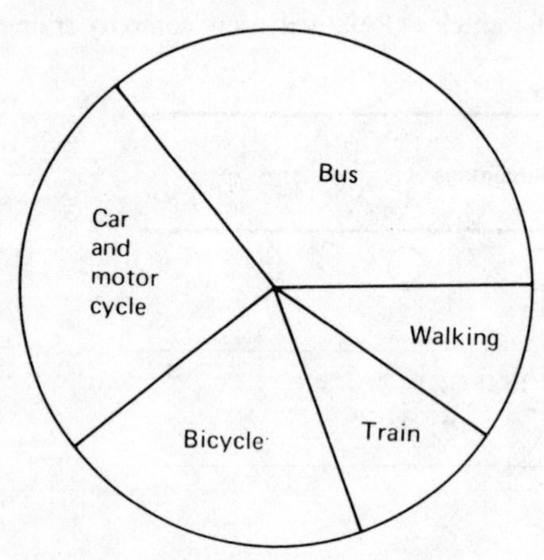

Fig. 12.2 Pie diagram showing mode of travel to work for employees of a firm

Bar charts

In a bar chart, data is represented by a series of bars or strips, the width of each bar being unimportant but the length of the bar being proportional to the quantity being represented. The bars can be drawn either horizontally or vertically. Figure 12.3 shows the data concerning mode of travel shown by means of a vertical bar chart, whereas Fig. 12.4 shows the same data presented as a horizontal bar chart.

As with graphs, care must be taken to avoid false zeros where possible, or a misleading representation can result.

Percentage component bar charts

These are usually used to show changing trends with time. Table 12.2 shows how the mode of travel of employees changed over a three-year period. To represent this data on a percentage component bar chart, a rectangle of any width but whose height corresponds to 100% is constructed. The height is sub-divided to represent the data given in Table 12.2. For just one set of data, a percentage component bar chart does not have a great deal of value, but if these bars are produced at various time intervals, changing trends can be seen readily.

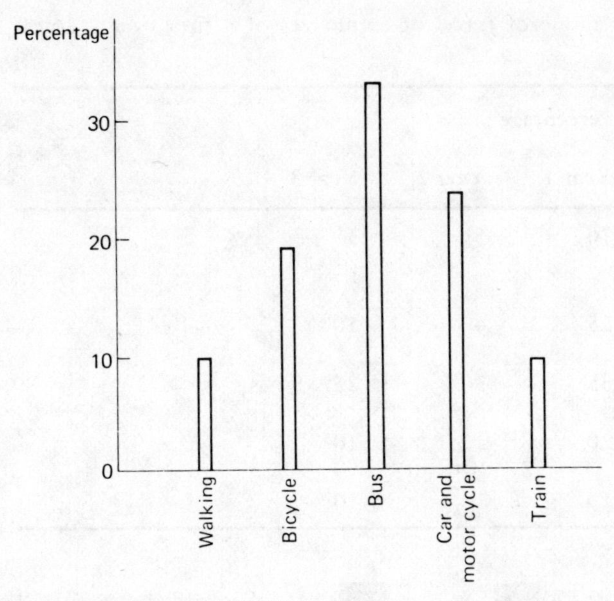

Fig. 12.3 Vertical bar chart showing mode of travel to work for employees of a firm

It can be seen from Fig. 12.5 that the number of people walking to work is remaining constant, more employees are bringing cars and motor-cycles at the expense of those cycling or travelling by public transport.

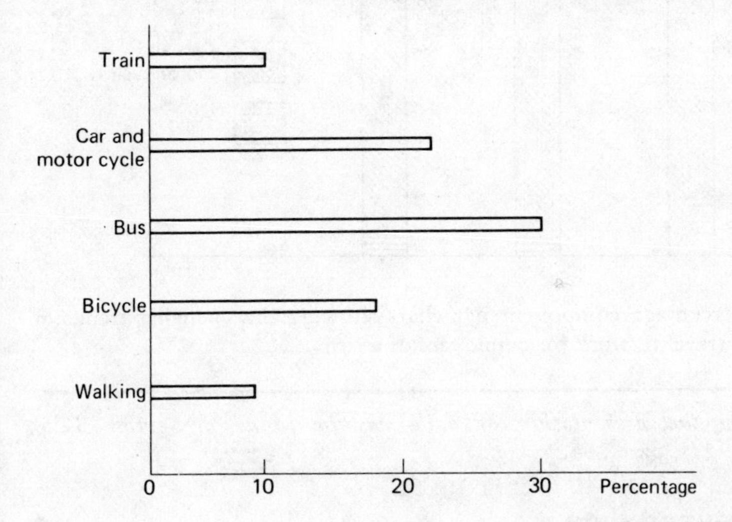

Fig. 12.4 Horizontal bar chart showing mode of travel to work for employees of a firm

Table 12.2 To show mode of travel of employees of a firm over a 3-year period.

Mode of travel	Percentage Year 1	Year 2	Year 3
Train	10	5	5
Car and motor cycle	25	40	50
Bus	35	30	25
Bicycle	20	15	10
Walking	10	10	10

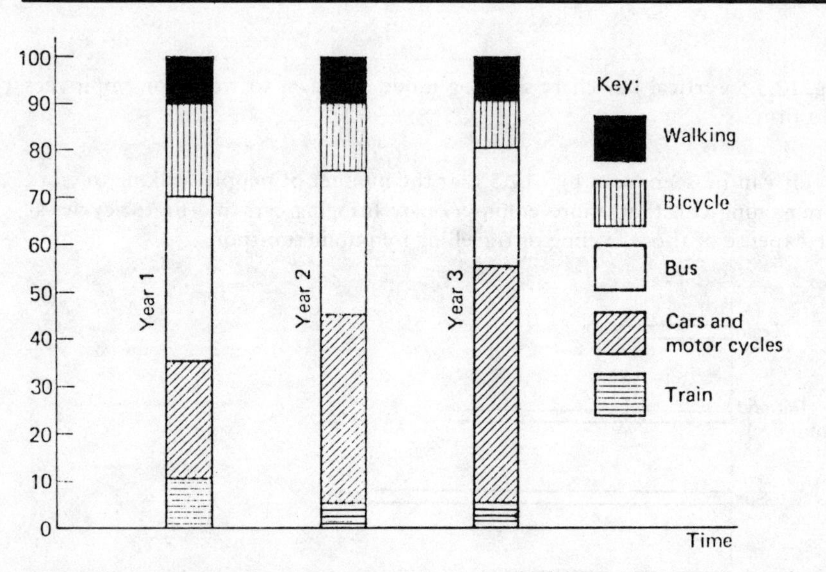

Fig. 12.5 Percentage component bar chart showing the changing trends in the mode of travel to work for employees of a firm

Problems on the presentation of data may be found in Section 12.5 (Problems 5–11), page 384.

12.4 Grouped data

When a set has a large number of members, it is not practical to use the methods introduced in Section 12.3 to depict the information given. Consider the set containing the heights of fifty children of the same age group, measured correct to the nearest centimetre, as shown.

171	170	172	170	171	166	169	169	169	168
175	168	167	168	168	172	168	173	168	170
169	173	169	173	170	170	177	167	167	172
173	171	174	168	174	171	166	169	167	165
167	174	170	168	163	169	173	174	170	168

This data is called **raw** or **ungrouped data.** The data is presented in a haphazard form and no conclusions can be drawn easily from such data. To present and analyse the information given in the set, a **frequency distribution** can be formed and the data is then said to be **grouped.** To form a frequency distribution, the **range** of the set is found by taking the smallest value from the largest; for this set, the range is 177–163, that is, 14 centimetres. It is normal to group the information together to give about six to eight **classes.**

Table 12.3 A frequency distribution showing the heights of children of the same age group

Class (centimetres)	**Tally**	**Frequency**
162–164	1	1
165–167	~~1111~~ 111	8
168–170	~~1111~~ ~~1111~~ ~~1111~~ ~~1111~~ 111	23
171–173	~~1111~~ ~~1111~~ 11	12
174–176	~~1111~~	5
177–179	1	1
	Total	50

The information shown in Table 12.3 has been selected to give six classes with no overlap between any two numbers forming them. For example, the first class ends at 164 centimetres and the second class starts at 165 centimetres. The class shown as 168–170 has **class limits** of 168 and 170 and 168–170 is called the **class interval.** Figure 12.6 shows that this class actually extends from 167.5 to 170.5 centimetres since the boundaries are halfway between 167 and 168, and 170 and 171. These two boundaries are called the **lower** and **upper class boundaries** respectively. The midpoint between these two boundaries is called the **class midpoint** and for the 168–170 class it is

$$\frac{167.5 + 170.5}{2} = 169 \text{ cm}.$$

Selecting each height in turn from the ungrouped data, a mark is made in the appropriate row corresponding to the class required, thus for 171 centi-

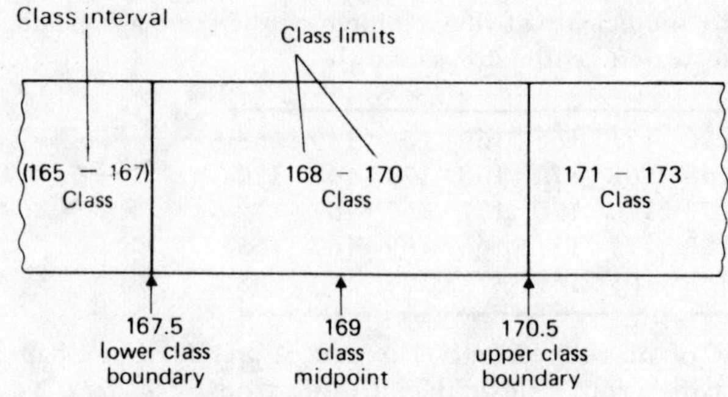

Fig. 12.6 A class of grouped data showing the terminology used

metres, the class selected will be the 171–173 class. To assist with counting, a tally system is used, as introduced in section 12.2. By adding the tally, the frequency of each class is obtained. A table containing class intervals together with their frequencies is called a frequency distribution. Grouped data is usually presented pictorially either by a **histogram** or a **frequency polygon.**

Histograms

A histogram is probably one of the most widely used methods of depicting grouped data. Rectangles are drawn vertically, the width of the rectangles representing the class intervals and the height of the rectangles the class frequencies. Since the area is equal to class interval multiplied by frequency, it is the area which is proportional to the number of observations in the class

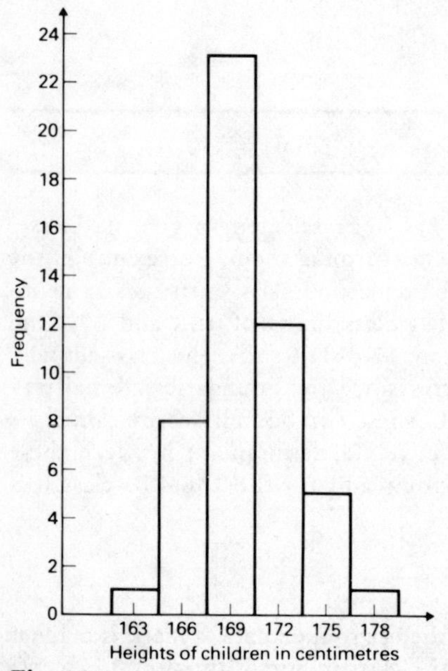

Fig. 12.7

concerned. When class intervals are equal in size, the heights of the rectangles also represent the frequencies.

A histogram depicting the frequency distribution given in Table 12.3 is shown in Fig. 12.7. Class midpoint values are used on the horizontal axis and rectangles having heights directly proportional to the class frequencies are constructed vertically as shown.

For unequal class intervals, the heights of the rectangles must be adjusted so that the area is proportional to the frequency of the class. The height of a rectangle is proportional to the frequency divided by the class interval, an illustration of this being shown in Table 12.4. A histogram depicting the data given in Table 12.4 is shown in Fig. 12.8.

Table 12.4 Determining the height of histogram rectangles for unequal class intervals

Class intervals	Frequency	Height of rectangle of histogram
0–9	15	$\frac{15}{10} = 1.5$ units
10–19	40	$\frac{40}{10} = 4.0$ units
20–39	70	$\frac{70}{20} = 3.5$ units
40–99	36	$\frac{36}{60} = 0.6$ units
100–199	20	$\frac{20}{100} = 0.2$ units

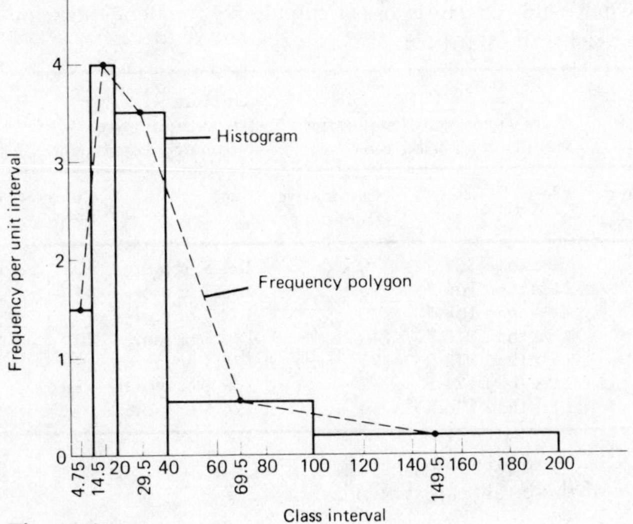

Fig. 12.8

Class boundaries or class midpoints are used for marking the horizontal scale. It is usually easier to mark the class midpoints and then to draw the bases of the rectangles symmetrically about the midpoints. Frequency is used for the vertical scale for equal class intervals or frequency per unit class interval for unequal class intervals.

Frequency polygons

By connecting the midpoints of the tops of the rectangles of a histogram together by straight lines and joining the midpoint of the top of the first rectangle to the base at the centre of the preceding class interval and the midpoint of the top of the last rectangle to the base at the centre of the next class interval, a figure called a **frequency polygon** is formed. Frequency polygons for data having unequal class intervals cannot be 'anchored', because the class midpoint of the next class interval is not known. A frequency polygon for the data given in Table 12.4 is shown in Fig. 12.8 by a broken line.

Ogives

The curve obtained when a cumulative frequency is plotted as a graph is called an **ogive** (pronounced o-jive) or **cumulative frequency curve.** A frequency distribution can be presented as an ogive in two ways. By taking upper class boundaries, frequencies which are 'less than' based can be determined. Alternatively, by taking the lower class boundaries, an 'or more' based cumulative frequency distribution can be determined, sometimes called a reverse cumulative frequency distribution. Both 'less than' and 'or more' based cumulative frequency distributions are shown in Table 12.5, the frequency distribution relating to the heights of 60 people measured at random.

Table 12.5 'Less Than' and 'Or More' based cumulative frequency distributions, based on the heights of 60 people

Frequency distribution		'Less Than' based cumulative frequency distribution		'Or More' based cumulative frequency distribution	
Class intervals	**Frequency**	**Class**	**Cumulative frequency**	**Class**	**Cumulative frequency**
		Less than 164.5	0	164.5 or more	60
165–166	6	Less than 166.5	6	166.5 or more	54
167–168	14	Less than 168.5	20	168.5 or more	40
169–170	20	Less than 170.5	40	170.5 or more	20
171–172	11	Less than 172.5	51	172.5 or more	9
173–174	6	Less than 174.5	57	174.5 or more	3
175–176	3	Less than 176.5	60	176.5 or more	0

A graph or the ogive of these values is shown in Fig. 12.9.

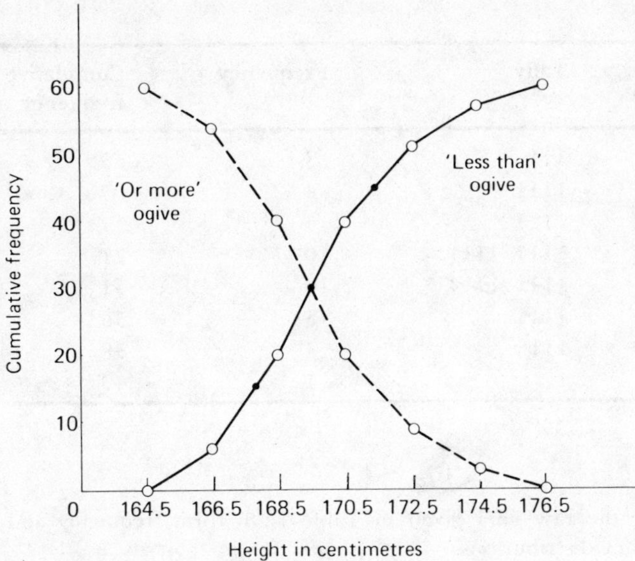

Fig. 12.9

Worked problems on grouped data.

Problem 1. Form a frequency distribution and cumulative frequency distribution of the data given in Table 12.6 which refers to the diameter of components produced by a machine, measured correct to the nearest millimetre.

Table 12.6 Measurements of 40 components produced by a machine correct to the nearest millimetre.

1.37	1.38	1.34	1.30	1.39	1.36	1.28	1.33
1.27	1.30	1.37	1.42	1.29	1.35	1.24	1.31
1.38	1.33	1.33	1.35	1.45	1.43	1.40	1.38
1.35	1.32	1.25	1.38	1.35	1.34	1.28	1.36
1.41	1.34	1.26	1.38	1.36	1.42	1.40	1.39

Each member of the set is examined to find the largest and smallest members. This gives the range and is (1.45–1.24) or 0.21 mm. Classes are selected to give, say, a total of eight classes. These are shown in Table 12.7. Each member of the set is placed in its appropriate class, using a tally system, and hence the frequency of each class is determined. By adding each frequency to the sum of the previous frequencies the cumulative frequencies are determined.

Table 12.7

Class intervals	Tally	Frequency	Cumulative frequency
1.24–1.26 mm	111	3	3
1.27–1.29 mm	1111	4	7
1.30–1.32 mm	1111	4	11
1.33–1.35 mm	~~1111~~ ~~1111~~	10	21
1.36–1.38 mm	~~1111~~ ~~1111~~	10	31
1.39–1.41 mm	~~1111~~	5	36
1.42–1.44 mm	111	3	39
1.45–1.47 mm	1	1	40

Problem 2. Using the raw data given in Table 12.8, form frequency and cumulative frequency distributions.

Table 12.8

47	18	35	8	12	182	19	27	15	49
73	25	3	77	29	53	71	6	82	7
6	58	36	38	5	6	9	42	57	28
104	22	44	16	30	13	27	137	11	64
87	45	5	15	23	10	79	12	2	40
17	44	123	9	43	7	14	37	35	161
36	94	14	20	97	10	67	32	59	18

Inspection of the data given in Table 12.8 shows that the range is (182–2) or 180. However, it is immediately apparent that there are very few numbers larger than 100. A quick count, to give some idea of the distribution of the numbers, shows that:

Class intervals	Frequency
0–49	51
50–99	14
100–149	3
150–199	2

This shows that to overcome the disadvantage of having most of the data in very few classes, unequal class intervals should be selected. A second count in the range 0–100 shows that:

Class intervals	Frequency
0–19	27
20–39	16
40–59	12
60–79	6
80–99	4

The majority of numbers are less than 40. There is no hard and fast rule to help determine the final selection of class intervals and one possible selection is shown in Table 12.9. Many other arrangements of class intervals could have been selected and would have been equally correct.

Table 12.9

Class intervals	Tally	Frequency	Cumulative frequency
1–4	11	2	2
5–9	~~1111~~ ~~1111~~	10	12
10–19	~~1111~~ ~~1111~~ ~~1111~~	15	27
20–49	~~1111~~ ~~1111~~ ~~1111~~ ~~1111~~ 1111	24	51
50–99	~~1111~~ ~~1111~~ 1111	14	65
100–199	~~1111~~	5	70

Problem 3. Draw a histogram, frequency polygon and ogive for the data given in Table 12.10.

Table 12.10

Class intervals	Frequency	Cumulative frequency
1.24–1.26 mm	2	2
1.27–1.29 mm	4	6
1.30–1.32 mm	4	10
1.33–1.35 mm	10	20
1.36–1.38 mm	11	31
1.39–1.41 mm	5	36
1.42–1.44 mm	3	39
1.45–1.47 mm	1	40

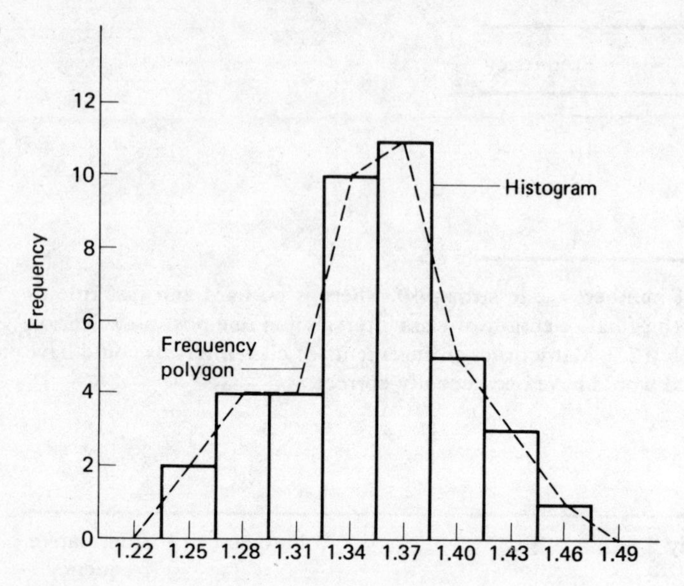

Fig. 12.10

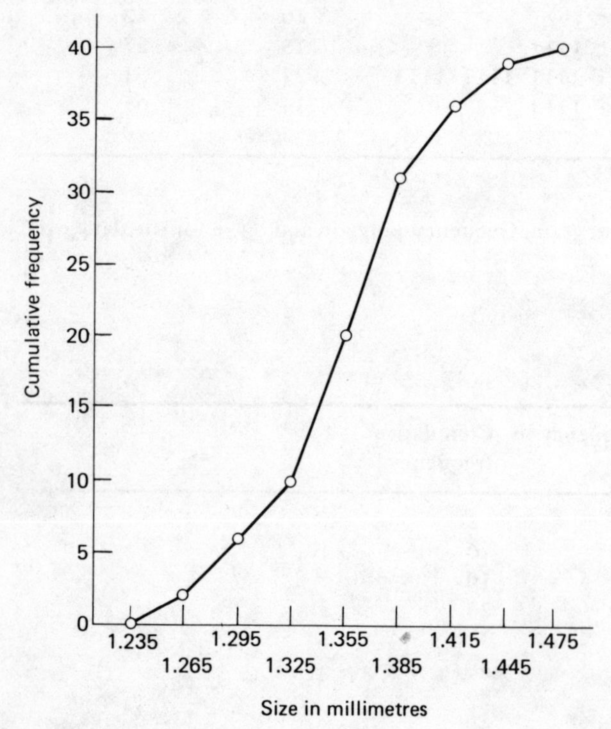

Fig. 12.11

The class midpoints are selected for the horizontal scale of the histogram and frequency for the vertical scale. Rectangles are drawn as shown in Fig. 12.10 so that they symmetrically span the class midpoints and touch one another at the vertical common line. The frequency polygon is also shown in Fig. 12.10 by the broken line and is produced by joining the central points of the tops of the rectangles together by straight lines. The polygon is 'anchored' at the start by drawing a straight line from the top centre of the first rectangle to the zero frequency of the midpoint of the class preceding it. It is 'anchored' at the other end by drawing a straight line from the top centre of the last rectangle to the zero frequency of the next class after it.

Because the ogive is not specified as having a 'less than' or an 'or more' basis, the 'less than' basis is assumed. Upper class boundary values are selected for the horizontal scale and cumulative frequency as the vertical scale. The ogive is shown in Fig. 12.11.

Problem 4. Using the data given in Table 12.11, construct a histogram, frequency polygon and ogive.

Table 12.11

Class intervals	Frequency
1–4	2
5–9	10
10–19	15
20–49	24
50–99	14
100–199	5

The class intervals in this distribution are unequal and the heights of the rectangles of the histograms have to be calculated using the relationship:

$$\text{height of histogram rectangle} = \frac{\text{class frequency}}{\text{class interval}}$$

The data from Table 12.11 together with calculated data is given in Table 12.12.

For the horizontal axis of the histogram, the class midpoints are $\frac{0.5 + 4.5}{2}$ or 2.5, $\frac{4.5 + 9.5}{2}$ or 7 and so on, and these are used as the central points of the rectangles. For histograms having unequal class intervals, the heights of the rectangles are not proportional to frequency. The units of the vertical scale are $\frac{\text{frequency}}{\text{class intervals}}$ and are shown as frequency per unit class interval. In this way, the area of the histogram is kept proportional to frequency. The histo-

 Table 12.12

Class intervals	Frequency	Upper class boundary	Cumulative frequency	Height of histogram rectangle
		Less than 0.5	0	
1–4	2	Less than 4.5	2	$\frac{2}{4} = 0.5$ units
5–9	10	Less than 9.5	12	$\frac{10}{5} = 2$ units
10–19	15	Less than 19.5	27	$\frac{15}{10} = 1.5$ units
20–49	24	Less than 49.5	51	$\frac{24}{30} = 0.8$ units
50–99	14	Less than 99.5	65	$\frac{14}{50} = 0.28$ units
100–199	5	Less than 199.5	70	$\frac{5}{100} = 0.05$ units

gram is shown in Fig. 12.12. By joining the central points of the tops of the rectangles together by straight lines, the frequency polygon is produced, shown by the broken line in Fig. 12.12. It is not possible to anchor the end points because the midclass points of the classes on either side of the polygon are not known.

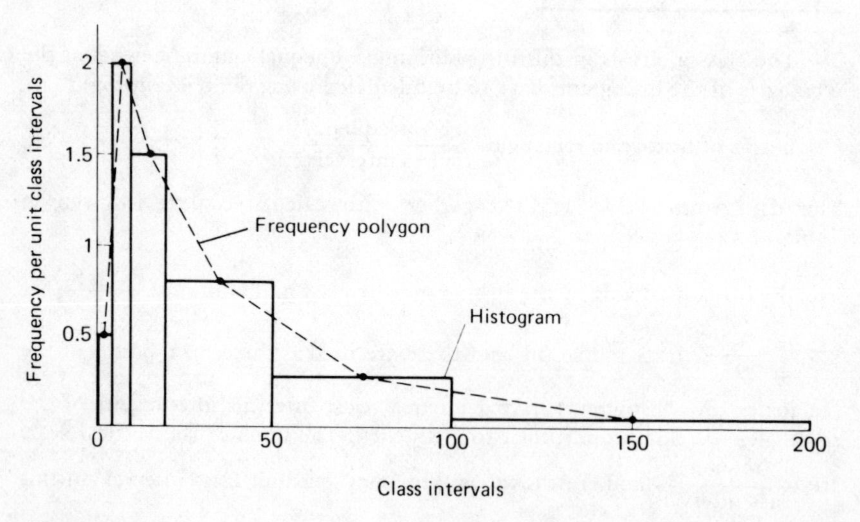

Fig. 12.12

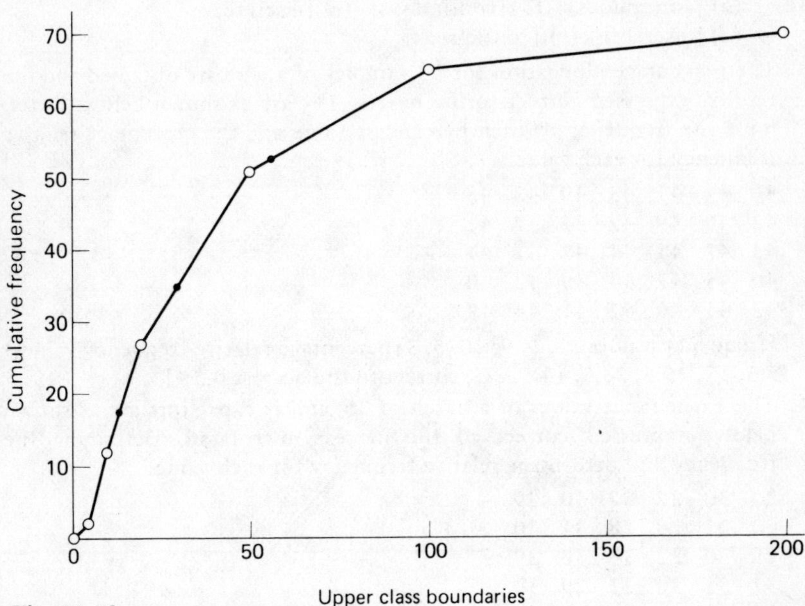

Fig. 12.13

The ogive is shown in Fig. 12.13 and is produced by plotting the values of the cumulative frequencies against the values of the upper class boundaries.

Further problems on grouped data may be found in Section 12.5 following (Problems 12–21), page 386.

12.5 Further problems

Data classification

1. State whether the data obtained on the following topics is likely to be discrete or continuous:
   (*a*) the diameter of a bolt produced by a machine tool,
   (*b*) the number of litres of water in a washing machine,
   (*c*) the number of books in a library, and
   (*d*) the diameter of a cylinder.

   (*a*) [continuous] (*b*) [continuous] (*c*) [discrete]
   (*d*) [continuous]

2. Classify the data obtained on the following items as discrete or continuous:
   (*a*) the rainfall measured daily in a weather station,
   (*b*) the speed of a car measured in km h^{-1},
   (*c*) the number of £1 notes in circulation,
   (*d*) the number of students in classes in a college, and
   (*e*) the time of flight of an aircraft.

(*a*) [continuous] (*b*) [continuous] (*c*) [discrete]
(*d*) [discrete] (*e*) [continuous]

3. The percentage elongation for 35 samples of a wire are obtained and the results, expressed correct to the nearest 1%, are as shown below. Determine the frequency of each percentage value and the percentage relative frequency for each value.

   40 41 41 43 40 42 43
   42 40 39 42 41 42 41
   43 42 41 41 42 41 44
   40 43 42 44 39 43 40
   42 41 40 42 41 44 42

   [frequency values: 2, 6, 9, 10, 5, 3; percentage relative frequency values: 5.5, 17, 25.5, 28.5, 14.5, 8.5, correct to the nearest 0.5%]

4. The capacitance values of a batch of 36 similar capacitors are as shown below, expressed correct to the nearest microfarad. Determine the frequency and percentage relative frequency for each value:

   29 30 32 32 30 29
   30 31 29 28 31 30
   29 30 32 31 30 29
   31 31 30 32 29 30
   30 32 28 30 31 29
   31 30 31 28 30 28

   [frequency values: 4, 7, 12, 8, 5; percentage relative frequency values: 11, 19.5, 33.5, 22, 14, correct to the nearest 0.5%]

Presentation of data

5. The proportions of raw materials produced in certain areas are given below:

   Area A – 40 units
   B – 16 units
   C – 16 units
   D – 12 units
   E – 8 units
   F – 8 units

   Represent this data on (*a*) a pictogram, (*b*) a pie diagram, (*c*) a horizontal bar chart and (*d*) a vertical bar chart.

6. A company produces five products in the following proportions:

   Product P 25
   Product Q 17
   Product R 15
   Product S 10
   Product T 5

   Draw (*a*) a pie diagram, (*b*) a horizontal bar chart and (*c*) a vertical bar chart to present this data visually.

7. An analysis of 100 items produced in a factory gave the following results:

Perfect	60
Defect due to poor finish	10
Defect due to being oversized	15
Defect due to being undersized	5
Defect due to mal-operation	10

Represent this data visually in three different ways.

8. The water used by the chemical industry in the United Kingdom may be divided into classes as follows:

Pure uncontaminated water	77.8%
Contaminated water without pretreatment	3.5%
Contaminated water with pretreatment	18.7%

Express these results as a pie chart.

9. The Gas Industry on assessing quality and reliability costs obtained the following values (Column 1) and considered that by involving management and workers in the quality control process this could be improved (Column 2).

	Column 1	Column 2
Costs to prevent defects	5%	10%
Cost of appraising output quality	30%	20%
Cost from defect or quality failure	65%	35%

Express these results as pie charts, showing clearly the savings involved in the second process.

10. The percentage sales in four departments of a company over a period of 5 years are as follows:

Year	1	2	3	4	5
Department A	10	15	17	16	8
Department B	27	22	20	21	29
Department C	34	38	42	47	50
Department D	29	25	21	16	13

Represent this data visually by means of a percentage component bar chart.

11. Over a three-month period, the six components produced by a factory varied as shown:

Month	1	2	3
Component P	9	10	8
Component Q	4	2	2
Component R	25	30	35
Component S	27	22	15
Component T	22	26	30
Component U	13	10	10

Represent this data by a visual diagram method.

12. Draw a histogram showing the haemoglobin levels in a number of children from the following data:

Level	60–64	65–69	70–74	75–79	80–84	85–89	90–94
Number	2	3	32	91	165	248	200
Level	95–99	100–104	105–109	110–114			
Number	156	98	23	9			

13. The lengths of components in centimetres cut off by an automatic guillotine machine are given below. Group this data into a frequency distribution.

5.38	5.36	5.26	5.37	5.35	5.36	5.39	5.45
5.29	5.41	5.39	5.28	5.24	5.33	5.35	5.36
5.43	5.35	5.34	5.37	5.33	5.42	5.40	5.28
5.40	5.31	5.27	5.36	5.42	5.38	5.25	5.33
5.34	5.32	5.33	5.30	5.32	5.30	5.39	5.34

[The distributions obtained will depend on the class intervals selected.]

14. The annual incomes of 50 employees of a company are listed below, the amount shown being correct to the nearest £10. Produce a frequency distribution for this data.

1 500	2 650	3 220	930	2 900	9 500	2 120	2 800
1 000	3 170	3 600	1 750	490	2 550	4 700	750
2 750	3 000	2 000	2 550	2 900	4 250	2 500	6 750
2 940	2 360	3 050	1 250	3 480	4 750	3 110	980
3 150	1 450	3 170	2 180	2 900	3 000	480	2 700
3 050	1 800	3 350	2 750	2 450	3 127	2 950	3 330
2 400	3 420						

[A distribution having unequal class intervals should be selected. The distribution obtained will depend on the class intervals selected.]

15. A batch of transistors were tested to destruction by operating them on overload conditions. The number of hours they functioned satisfactorily before failure occurred are shown below. Form frequency and cumulative frequency distributions for this data.

63	94	76	73	84	60	75	89	83	56
63	76	77	82	88	62	80	75	73	88
93	54	78	77	97	93	57	96	60	90
74	76	86	76	77	73	95	87	91	71
81	78	79	53	78	71	95	74	59	82
60	75	61	85	75	85	75	74	72	75

[The distributions obtained depend on the class intervals selected.]

16. The percentage elongation for samples of a wire are obtained and the results are as shown. Produce a frequency distribution and a cumulative frequency distribution for this data.

43	39	37	40	36	40	37	35	39	38
43	41	40	42	38	42	41	39	40	41
39	35	38	42	40	37	38	39	42	37
36	43	39	35	39	38	39	36	39	40
40	39	37	40	38	41	39	42	38	39
37	41	36	38	40	36	41	38	41	39

[The distributions obtained depend on the class intervals selected.]

17. The length in millimetres of a sample of bolts is shown below. Produce a frequency distribution, an 'or more' ogive and a 'less than' ogive for this data.

Height (cm)	165	166	167	168	169	170	171
Frequency	5	14	18	28	36	29	29
Height (cm)	172	173	174	175	176	177	
Frequency	24	19	15	6	3	2	

18. The annual incomes of a group of employees are given below. Draw a histogram, frequency polygon and ogive for this data.

Income (£)	500–999	1 000–1 499	1 500–1 999	2 000–2 499
Frequency	12	38	40	30
Income (£)	2 500–2 999	3 000–3 499	3 500–4 999	5 000–9 999
Frequency	19	12	8	4

(Note that the class intervals are unequal.)

19. Draw a histogram and ogive for the data given below, which refers to the number of hours a sample of electric light bulbs burned before failure.

Bulb life (hours)	Under 499	500–699	700–899	900–999
Frequency	14	23	36	40
Bulb life (hours)	1 000–1 099	1 100–1 299	1 300--1 499	Over 1 500
Frequency	41	34	25	12

(Note that the class intervals are unequal.)

20. The data given below refers to the intelligence quotient of a group of children. Draw a histogram, frequency polygon and ogive for this data.

I.Q.	60–79	80–89	90–94	95–99	100–104	105–114	115–129	130–149
Frequency	9	29	51	57	61	53	31	9

21. The percentage of chlorine in a chloro hydrocarbon was estimated 80 times with the following results:

21.83	21.84	21.85	21.86	21.87	21.88	21.89	21.90	21.91
1	1	2	6	8	10	28	13	6

21.92	21.93
3	2

Construct a histogram and frequency polygon of these results.

Index